HACCP in the meat industry

Related titles from Woodhead's food science, technology and nutrition list:

Lawrie's meat science Sixth edition (ISBN: 1 85573 395 1)
R A Lawrie

This book remains a standard for both students and professionals in the meat industry. It provides a systematic account of meat science from the conception of the animal until human consumption.

'Overall this is one of the best books available on the subject of meat science, and is ideal for all students of food science and technology.' *Chemistry in Britain*

Principles and practices for the safe processing of food (ISBN: 1 85573 362 5)
H J Heinz Company

This food industry handbook is a practical, science-based guide to food safety in food processing operations. The text is organised for easy reference, illustrated with numerous schematics and includes important reference data tables.

'... readers will want to have this book, not just because it is such a comprehensive text on safe processing, but because it is so full of sound advice. For food companies working on HACCP programmes, this book is a must-have.' *Food Engineering*

Chilled foods Second edition (ISBN: 1 85573 499 0)
Edited by Michael Stringer and Colin Dennis

The first edition of this book rapidly established itself as the standard work on the key quality issues in one of the most dynamic sectors in the food industry. This new edition has been substantially revised and expanded, and now includes three new chapters on raw material selection for chilled foods.

'This book lives up to its title in reviewing a major section of the food industry.' *International Food Hygiene*

Details of these books and a complete list of Woodhead's food science, technology and nutrition titles can be obtained by:

- visiting our web site at www.woodhead-publishing.com
- contacting Customer Services (e-mail: sales@woodhead-publishing.com; fax: +44 (0)1223 893694; tel: +44 (0)1223 891358 ext. 30; address: Woodhead Publishing Ltd, Abington Hall, Abington, Cambridge CB1 6AH, England)

If you would like to receive information on forthcoming titles in this area, please send your address details to: Francis Dodds (address, tel. and fax as above; e-mail: francisd@woodhead-publishing.com). Please confirm which subject areas you are interested in.

HACCP in the meat industry

Edited by
Martyn Brown

CRC Press
Boca Raton Boston New York Washington, DC

WOODHEAD PUBLISHING LIMITED
Cambridge England

Published by Woodhead Publishing Limited
Abington Hall, Abington
Cambridge CB1 6AH
England
www.woodhead-publishing.com

Published in North and South America by CRC Press LLC
2000 Corporate Blvd, NW
Boca Raton, FL 33431
USA

First published 2000, Woodhead Publishing Limited and CRC Press LLC

© 2000, Woodhead Publishing Limited
The authors have asserted their moral rights.

Conditions of sale
This book contains information obtained from authentic and highly regarded sources. Reprinted material is quoted with permission, and sources are indicated. Reasonable efforts have been made to publish reliable data and information, but the authors and the publishers cannot assume responsibility for the validity of all materials. Neither the authors nor the publishers, nor anyone else associated with this publication, shall be liable for any loss, damage or liability directly or indirectly caused or alleged to be caused by this book.

Neither this book nor any part may be reproduced or transmitted in any form or by any means, electronic or mechanical, including photocopying, microfilming and recording, or by any information storage or retrieval system, without permission in writing from the publishers.

The consent of Woodhead Publishing Limited and CRC Press LLC does not extend to copying for general distribution, for promotion, for creating new works, or for resale. Specific permission must be obtained in writing from Woodhead Publishing Limited or CRC Press LLC for such copying.

Trademark notice: Product or corporate names may be trademarks or registered trademarks, and are used only for identification and explanation, without intent to infringe.

British Library Cataloguing-in-Publication Data
A catalogue record for this book is available from the British Library.

Library of Congress Cataloging in Publication Data
A catalog record for this book is available from the Library of Congress.

Woodhead Publishing Limited ISBN 1 85573 448 6
CRC Press ISBN 0-8493-0849-6
CRC Press order number: WP0849

Cover design by The ColourStudio
Project managed by Macfarlane Production Services, Markyate, Hertfordshire
Typeset by MHL Typesetting Limited, Coventry, Warwickshire
Printed by T J International, Padstow, Cornwall

Contents

Preface xi
List of contributors xiii

Part 1   General issues

1   Introduction 3
T. H. Pennington, University of Aberdeen
1.1   E. coli O157 4
1.2   HACCP and food safety 6
1.3   The successful implementation of HACCP 8
1.4   References 9

2   The regulatory context in the EU 11
M. Fogden, Meat and Livestock Commission
2.1   Introduction: the international context 11
2.2   EU food policy and HACCP 12
2.3   EU meat hygiene legislation and HACCP 14
2.4   Fishery products 20
2.5   Future trends 22
2.6   Sources of further information and advice 25
2.7   References 26

3   HACCP in the United States: regulation and implementation 27
L. Crawford, Center for Food and Nutrition Policy,
Georgetown University
3.1   Introduction: the regulatory background 27

3.2	Development of HACCP in the United States	28
3.3	HACCP implementation in practice	30
3.4	Beyond HACCP	31
3.5	Bibliography	33

Part 2 HACCP on the farm and in primary processing

4 HACCP and farm production 37
A. M. Johnston, Royal Veterinary College, University of London

4.1	Introduction	37
4.2	Hazard analysis in animal rearing	38
4.3	Setting up the HACCP system	43
4.4	HACCP plans for cattle	45
4.5	HACCP plans for sheep and goats	61
4.6	HACCP plans for a poultry unit	62
4.7	HACCP plans for a pig unit	65
4.8	Summary: the effectiveness of HACCP on the farm	67
4.9	References	76

5 HACCP in primary processing: red meat 81
C. O. Gill, Agriculture and Agri-Food, Canada

5.1	Introduction	81
5.2	Microbiological data: collection and analysis	83
5.3	HACCP implementation: the general approach	87
5.4	Stock reception	92
5.5	Slaughter and predressing	94
5.6	Carcass dressing	96
5.7	Collection and cooling of offals	104
5.8	Carcass cooling	107
5.9	Carcass breaking; equipment cleaning	110
5.10	Smaller plants	113
5.11	Microbiological criteria	115
5.12	References	117

6 HACCP in primary processing: poultry 123
G. C. Mead, Royal Veterinary College, University of London

6.1	Introduction	123
6.2	Hazard analysis in the slaughter process	127
6.3	Establishing CCPs	134
6.4	Other processing operations	142
6.5	Future trends	145
6.6	Decontamination of carcasses	147
6.7	Sources of further information and advice	149
6.8	References	150

Contents vii

Part 3 HACCP tools

7 Microbiological hazard identification in the meat industry 157
P. J. McClure, Unilever Research, Sharnbrook
7.1   Introduction 157
7.2   The main hazards 158
7.3   Analytical methods 171
7.4   Future trends 173
7.5   Sources of further information and advice 174
7.6   References 174

8 Implementing HACCP in a meat plant 177
M. H. Brown, Unilever Research, Sharnbrook
8.1   Introduction 177
8.2   The elements requiring implementation 178
8.3   The implementation process 187
8.4   The differences between large and small businesses 188
8.5   Where to start with implementation 189
8.6   Explanation of the reasons for HACCP 190
8.7   Review of food safety issues 190
8.8   Planning for implementation 191
8.9   Allocation of resources 193
8.10 Selecting teams and activities 193
8.11 Training 193
8.12 Transferring ownership to production personnel 195
8.13 Tackling barriers 197
8.14 Measuring performance of the plan 198
8.15 Auditing and review 199
8.16 Conclusions 199
8.17 References 199

9 Monitoring CCPs in HACCP systems 203
J. J. Sheridan, TEAGASC (The National Food Centre), Dublin
9.1   Introduction 203
9.2   Establishing criteria 206
9.3   Determination of critical limits 207
9.4   Setting up monitoring systems 208
9.5   Verification of HACCP systems 213
9.6   Validation of the HACCP plan 222
9.7   Identifying problem areas 223
9.8   Feedback and improvement 224
9.9   Future trends 224
9.10 References 226

10 Validation and verification of HACCP plans ... 231
M. H. Brown, Unilever Research, Sharnbrook
- 10.1 Introduction ... 231
- 10.2 The background to validation and verification of HACCP ... 238
- 10.3 How far along the supply chain should a HACCP study extend? ... 241
- 10.4 The importance of Good Manufacturing Practice (GMP) ... 242
- 10.5 Decision making within a HACCP-based QA system ... 242
- 10.6 Monitoring ... 243
- 10.7 Validation, microbiological and other hazards ... 244
- 10.8 Introducing validation and verification ... 247
- 10.9 Validation – is it the right plan? ... 249
- 10.10 Verification – are we doing it correctly? Is it working? ... 253
- 10.11 Reporting conclusions and agreeing an action plan ... 261
- 10.12 Specific additional requirements for the meat industry ... 262
- 10.13 Involvement of plant management in validation and verification ... 263
- 10.14 Involvement of the HACCP team in validation and verification ... 263
- 10.15 How to validate a new HACCP study ... 264
- 10.16 How to validate an implemented HACCP plan ... 264
- 10.17 Sampling plans for validation ... 265
- 10.18 Sampling plans for verification ... 265
- 10.19 Output from validation and verification ... 267
- 10.20 Conclusions ... 267
- 10.21 References ... 269

11 Auditing HACCP-based QA systems ... 273
N. Khandke, Unilever Research, Sharnbrook
- 11.1 Introduction ... 273
- 11.2 HACCP and quality systems ... 275
- 11.3 Establishing benchmarks for auditing ... 277
- 11.4 What the auditor should look for ... 287
- 11.5 Future trends ... 288
- 11.6 References ... 290

12 Moving on from HACCP ... 293
J.-L. Jouve, Ecole Nationale Veterinaire de Nantes
- 12.1 Introduction ... 293
- 12.2 Future trends ... 295
- 12.3 Development of a risk-based food safety strategy ... 298
- 12.4 The Food Safety Programme ... 301

12.5	HACCP revisited: introduction of risk assessment techniques		310
12.6	Summary		316
12.7	References		317
Index			321

Preface

Over the past ten years or so, HACCP has become the preferred tool for the management of microbiological food safety. As its use has spread and reliance on it as the main means of ensuring the safe production of food has increased, so the need for checking its effectiveness has also increased. This same period has also seen dramatic changes in the processing, storage and distribution of meat products; and new pathogens, e.g. *E. coli* O157, have challenged both existing supply chain controls and also consumer and regulatory views of safety requirements. Extensive reliance on the use of HACCP for managing microbiological safety has also been questioned and it is hoped that this book will show that HACCP is the most powerful management tool available for ensuring product safety. In this context it is important to remember that the retrospective nature of microbiological testing makes it unsuitable for supporting day-to-day decisions on product release, or for monitoring CCPs, as products are likely to be out of the control of the producer by the time the results of testing are available. Therefore the 'real time' control available from a well designed and implemented HACCP plan offers manufacturers and consumers of meat products the best protection. However microbiological testing in a plant with a HACCP-based QA system still has a role to validate and verify the effectiveness of the process control and hygiene measures in place.

The purpose of this book is to present chapters written by experts on particular aspects of HACCP in the meat industry. The structure should provide the reader with chapters that taken singly give a clear account of one aspect or cover one type of material. Taken together they present a practical and coherent guide to HACCP for the industry. The chapters are divided into 'General Issues' giving the current (1999) expert view and the legislative context of HACCP in Europe and the United States. 'HACCP on the farm and in primary processing' sets out issues related to product and raw material groups and provides an outline of considerations specific to farm production, red and poultry meat. 'HACCP tools', the last chapters, provide a guide to the tools and information available for developing, implementing and managing HACCP-based QA systems.

It is important to appreciate that each chapter represents current thinking and current techniques; the challenges addressed and the sticking points may change as the industry develops, new pathogens emerge or as knowledge and familiarity with the techniques improves. Implemented and validated HACCP plans provide the meat industry with the best tool to manage food safety reliably and demonstrate how the quality of its raw materials, its standards of hygiene and process control lead to safe, high quality products. Its importance for managing food safety and ensuring free trade is reflected by support for HACCP in the SPS parts of the GATT agreement. Familiarity with the principles will help those actively involved in HACCP study teams produce reliable, soundly based actions and requirements. Similarly, less detailed knowledge will help any manufacturer or enforcement officer, not actively involved with HACCP, decide whether the QA principles and practices proposed are effective and later if the HACCP plan is scientifically valid and working. Using the 'primary processing' chapters they should be able to decide on the relevance of pathogens or toxins to their raw materials or products, or the likelihood of faulty manufacturing practices or controls prejudicing product safety.

I would like to thank the chapter authors for their co-operation in the preparation of their contributions and my colleagues at Colworth House for their help and advice.

Martyn Brown

Contributors

Chapter 1

Professor T. H. Pennington
University of Aberdeen
Department of Medical Microbiology
Medical School Buildings
Foresterhill
Aberdeen AB25 2ZD

Tel: +44 (0)1224 681818
Fax: +44 (0)1224 685604 Ext 52867
E-mail: mmb036@abdn.ac.uk

Chapter 2

Mr Michael Fogden
3 Kipling Drive
Towcester
Northamptonshire NN12 6QY

Answerphone: +44 (0)1327 359673
Fax: +44 (0)1327 359673
E-mail: fogden@medvek.fsnet.co.uk

Chapter 3

Professor Lester Crawford
Center for Food and Nutrition Policy
Georgetown University
3240 Prospect Street NW
Washington DC 2007-2197
USA

Tel: +1 202 965 6400
Fax: +1 202 965 6444
E-mail: ceres@erols.com

Chapter 4

Professor A. M. Johnston
Royal Veterinary College
University of London
Department of Farm Animal & Equine
Medicine & Surgery
Boltons Park
Hawkshead Road
N Mymms
Herts AL9 7TA

Tel: +44 (0)1707 666277
Fax: +44 (0)1707 660671
E-mail: Johnston@rvc.ac.uk

Chapter 5

Dr Colin Gill
Agriculture & Agri-Food Canada
Lacombe Research Centre
6000 C & E Trail
Lacombe
Alberta T4L 1W1
Canada

Tel: +1 403 782-8113
Fax: +1 403 782-6120
E-mail: gillc@cm.agr.ca

Chapter 6

Professor G. C. Mead
Royal Veterinary College
University of London
Boltons Park
Hawkshead Road
Potters Bar
Herts EN6 1NB

Chapter 7

Dr P. J. McClure
Unilever Research
Colworth House
Sharnbrook
Bedford MK44 1LQ

Tel: +44 (0)1234 222010
Fax: +44 (0)1234 222277
E-mail: Peter.McClure@unilever.com

Chapters 8 and 10

Professor Martyn Brown
Unilever
Microbiology Department
Unilever Research Colworth Laboratory
Colworth House
Sharnbrook
Bedford MK44 1LQ

Tel: +44 (0)1234 222351
Fax: +44 (0)1234 222277
E-mail: martyn.brown@unilever.com

Chapter 9

Dr James Sheridan
Food Safety Department
TEAGASC
National Food Centre
Dunsinea
Castleknock
Dublin 15
Ireland

Tel: +353 (1) 805 9500
Fax: +353 (1) 805 9550
E-mail: a.shalloo@teagasc.ie

Chapter 11

Dr Neil Khandke
Ice Cream Technology Unit
Unilever Research
Colworth House
Sharnbrook
Bedford MK44 1LQ

Fax: +44 (0)1234 222000
E-mail: Neil.Khandke@unilever.com

Chapter 12

Professor J.-L. Jouve
Ecole Nationale Veterinaire de Nantes
Atlanpole-La Chanterie
BP 40706
44307 Nantes Cedex 03
France

Part 1

General issues

1

Introduction

T. H. Pennington, University of Aberdeen

Even in a new millennium we can be certain that myths will continue to play an important role in people's lives. A longstanding and pervasive one is that the only general spin-off from space travel and rocket science has been the non-stick frying pan. Far more important, of course, was the development of HACCP by NASA, Pillsbury and others. What a debt we owe to those who addressed the need to protect space missions from food poisoning and the appalling prospect of diarrhoea in zero gravity!

As a medical microbiologist specialising in the molecular typing of human pathogens my involvement with HACCP was, until recently, remote and indirect. This changed suddenly and dramatically at the end of 1996, when Central Scotland suffered one of the largest outbreaks of *E.coli* O157 food poisoning ever recorded with more than 500 cases and 21 associated deaths. It centred on a butchery business.[1] Like the 1993 Jack-in-the-Box hamburger chain outbreak in the United States,[2] it had a profound impact on politicians as well as public opinion. While it gave red meat – yet again – a negative role as a vector of disease, it also created a window of opportunity for driving forwards improvements in food safety. Early in the outbreak I was asked by the Secretary of State for Scotland to chair an Expert group 'to report on the circumstances leading to the outbreak, the implications for food safety, and the lessons to be learned.'

In the deliberations which led to our final report[3] we tried to identify measures which would help to reduce the incidence of future infections with *E.coli* O157 and, in particular, outbreaks of the scale involved in Central Scotland. We were also determined, in considering food safety legislation, guidance and practices that, in coming to our views, public health considerations should be regarded as paramount in the handling of potential and actual outbreaks of food poisoning. We were persuaded of the overriding need to tackle the dangers which *E.coli* O157 presents and to reinforce public health considerations in the area of food safety. This overarching principle guided our work. Moreover, while we believed that the measures proposed were justified with reference to the circumstances of the outbreaks examined, we also acknowledged the influence of more general concerns about the growing incidence of food poisoning cases, and their economic and social costs, in supporting the precautionary and preventive approach adopted.

A 'public health' approach concerns itself primarily with prevention. So does HACCP, with both its philosophy and its practice centring on critical control points. This is why the principles of HACCP were central to our deliberations. We sought to identify the critical points in the process of food production 'from farm to fork' at which, based on our examination of the circumstances of recent outbreaks, there seems to be most risk of contamination.

It was brought home to us early in our investigation of the Central Scotland outbreak that the successful introduction and implementation of HACCP is not a trivial undertaking. A prerequisite for these is an understanding by management and workforce of the hazards and risks that underpin Good Hygiene practice, and the effective operation of the latter. All these things were lacking in John Barr's, the butchery business that was the source of the outbreak. Thus at the time of the outbreak there was no training programme for its staff, no cleaning schedule for its equipment or premises, no temperature monitoring of cookers or refrigerators, and neither soap nor drying facilities at the inadequate number of wash hand basins. There were, on the other hand, more than 30 points at which there was a high risk of cross-contamination. The sheriff principal who conducted the Fatal Accident Inquiry into the 21 deaths associated with the outbreak summarised the problem succinctly: 'I have no doubt Mr John Barr liked a clean shop and maintained a clean shop. What he failed to do was to maintain a safe shop and the main ingredient of his failure was ignorance of the requirements which would produce that result.'

1.1   *E.coli* O157

As a test for food safety systems *E.coli* O157 is unparalleled. This is because of its propensity to be transmitted to people at any point in the food chain, because of other properties like its ability to survive well in hostile environments and its low infectious dose, and because of its nastiness as a pathogen. These things make it an important public health problem and a serious challenge to the meat industry. It cannot be bettered as a focal point on which to centre considerations of HACCP. So it is worth considering the biology and natural history of the organism in some detail.

E.coli O157 exists in a wide range of animals (wild, farmyard and domestic) and even birds. It is generally accepted that its main reservoir is in the rumens and intestines of cattle and, possibly, sheep. The organism can be excreted and may therefore exist in animal manure or slurry, which could be a source of environmental or water contamination, or direct contamination of food such as vegetables. (Most of the evidence for this is, however, circumstantial.) It seems likely that there can be animal to animal infection/reinfection. There is good evidence that it is transferred to animal carcasses through contamination from faecal matter during the slaughter process. Many early outbreaks were associated with the consumption of hamburgers. There have also been documented cases attributed to meat, meat products and other foods such as milk, cheese and apple juice. In the very large Japanese outbreak, radishes were identified as a possible source of the infection. The vehicle for most cases of infection, however, remains unknown. The organism survives well in frozen storage and freezing cannot be relied upon to kill it. It is killed by heating but can survive if food is not properly cooked. If appropriate hygiene measures are not taken, there can also be cross-contamination between raw meat carrying the organism and cooked or ready to eat foods. *E.coli* O157 appears to be relatively tolerant to acidic conditions (compared, for example, to *Salmonella*).

Human infection may occur as a result of direct contact with animals carrying the organism, from contamination from their faeces, or through consumption of contaminated food or water. It may also spread directly from person to person as a result of poor hygiene practices which allow faecal–oral spread. The latter is, obviously, a particular potential problem in institutions such as nursing homes, day-care centres or hospitals and in places where pre-school children meet, and underlines the need for good personal hygiene and meticulous attention to procedures designed to prevent cross-infection. Cases may be related to outbreaks or may be sporadic (i.e. isolated and apparently unrelated to other cases). The role of asymptomatic food handlers in outbreaks is unclear but may be important in light of the low infectious dose.

Infection with *E.coli* O157 is potentially very serious for vulnerable groups, particularly the elderly and the very young. There is no specific treatment available for infection or to prevent complications. These include haemorrhagic colitis (bloody diarrhoea), the haemolytic uraemic syndrome (HUS) and thrombotic thrombocytopaenic purpura (TTP). The latter two complications are much less common but can be very serious, causing kidney and other problems and, in the most severe cases, even death. Infection with *E.coli* O157 and associated HUS is the most common cause of acute renal failure in children in the UK. Morbidity for the vulnerable groups is particularly high compared to other forms of foodborne illness.

Despite improvements in surveillance and testing techniques, the organism remains more difficult to detect and identify accurately than most other important foodborne bacterial pathogens. *E.coli* O157 does not generally cause illness in animals other than, at worst, transient diarrhoea in very young animals. There is, therefore, no reason for farmers to seek to identify the presence of the organism in their animals.

The very few organisms that are required to cause harm in humans can, under present rules and practices, easily escape detection and pass along the food chain, whether from animal faeces, carcasses, meat, equipment or humans. *E.coli* O157 has been difficult to identify in foods and, although techniques have improved over the years, rates of detection are still unsatisfactory. This is due in part to the low levels of the organism which appear to occur in food. The most sensitive techniques for identifying the organism (particularly, but not only, in food) are complex and sophisticated, requiring specialised equipment and expertise that is not generally available.

1.2 HACCP and food safety

Clearly, all these things make *E.coli* O157 a formidable challenge. No single immediately and universally applicable technical fix is available to eliminate it from the food chain. Eradication from its ruminant hosts is not a practical proposition at the present time. This is why my expert group spent a lot of time considering the HACCP system. This was not just because it is the overarching system which governs the UK's (and indeed the emerging global) approach to tackling food safety issues, but because of its applicability at many parts of the food chain. This derives from its nature – a structured approach to analysing the potential hazards in an operation; identifying the points in the operation where the hazards may occur; and deciding which points are critical to control to ensure consumer safety. These critical control points are then monitored and remedial action, specified in advance, is taken if conditions at any point are not within safe limits. Thus HACCP is both a philosophy and a practical approach to food safety.

European Union (EU) food law places the responsibility for ensuring the safety and protection of the consumer very firmly with individual food businesses. HACCP-based principles, some of which are enshrined in much of this law, provide the tool for food businesses to address this responsibility, and these principles are backed up in law by prescriptive requirements and provisions requiring enforcement. The advantages of the HACCP approach are now internationally recognised, through the Codex Alimentarius Commission, where it is agreed that HACCP is based on seven principles:

1. Conduct a hazard analysis. Identify the potential hazards associated with food production at all stages up to the point of consumption, assess the likelihood of occurrence of the hazards and identify the preventive measures necessary for their control.
2. Determine the critical control points (CCP). Identify the procedures and operational steps that can be controlled to eliminate the hazards or minimise the likelihood of their occurrence.
3. Establish critical limit(s). Set target levels and tolerances which must be met to ensure the CCP is under control.
4. Establish a system to monitor control of the CCPs.
5. Establish the corrective actions to be taken when monitoring indicates that a particular CCP is not under control.

6. Establish procedures for verification to confirm that the HACCP system is working effectively.
7. Establish documentation concerning all procedures and records appropriate to these principles and their application.

How did my group apply these principles to *E.coli* O157?

After the farm, slaughterhouses represent the second critical point in the food production process. My expert group believed that HACCP should apply to the slaughter process. In this context, we took cognisance of the Australian approach where slaughterhouses have adopted HACCP in full and accept their responsibility for food safety. These measures appear to have largely eliminated the problem of faecal contamination of carcasses. However, even starting from the base of high standards necessitated by the demands of export markets, it took at least five years for Australia to reach this position. My group identified a range of issues relating to slaughterhouses and the potential for cross-contamination at various key stages in the slaughter process. These included:

- the presentation of animals in a clean and dry condition suitable for slaughter;
- processes relating to removal of the hide and the intestines of the animal;
- the need to consider and evaluate end-of-process treatments;
- issues related to the transportation of carcasses and meat; and
- more general issues related to the achievement and enforcement of good hygiene standards within abattoirs.

My group also concluded that HACCP principles and the need for the highest hygiene standards should apply to the transportation of carcasses and meat. We felt it to be pointless promoting hygiene within abattoirs and butchers if meat was permitted to become cross-contaminated during transportation to or from cutting plants or butchers. We indicated that vital importance of preventing, for example, unwrapped meat from touching the sides of transport vehicles during loading, carriage and unloading and that HACCP principles needed to be understood by transport interests and reflected in regulations and subsequent enforcement in this area.

The potential for cross-contamination of foods points to the critical nature of meat production and butchers' premises in the food chain. Even with measures taken earlier in the chain to help prevent contamination, it is inevitable that from time to time meat will enter the premises contaminated with *E.coli* O157. All raw meat, therefore, needs to be treated as though it is potentially contaminated and appropriate handling and hygiene standards adopted with HACCP as the universal approach. Clearly, if an effective HACCP had been in place at the butchery business responsible for the Central Scotland outbreak, the large amount of raw and ready-to-eat meats being handled there daily – with a very high cross-contamination potential – would have figured high in the list of critical control points and for action. Many lives would have been saved.

1.3 The successful implementation of HACCP

The successful application of HACCP requires the full commitment of management and the workforce. It also requires a multi-disciplinary approach. A prerequisite to implementation is knowledge, understanding and expertise in identifying the hazards and assessing the risk involved in an operation. Introduction of a new system requires structured implementation. From information and evidence that my expert group collected during the course of its work, we identified a number of concerns about the current position of HACCP in the UK.

- The scheme relies primarily on businesses themselves, albeit with external expert advice and assistance as appropriate, to identify potential hazards and critical control points within their own operations.
- Businesses require expertise and training for successful implementation.
- Many businesses have yet to discover HACCP, or to put it into practice. The concept is sound, but it is relatively new and as yet insufficiently well known or understood – across the spectrum of issues involved or of business.
- The period over which HACCP principles can effectively be introduced is lengthy (in the UK senior environmental health officers with first-hand knowledge and experience of food premises, and individuals involved in education and training in food safety, suggest that this will take up to five years or even longer, regardless of the risks inherent in a particular business). As recent events have shown, there is the potential for many serious outbreaks of food poisoning over that period.

E.coli O157 is of course not the only hazard that challenges the meat industry. Other *E.coli* serogroups like O111 and O26 behave in a similar way, and *Salmonella* is still with us. However, I have focused on it in this introduction for four main reasons. First, its versatility and nastiness as a food poisoning organism makes it an unremitting and particularly severe – and therefore good – test for food safety systems. Second, its propensity to cause dramatic and severe outbreaks means that in addition to its direct effects on those who suffer disease – devastating as these often are – it also has a broad range of negative impacts of a general kind. Thus in addition to ruining businesses, its impact diminishes public confidence in food safety. Third, its public impact can drive public policy. Thus on both sides of the Atlantic major outbreaks have led to an acceleration and an increase in rigour in the development of HACCP programmes.[4,5,6]

Finally, and important for HACCP, as a new and emerging pathogen with distinct properties *E.coli* O157 has reminded us that even the best HACCP relies on past information in its identification and management of critical control points. Continued programmes of research are needed to keep up with the evolution of pathogens as well as the lessons from outbreaks, which are still occurring and still giving new insights into pathogen behaviour.

The prevention of food poisoning by HACCP is not, of course, an issue restricted to the meat industry. The Central Scotland outbreak highlighted the importance of food hygiene at the point of consumption – eight of those who died were infected by food served in a church hall – and one of the recommendations of my Expert Group was 'that steps should be taken by local authorities to encourage the adoption of HACCP principles in non-registered premises where there is catering for functions for groups of people involving the serving of more than just tea, coffee and confectionery goods.'

Fortunately, *E.coli* O157 infections are still relatively rare. It is an unfortunate and depressing fact, however, that when outbreaks are studied in detail it turns out that for many of them their root cause was ignorance or disregard of well-understood safety principles, with failure at management levels being key. Even though *E.coli* O157 is a relatively new organism, having emerged as a problem only in the last 20 years, its critical control points are in principle the same as for other meat-borne pathogens.

For all these reasons, it is abundantly clear that the solution to these problems lies in the effective implementation of HACCP. The authoritative chapters which follow show how this can be done.

1.4 References

1. AHMED S and DONAGHY M, 'An outbreak of *Escherichia coli* O157:H7 and other shiga toxin-producing *E.coli* strains', in *Escherichia coli O157:H7 and other Shiga Toxin-producing E.coli Strains*, ed KAPER J B and O'BRIEN, A D, pp 59–65, Washington DC, ASM Press, 1998.
2. TUTTLE J, GOMEZ T, DOYLE M P, WELLS J G, ZHAO T, TAUXE R V and GRIFFIN P M, 'Lessons from a large outbreak of *E.coli* O157:H7 infections: insight into the infectious dose and method of widespread contamination of hamburger patties', *Epidemiol Infect*, 1999 **122** 185–92.
3. PENNINGTON T H, 'The Pennington Group. Report on the circumstances leading to the 1996 outbreak of infection with *E.coli* O157 in Central Scotland, the implications for food safety and the lessons to be learned'. Edinburgh, The Stationery Office, 1997.
4. PENNINGTON T H, 'Factors involved in recent outbreaks of *E.coli* O157:H7 in Scotland and recommendations for its control', in *Food Safety: the Implications of Change from Producerism to Consumerism*, ed SHERIDAN J J, O'KEEFE M and ROGERS M, pp 127–35, Food and Nutrition Press, Trumbull, CT, 1998.
5. The UK government has accepted the recommendations of the Pennington Report, including those which said 'HACCP (i.e. the Codex Alimentarius approach and the seven principles) should be adopted by all food businesses to ensure food safety' and 'The Government should give a clear policy lead on the need for the enforcement of food safety measures and the accelerated implementation of HACCP.'
6. USDA Food Safety and Inspection Service, 'Pathogen reduction; hazard analysis and critical control systems (HACCP); final rule', Federal Register **144** July 25 1996.

2

The regulatory context in the EU

M. Fogden, Meat and Livestock Commission*

2.1 Introduction: the international context

2.1.1 The European Union and its Member States
The attitude of the EU to risk assessment and management in the food chain, in particular to HACCP, is extremely important within its 15 component states, and also in most other European territories, for it has a significant influence on its neighbours. In particular, countries within or applying to join the EU must implement its regulatory systems. However, that attitude is also important to global trading partners ('third countries'), which must comply with its rules to import food into any Member State, since the EU applies its internal regulatory control processes to trade with other nations so that the standards applicable within its boundaries are adequate and reasonably uniform.

2.1.2 Global policy
There is a powerful trend towards improved food hygiene throughout the world, largely driven by consumer pressures, particularly noticeable where advanced processing, storage and handling techniques may result in serious problems should the control system fail. Food chain controls are expected to ensure food safety in all reasonably foreseeable circumstances, including potential abuse.

Codex Alimentarius Commission (Codex)
This organisation was set up by the Food and Agriculture Organisation (FAO) and World Health Organisation (WHO) to implement their joint food standards

---

* This chapter expresses the personal views of the author and must not be attributed to MLC.

programme. It works to harmonise international food standards and hygiene requirements and to protect public health, and publishes standards and other guidance in the 'Codex Alimentarius'. In June 1997, Codex adopted three texts on food hygiene, including a Recommended International Code of Practice 'General Principles of Food Hygiene' to which was annexed 'Hazard Analysis and Critical Control Point (HACCP) System and Guidelines for its Application'.

The EU is not a full member of Codex, although it has observer status and has sought membership. To progress this, it must establish with its Member States where it has the legal competence to negotiate and vote and where this remains with the individual states. The application of HACCP, following these Codex texts, is as beneficial within the meat and meat products sector as elsewhere in the food industry. The EU and its Member States accept that the Codex texts are to be taken into consideration when they are developing their own control measures.

World Trade Organisation (WTO)
This is the successor to the General Agreement on Tariffs and Trade (GATT), which included the Sanitary and Phytosanitary (SPS) Agreement among its achievements. This incorporates the basic principle that its requirements must be based on sound science, essential to avoid the creation of adverse trade barriers, and provides for the Codex Alimentarius standards to be given full consideration. The SPS Agreement applies to all relevant measures that may affect international trade, prohibiting measures having an overt or covert protectionist effect unless they are justifiable and based on sound science. Article 5 clarifies this, requiring sanitary and phytosanitary measures to be based on an assessment of risk, with particular internationally accepted assessment and control techniques being taken into consideration. HACCP is such a technique, applicable in the realm of food safety [Annex A (3)(a)].

These texts do of course require authoritative interpretation, which has developed in recent decisions following challenges to the European Union's prohibition on the importation of cattle treated with growth-promoting hormones, by Canada and the United States. The EU and its Member States and therefore producers have agreed to comply with the WTO rules.

2.2 EU food policy and HACCP

2.2.1 EU food policy
The EU has a policy of providing a high level of consumer protection, including food safety measures, as shown in the Treaty of Rome (as amended):

- Article 95 declares that the Commission, in approximating laws concerning health, safety, environmental protection and consumer protection affecting the establishment or functioning of the common market, 'will take as a base a high level of protection, taking account in particular of new developments

based on scientific facts'. The European Parliament and Council will seek to do likewise.
- Article 152 declares that 'a high level of human health protection shall be ensured in the definition and implementation of all Community policies and activities', and that 'Community action, which shall complement national policies, shall be directed towards improving public health, preventing human illness and diseases, and obviating sources of danger to human health'.
- Article 153 declares that 'to promote the interests of consumers and to ensure a high level of consumer protection, the Community shall contribute to protecting the health ... of consumers'.

The Directorate-General of the European Commission responsible for consumer health and consumer protection (DG SANCO, formerly DG XXIV) has recently been much strengthened. It works with colleagues dealing with, for example, agriculture, the development of the internal market and trade policy to develop appropriate legislation. It is responsible for the provision of independent scientific advice and for monitoring the implementation of consumer-oriented legislation.

2.2.2 Internal market development of hygiene legislation

Within the EU, great steps have been taken since the mid-1980s to harmonise food legislation, building on previous regulatory requirements to construct a single control system. A programme to develop a comprehensive set of food hygiene controls was essentially completed by 1994, although this has been modified since then. Further, a review of these controls has recently taken place (see Section 2.5), and draft proposals to reconstruct the regulatory legislative situation are currently under consideration. They provide some simplification of the current legislative texts. One important feature is the status of HACCP.

The legislation is currently entirely in the form of directives, a format which requires each Member State to introduce legislation to reach the agreed objectives contained in the EU texts according to the national legal culture. This has the result that each state also tends to modify the technical requirements to suit its cultural preferences, causing diversity in the requirements applicable in the various territories, which can in certain cases be reconciled with the original intention only with some difficulty. The European Commission attempts to restrain such divergences and to ensure that each state system matches the EU ideal. It would prefer to legislate in the form of regulations, the texts of which would be immediately applicable in each Member State without the domestic legislators having the opportunity to introduce a national slant. Even with regulations, in the format of the present draft proposals, national interpretations are likely to vary.

Food businesses are put under varying obligations in each of the directives, which are intended to give assurance that the foods that they produce are processed hygienically and in accordance with the provisions made in the relevant legislation, with sufficient monitoring being undertaken to confirm this. These provisions may be part of, or accompany, critical control point systems. In these and other directives, operators are put under duties generally, or specifically, or as an explicit or implied condition of approval of the premises and activities therein. The ultimate responsibility for the safety aspects of food in his control always lies with an operator, not with the competent authority that monitors and permits his activities. The authority's responsibility lies in ensuring that public health is not put at risk, and not directly in the practical aspects of the control measures effected in individual premises to achieve this, although the distinction is subtle and there is a very large overlap of interest.

Influence of HACCP
While HACCP has voluntarily played a significant part in food safety control in larger businesses for some years, its formal use by smaller operators has been very limited. However, the informal and unknowing use of at least some of its principles has been present in most businesses, albeit not always fully effectively, because it is natural and in his own interest for an operator to consider where things may go wrong and to try to prevent this from happening. Some recent EU regulatory texts insist on more systematic implementation of risk assessment and control based on the principles of HACCP, but this is by no means universally applied and enforcement authorities have not yet succeeded in ensuring compliance with these requirements. Nevertheless, progress is gradually being made, and this is encouraged by the need to respond to public outcries following food safety incidents that might have been prevented had systematic control procedures been properly in place. For example, in the UK, the fatal outbreaks of *Escherichia coli* O157:H7 in Scotland have resulted in significant national expenditure on specific training and enforcement initiatives to ensure that butchers' shops apply full HACCP systems; these will be mandatory where raw meat and ready-to-eat foods are handled in the same premises, before such premises are licensed. Experience gained here can be offered to other states, and will be able to be applied in other sectors, such as the catering industry.

2.3 EU meat hygiene legislation and HACCP

It is unnecessary to consider non-HACCP controls in great detail, but because EU food hygiene directives (see Section 2.7) are generally not based on HACCP, it is important to note the breadth and detail of the prescriptive controls they do contain so that the possibility of replacing or supplementing these with HACCP requirements can be considered. A range of controls provide hygiene assurance about raw materials, their receipt, storage, processing, packaging and handling throughout the food chain. There are also measures applying checks and controls in primary production, on the production of the live animals entering the chain.

Horizontal and vertical controls
Horizontal food legislation applies general or specific controls across a range of foods; this is the case with the general food hygiene directive, whereas the other directives under consideration are vertical measures, applying controls to particular categories of foods.

2.3.1 General food hygiene directive (93/43/EEC)
To ensure food safety, this directive requires food business operators to comply with general hygiene rules, and there are also limited detailed requirements in an Annex. It applies to meat and foods made from meat in those cases not controlled by specific vertical legislation. In particular, the latter do not apply generally in retail or catering premises, or elsewhere that food is supplied to the consumer where it is prepared, or to foods that are not of animal origin.

The directive obliges food businesses to operate in a hygienic way. It requires all stages of production to be carried out hygienically, with hazard assessment and control procedures being implemented by food business operators to ensure that adequate food safety is obtained. This provision will vary from business to business, for example because of the nature of the foods handled, the hazards that are present because of the food type or as a result of structural and operating procedures, and the resources available.

The control procedures must be developed and applied in accordance with the principles used to develop the HACCP system, although that system is not currently required to be employed in full. These principles are specified, in Article 3(2), as:

- analysing the potential food hazards in a food business operation,
- identifying the points in those operations where food hazards may occur,
- deciding which of the points identified are critical to food safety – the 'critical points',
- identifying and implementing effective control and monitoring procedures at those critical points, and
- reviewing the analysis of food hazards, the critical control points and the control and monitoring procedures periodically and whenever the food business operations change.

This provides the fundamental practical requirement of this directive, for the adequate and systematic control of potential food hazards based on risk assessment and management. The food industry tends to prefer flexible regulatory provisions, rather than a rigid approach. Such principles must of course be able to ensure food safety and, for fishery products, they have been further elaborated (see Section 2.4).

Provisions permit individual Member States to 'maintain, amend or introduce national hygiene provisions that are more specific' than those laid down by this Community legislation. Any such rules must be at least as stringent as those contained in this directive, and must not restrict, hinder or bar trade in foodstuffs produced in accordance with this directive. Unsurprisingly, this can result in barriers being erected.

The directive also requires other specific provisions to be met, in an Annex. Structures, areas, tools and other equipment should be kept clean and disinfected as appropriate. Food must be cleaned where necessary, and be placed and protected to minimise the risk of contamination so far as is reasonably practicable. Broad protection is required 'against any contamination likely to render the food unfit for human consumption, injurious to health or contaminated in such a way that it would be unreasonable to expect it to be consumed in that state'.

The Annex provides more specific controls: for example, it prohibits a food business from accepting raw materials which are known to be, or might reasonably be expected to be, so contaminated that normal procedures hygienically applied would be inadequate to make them fit for human consumption. This realistically prevents the general introduction of raw materials that are not fit for human consumption into the establishment, but allows them to be accepted if they can readily be rendered fit for human consumption. Raw materials must be stored in appropriate conditions to prevent harmful deterioration and to protect them from contamination; perishable raw materials must be kept at temperatures that would not result in a risk to health. Similarly, pests, waste and refuse must be controlled.

Appropriate temperature controls must guard against microbiological hazards and the formation of toxins; these apply to ingredients and products, including intermediaries, and are to allow where necessary for limited uncontrolled periods, provided always that this is consistent with food safety. Temperatures of finished foods must be controlled to the extent necessary to prevent a food safety risk. Food must be cooled as quickly as possible to a temperature that avoids health risks after the final heating stage, or following the final stage of preparation if no heat process is applied, if is to be held or served at chill temperatures.

A general food safety hazard analysis must be considered also by the competent authority during its inspections, which must include a review of the business's critical control point procedures.

Thus, this directive requires hygienic handling of food throughout its storage, transportation, distribution, handling and offering for sale or supply, using hazard assessment and control techniques based on the specified principles and prescriptive controls. These requirements apply whenever food is in the possession of food businesses, unless these horizontal rules are supplanted by more specific ones contained in the vertical directives.

But the vertical directives are generally less flexible in their approach, especially those relating to the production of meat itself. There are no provisions relating to the principles of HACCP in those controlling fresh red meat, poultrymeat, wild game meat, and rabbit and farmed game meat (see Section 2.3.2), nor in the control of waste materials. However, such provisions are included in directives concerning minced meat and meat preparations (Section

2.3.3) and meat products (Section 2.3.4). They are notably present in controls on fishery products (Section 2.4).

2.3.2 Fresh meat hygiene directives (Section 2.7)

Red meat (64/433/EEC and 91/497/EEC)
There are no references to risk management based on HACCP or any similar system in these directives. However, they are not needed because the prescriptive controls on the raw materials for the production of fresh meat are exceptionally extensive and detailed. They apply initially to the production of carcasses and of part-carcasses in approved premises, from animals that have been inspected before and after slaughter. Controls are similar as carcasses are cut into smaller pieces, and then other legislation applies as they are comminuted or converted into meat preparations (Section 2.3.3) or processed into meat products (Section 2.3.4).

There are comprehensive requirements to ensure hygiene, including structural and storage provisions, as well as specific controls, for example on veterinary residues. These restrictions can be compared with the limited, although presumably adequate, rules applicable to most other foods.

The directive requires meat unfit for human consumption to be clearly distinguished from meat fit for human consumption and to be treated according to the requirements of the animal waste directive. This is important to ensure the hygiene of meat that is to be consumed as such or used as a raw material for meat preparations or meat products. Intense veterinary inspection procedures are detailed. Carcasses passed fit for human consumption under such veterinary control must be stamped in ink or branded with a health mark in a prescribed manner. Cut meat and offal must be treated similarly.

Fresh meat must be chilled immediately after post-mortem inspection and kept constantly at specified internal temperatures, subject to derogations for transportation to cutting plants or butchers' shops near the slaughterhouse. Fresh meat can be frozen only where it was slaughtered or cut, or in an approved cold store to which it has come directly from such premises. Freezing must be carried out without delay, to below $-12°C$, although such immediacy of freezing cannot be justified in the general case on hygiene grounds.

Fresh meat must be wrapped or packaged hygienically. In general, cut meat and offal must be wrapped, unless it is to be suspended throughout its transport, and wrapped meat must be packaged unless the wrapping itself provides the protection that would be afforded by packaging. Conditions are laid down to ensure hygienic storage and transportation, including conditions for the approval of cold stores and rules relating to documentation. Measures must be taken to avoid contamination or other adverse effects on the hygiene of meat during loading and transportation.

It is thought that many of these provisions are unnecessarily ponderous and rigid, being replaceable quite adequately by a risk assessment procedure based on HACCP as in the General Food Hygiene directive.

Poultrymeat (71/118/EEC and 92/116/EEC)
Much as for red meat (above), controls on poultrymeat omit risk assessment and management provisions based on HACCP. They are instead again detailed and intensive.

The relevant directive provides detailed prescriptive requirements for carcasses, part-carcasses, boned meat and offal, demanding that they come from approved slaughterhouses, from birds considered to be suitable for human consumption as the result of an ante-mortem inspection, usually on-farm. The directive requires veterinary supervision of farms delivering poultry to slaughterhouses.

This concentration on hygiene assessment at the farm of origin is not mirrored in the controls applicable to larger animals (red meat and farmed game), which are principally carried out at the slaughterhouse. However, this does not indicate any lack of concern about farm hygiene and disease control for meat derived from those larger animals. It is rather a matter of seeking practical and convenient means to achieve, monitor and ensure an acceptable standard of health.

The directive provides structural, inspection and other rules, much as in the case of the red meat controls. Again as for fresh meat, it requires birds to be slaughtered in accordance with prescriptive requirements in an approved abattoir, then cut up or boned in approved premises under temperature-controlled conditions. The carcasses, poultrymeat and offal must be handled, wrapped, packaged, stored and transported hygienically, largely similarly to red meat. Rules on packaging and wrapping include segregation of packaged fresh poultrymeat from unpackaged fresh meat during storage.

Again, it is argued that flexible and self-controlled risk management control systems could replace some of the detailed controls.

Rabbit and game meat (91/495/EEC and 92/45/EEC)
Control mechanisms applied using the principles of HACCP are also not incorporated in these directives.

Rabbit meat must be obtained in establishments that fulfil the general conditions of the poultrymeat directive with the source animals being similarly checked for their health status. The requirements for cutting, handling, storing, transporting and supplying rabbit meat are also related to poultrymeat provisions. Similar provisions are applicable to farmed game birds, whereas red meat directive controls form the basis for the control of farmed game meat obtained from cloven-hoofed wild land mammals.

Wild game must be killed for human consumption in a hunting area that is not subject to restrictions resulting from animal health considerations or from the presence of contaminants found in the environment. The controls are less stringent than for farmed animals, although adequate opportunities should be available for appropriate hygiene checks to be performed. The killed game must be prepared in accordance with this directive, and processed under specified conditions into meat in special approved premises, or as appropriate in red meat or poultrymeat approved premises. It must then be handled, stored and transported hygienically, much as under the appropriate (red meat or poultrymeat) directives.

Wild game meat is prohibited from use for human consumption where the animal was diseased or suffered changes during killing that make its meat dangerous to human health or it is otherwise unfit for the purpose.

2.3.3 Minced meat and meat preparations (94/65/EC)

This directive applies controls to the production and supply of minced meat and meat preparations, the latter being meat-based foods (such as burgers and breakfast sausages) that have not been treated in ways that make them 'meat products'. This relates to a defining requirement that any treatment applied must have been insufficient to modify the internal cellular structure of the meat and thus to cause the characteristics of the fresh meat to disappear. The rules are understandably more stringent where comminuted meat is present in the food.

At a very late stage during the controversial adoption of this legislation (Fogden, 1991; Fogden and Taylor, 1995), an initiative to include self-regulatory provisions permitting controls to be based on the application of the principles of HACCP succeeded. Article 7(1), in requiring operators to take all necessary measures to comply with the directive's provisions, requires them to comply with Article 3 of the general food hygiene directive. Paragraph 1 of that Article 3 demands that hygienic practices are used, while paragraph 2 is the one introducing the requirements based on the principles of HACCP. Unfortunately this welcome inclusion somewhat lost its value when detailed prescriptive requirements were retained. It can be argued that comminuted meat presents a significant potential risk and that food safety assurance demands, for the time being, the parallel operation of the two control systems while competent authorities ascertain how well the modern system is implemented in practice. However, that argument is weakened by the absence of a provision allowing for gradual relaxation of the prescriptive rules where the HACCP-based system has demonstrably been effective in ensuring food safety.

The prescriptive requirements are detailed. There are stringent rules where the food is to be eaten raw or lightly cooked, intended to provide appropriate safeguards for all consumers: whatever their eating preferences, their safety must be protected. Limited derogations are available for minced meat and meat preparations that are to be cooked thoroughly before consumption, which are then confined to the national market. These derogations permit the use of traditional sources of meat and less-onerous operating practices.

Minced meat and meat preparations must be prepared in specially approved establishments from a restricted range of meat sources, and be inspected before being appropriately marked and labelled, wrapped and packaged, stored and transported. They may also have to meet microbiological and compositional standards.

The directive requires the competent authority to be notified when a health risk is perceived; further, in cases where an immediate human health risk exists, product withdrawal is required. It also requires certain own-checks to be carried out.

There seems to be limited purpose, except to provide minimal extra assurance, in requiring hazard management techniques to be used and then laying down a number of particular requirements that would almost inevitably result anyway from the hazard evaluation. However, the HACCP-based requirement was a late introduction into the text, which was eventually adopted hurriedly, and possibly the overlap with prescriptive rules was not realised fully. Further, the microbiological sensitivity of the foods controlled by the directive makes it realistic to accept that this step towards risk management control systems was probably made cautiously, and its successful implementation may permit a less rigid approach in future.

2.3.4 Meat products (77/99/EEC and 92/5/EEC)

These directives require operators to apply risk assessment and control procedures based on critical control point methodology much as in the general food hygiene directive, albeit without mentioning HACCP, including sampling for laboratory testing and record-keeping. This requirement thus permits increased flexibility in achieving hygiene, using techniques appropriate to the individual circumstances of each establishment. It replaces many of the detailed and rigid provisions that would have been included had the rigid approach in the meat hygiene directives been employed. Applying to foods which are meat-based and treated so that the meat content no longer resembles raw meat, some prescriptive controls are still required during their production and supply, but less so than for raw meats and meat preparations. This is reasonable because the risk should be less, provided processing treatments have been adequately applied and later contamination is appropriately controlled.

The specific rules include structural provisions, temperature requirements for cleaning tools, and during cutting, slicing and curing operations. As usual, preventative measures against contamination by other materials, other foods or the working environment are provided. The packaging of meat products that cannot be stored at ambient temperatures must bear an indication of the appropriate storage and transportation temperature, as well as the appropriate durability indication, for inspection purposes.

2.4 Fishery products

It is interesting to consider the clarification on the implementation of HACCP-based systems uniquely provided in fishery products' hygiene control measures. Directive 91/493/EEC is one of a set covering the hygienic production of food derived from aquatic animals; these were developed during the same period as those for food derived from land animals and have a similar structure.

Article 6 requires those responsible for the operation of establishments producing fishery products to carry out their own checks based on the following principles, which are not specifically related in the text to HACCP: identification of critical points in the processes used, establishment and implementation of methods for monitoring and checking these, taking samples to check in approved laboratories on cleaning, disinfection and compliance with various specified standards, and keeping records of these activities. In the event that there is, or may be, a health risk, appropriate action must then be taken.

The legislators followed up this measure with detailed rules in decision 94/356/EC, as required in the parent directive. Similar detail on the application of risk assessment and management has not yet been adopted in other EU food hygiene controls, even where HACCP principles have been specifically mentioned. This initiative may have been a tentative step towards the adoption of such rules throughout the hygiene control structure, or simply an attempt to provide additional controls on particular operators.

The decision, here indicated in general terms, does the following.

- It defines the scope of the checks as all actions necessary to ensure and demonstrate compliance with the directive, based on annexed general principles and requiring appropriate training of staff. These are clearly the full principles of HACCP although there is again, surprisingly, no mention of HACCP as such.
- It defines critical points, noting that these are specific to each establishment, and requires them to be identified in accordance with the annexed scheme. That recommends assembly of a sufficiently broad and expert multi-disciplinary team, leading to detailed description of the characteristics of the product, identification of its intended use, construction of a flow diagram of the manufacturing process, on-site verification during the operation of the plant, listing of hazards and control measures, identification of critical points using the supplied decision tree, and the design and implementation of effective control measures at each critical point where the hazard cannot otherwise be eliminated.
- The decision then states that monitoring and checking identified critical points includes all observations and/or measurements necessary to ensure the points are kept under control, but does not include verifying that the products comply with the standards laid down in the directive. The annexed recommendations on how to do this cover the establishment of critical limits and of their systematic monitoring and checking, together with a corrective action plan covering both loss of control and a trend towards this.
- Next, sampling for laboratory analysis as referred to in the directive is restricted to that intended to confirm that the critical control point system is operating effectively, but must also allow for validation and verification of the own-checks system, which also relates to compliance with the legislated standards. The annexed clarification of verification requirements again

includes a series of possible actions which will achieve this and incorporates the need for review of the system. Laboratories are to be approved by competent authorities, taking the European Standard EN 45001 into account for external laboratories, though lesser standards are acceptable for internal ones.
- Keeping a written record is indicated as requiring documentation of all information relating to the implementation and verification of own-checks, in two formats for submission to the competent authority. First, this comprises a detailed and comprehensive document describing the risk assessment for each product and the risk management system implemented; second, it includes a record of the observations and/or measurements obtained during operations, results of verification checks and reports on corrective actions.

This regulatory clarification of the meaning of HACCP without mentioning it is undoubtedly helpful in indicating what is expected and in enabling effective enforcement. It could usefully provide the basis for future development of regulatory risk management.

2.5 Future trends

Regulatory acceptance of risk assessment and management systems as the fundamental mechanism is currently largely absent from EU food law applicable to foods of animal origin, albeit tentative steps towards requiring such systems are present. That mechanism is recognised, however, in international agreements. What is the future of this approach within the EU?

2.5.1 Review of directives

The EU has reviewed the hygiene directives recently. New measures are under consideration and it seems likely that the influence of HACCP will be greater in future, although there are equally clear indications that it will run, at least for some years, in parallel with detailed prescriptive requirements. The following is based on document VI/1881/98-rev.2; III/5227/98-rev.4, of June 1999.

Consistency
The proposed regulation controlling the hygiene of foodstuffs would apply to all foods, with general measures being supplemented by specific provisions relating to particular products of animal origin, categorised much as before. Considerably more consistency would be achieved by introducing horizontal requirements across all, or groups of, foods.

Improvement of scientific basis, necessity and proportionality
It is stated in the preambles that deregulation is not permissible and the breadth and stringency of the hygiene controls is not substantially reduced. The

The regulatory context in the EU 23

necessity for certain measures has been challenged by observers (for example, Fogden, 1994–96), and it may be inappropriate and contrary to the principle of proportionality for this review to be based upon the premise that unnecessary provisions cannot be eliminated because deregulation is not an acceptable direction in which to proceed. Why are such measures present?

EU hygiene legislation has in the past frequently been adopted by different groups of officials and only after considerable negotiation, sometimes in a flurry of last-minute changes to achieve an acceptable compromise text. These conditions do not encourage logical or consistent controls.

EU hygiene legislation must be politically acceptable. Proportionality and scientific propriety can take second place to this need, and now Member State governments might find it difficult to explain to their electorates were existing prescriptive hygiene control measures, albeit unnecessary or disproportionate, to be eliminated.

Such changes would also admit past EU over-regulation, which has caused additional costs to industry and consumers, a suggestion made frequently and with passion (and rejected equally often), leading potentially to further unrest.

Elimination of other measures
A proposed directive, to be adopted prior to or approximately contemporaneously with the hygiene regulations, would repeal the existing hygiene directives. It is not entirely clear that this would be effective when the regulations would enter into force, but hopefully common sense would prevail and substantial enforcement of existing measures which were to be eliminated would not occur during any intervening period.

2.5.2 Outcome of review

HACCP
Article 3 in the proposed regulation on the hygiene of foodstuffs would require the whole food chain, from primary production to supply to the consumer, to operate hygienically in accordance with the regulations. Article 5 would require systematic controls based on principles of HACCP to be put in place for all food businesses other than those operating at the level of primary production, which covers all stages of animal production up to slaughter. Those principles are stated as:

(a) identification of any food safety hazards that must be prevented, eliminated or reduced to acceptable levels in order to ensure the production of safe food;
(b) identification of the critical control points at the step(s) where control is essential to prevent or eliminate a food safety hazard or reduce it to enable the objective of safe food to be met;
(c) establishment of critical limits at critical control points which separate

acceptability from unacceptability for the prevention, elimination or reduction of identified hazards;
(d) establishment and implementation of effective monitoring procedures at critical control points; and
(e) establishment of corrective actions when monitoring indicates that a critical control point is not under control.

The proposals would require the introduction of verification (including review) procedures, and businesses would have to establish documentation and records commensurate with their nature and size.

Member States would be required (Article 7) to ensure the existence of national guidelines to the application of the principles of HACCP within five years of the regulation coming into force, which is arguably unacceptably long, bearing in mind the wealth of such guidance already in existence. Although the consultations envisaged could be lengthy, there is no reason why development of guidance could not usefully begin prior to adoption of these rules. The guides would have to take account of the Codex Alimentarius Recommended International Code of Practice on the general principles of food hygiene (Section 2.1.2).

It is envisaged (Article 8) that Community guidance on the application of the principles of HACCP could be developed, although this would not necessarily supplant the national guidance.

The general introduction of systematic risk assessment and management based on the principles of HACCP would be most welcome, in the opinion of the author, as a measure that would enhance food safety assurance. However, this would only be effective with considerable education of food business operators; a culture change will be necessary for many.

A recent English initiative, confined to butchers, will have required over a year of concentrated effort using a significant proportion of relevant national technical resources to tackle this relatively simple sector alone. It is therefore obvious that simply requiring HACCP in legislation cannot result immediately in compliance. However, unless a regulatory impact strikes food businesses, and thereafter education and enforcement proceed in parallel for some years, it is likely that the EU will largely remain in its present 'unHACCPed' condition for the foreseeable future. Relevant education should commence as soon as possible, which will require substantial investment by governments, other authorities, technical specialists and above all by food businesses.

None of this would of course prevent food safety being compromised by consumers acting unwittingly or irresponsibly once food has been supplied to them, and consumer education in food safety should also be a priority.

Proportionality through self-regulation or prescription
It is clear that the vast majority of the controls have a basis in animal or human health control. Nevertheless some of them appear to have no clear relationship to hygiene.

It is imperative that health be appropriately protected, and probably wise to be cautious while ensuring that this does not result in unnecessary burdens on industry, leading to additional costs for consumers. General EU acceptance of risk assessment and control procedures for foods still sits rather uneasily with rigid and complex controls. The legislators could introduce further flexibility based on the implementation of HACCP and thus eliminate more prescriptive controls in establishments which had proved the consistency and effectiveness of their control systems. This might result in further lack of national congruence, but unwarranted stringency is expensive.

Confidence in industry management
It has become more acceptable to rely upon the operators of businesses, approved and monitored appropriately by the competent authority, to provide adequate hygiene controls within a framework of varying complexity, often based on critical control points. Inevitably at this stage in the general introduction of this type of control system, the EU feels obliged to parallel sophisticated elements with prescriptive obligations – but such precaution can properly be eliminated as businesses prove they can act responsibly.

Ease of enforcement
Article 10 of the proposed regulation would require competent authority staff to be adequately trained in food hygiene and safety, including the principles of HACCP. Nevertheless, it is more difficult in practice to enforce controls which may include a subjective element, although this suggests that in such situations a combination of education and compromise may be more effective than rigid enforcement. In the last resort, the courts will have to decide, but hopefully the need for this will be restricted to cases where serious food safety risks exist.

2.6 Sources of further information and advice

BROWN F L, *Hazard Analysis Critical Control Point Evaluations. A guide to identifying hazards and assessing risks associated with food preparation and storage*, World Health Organisation, Geneva, 1992.

CAMPDEN AND CHORLEYWOOD FOOD RESEARCH ASSOCIATION, *HACCP: a Practical Guide*, CCFRA, Chipping Campden, UK, 1997.

CODEX ALIMENTARIUS COMMISSION, *General Principles of Food Hygiene*, Rome, 1997.

CODEX ALIMENTARIUS COMMISSION, *Hazard Analysis and Critical Control Point (HACCP) System and Guidelines for its Application*, Rome, 1997.

CODEX ALIMENTARIUS COMMISSION, *Principles for the Establishment and Application of Microbiological Criteria for Foods*, Rome, 1997.

WORLD TRADE ORGANISATION/GENERAL AGREEMENT ON TARIFFS AND TRADE, *Agreement on the Application of Sanitary and Phytosanitary Measures*, Geneva, 1994.

2.7 References

Council directives of the European Community (note: many have been amended or updated substantially – other relevant instruments are listed in the Community list of Legislation in Force):

Number	OJ	Date	Page
General food hygiene			
93/43/EEC	L 175	19 July 1993	1
Fresh (red) meat			
64/433/EEC	121	29 July 1964	2012
91/497/EEC	L 268	24 September 1991	69
95/23/EC	L 243	11 October 1995	7
Fresh poultrymeat			
71/118/EEC	L 55	8 March 1971	23
92/116/EEC	L 62	15 March 1993	1
Rabbit meat and farmed game meat			
91/495/EEC	L 268	24 September 1991	41
Wild game meat			
92/45/EEC	L 268	14 September 1992	35
Minced meat and meat preparations			
94/65/EC	L 368	31 December 1994	10
Meat products			
77/99/EEC	L 26	31 January 1977	85
92/5/EEC	L 57	2 March 1992	1
Fishery products			
91/493/EEC	L 268	24 September 1991	15
92/48/EEC	L 187	7 July 1992	41
(Decision) 94/356/EC	L 156	23 June 1994	50
Animal waste (by-products)			
90/667/EEC	L 363	27 December 1990	51

FOGDEN M, 'European Community Minced Meat Legislation', *European Food Law Review*, (2) 150, Frankfurt am Main, Germany, 1991.

FOGDEN M, 'European Community Food Hygiene Legislation', *European Food Law Review*, Frankfurt am Main, Germany, 1994–96.

FOGDEN M and TAYLOR B, 'Minced meat and meat preparations: new EU hygiene legislation', *European Food Law Review*, (6), 177, Frankfurt am Main, Germany, 1995.

3

HACCP in the United States: regulation and implementation

L. Crawford, Center for Food and Nutrition Policy, Georgetown University

3.1 Introduction: the regulatory background

All three branches of the US government (legislative, executive and judicial) have a role in the development of legislation governing the food industry. Congress (the legislative branch) passes laws that establish general requirements and provide authority to regulating agencies to implement and enforce them. Once the President (the executive branch) signs the legislation, it becomes an official statute and is published in the United States Code (USC). The principal legislation governing safety in the meat industry is the Federal Meat Inspection Act and the Poultry Products Inspection Act which cover all products derived from domesticated animals, and the Processed Products Inspection Act. The first two Acts are administered by the Food Safety and Inspection Service (FSIS) of the United States Department of Agriculture (USDA). By agreement the Food and Drug Administration (FDA) has responsibility for foods containing less than 3% meat and 2% poultry and all closed meat-containing sandwiches. The language of these acts does not delineate the actual method of inspection, but requires that the foods covered be safe and unadulterated within the meaning of the legislation.

Because laws are broad and non-specific, the President is given the responsibility of implementing them through the various regulatory agencies by establishing regulations which provide detailed requirements and procedures. All such regulations must go through a public rule-making process which is mandated by the Administrative Procedure Act of 1946. This procedure is initiated by publishing an advance notice of proposed rule making (ANPR) in the Federal Register (FR), designed to alert interested parties that a new regulation is being considered and to solicit their views. The

FR is published by the government every working day of the year and allows all interested parties to comment on the provisions of the proposed regulation, and gives the appropriate agency the chance to respond. The second step is the publication of the proposed rule where, once again, interested parties may comment. Under US law all comments, however trivial, must be addressed and answered in written form in the FR as part of the Final Rule. If, as a result of comments or changing circumstances, the Proposed Rule must be substantially modified, a second Proposed Rule must be published. Once finalised, the regulation has the force and effect of law, unless reinterpreted by the courts (the judicial branch). Final regulations are published in the FR and, once a year, compiled into the Code of Federal Regulations (CFR). Final regulations in the FR include a preamble which discusses why the regulation is being proposed and the science base underpinning it. It also contains responses to comments received, a cost/benefit analysis (especially for the impact on small businesses), any potential environmental impacts, and an analysis of the paperwork required of those organisations affected by the regulation in question.

3.2 Development of HACCP in the United States

In the early 1970s it became generally accepted that the Hazard Analysis Critical Control Point (HACCP) system constituted an advanced and comprehensive system for producing safe food. A number of initiatives at this time anticipated or incorporated HACCP principles, notably the 1974 regulations governing low acid canned foods which applied to canned meats as well as other canned products. These prescribed a system designed to eliminate the threat of botulism and other microbiological hazards in the production of canned foods. Later partial and total quality control programs were developed as regulatory options by the USDA for use within food processing. The USDA also developed so-called streamlined inspection systems for meat plants as an alternative to traditional inspection regimes. These initiatives incorporated elements of HACCP philosophy.

1985 marked a turning point for HACCP. Two seminal reports by the National Academy of Sciences paved the way. The first of these was *Meat and Poultry Inspection, The Nation's Program*. This report firmly endorsed implementing HACCP systems as the key to safer meat and poultry products. The second report had a broader focus: *An Evaluation of the Role of Microbiological Criteria for Foods and Food Ingredients*. However, it also championed HACCP systems, especially for high-risk foods. As a result, the National Marine Fisheries Service (NMFS) began a pilot program designed to incorporate HACCP principles into the harvesting, production and processing of fish and fish products, and to explore how HACCP could be integrated into the regulatory and inspection process. The FSIS also produced a comprehensive response to the two National Academy of Science reports designed to adapt the agency to a new role of incorporating HACCP principles into meat and poultry processing and the way the industry was regulated.

Progress in incorporating HACCP principles into the regulatory framework, however, proved slow. Despite support from the two main trade organisations, the American Meat Institute and the National Fisheries Institute, consumer organisations were initially slow to champion HACCP as a concept, and the FSIS faced opposition from its inspectors' union. However, in January 1993 the new Clinton administration faced a major outbreak of food poisoning. A large number of cases of enterohemorrhagic *Escherichia coli* occurred in the Pacific Northwest, causing the deaths of some children who had consumed undercooked hamburgers. Renewed effort was put into developing a new HACCP-based regulatory system. Such a system was finally mandated for seafood plants in 1994 and for meat and poultry plants in 1996.

The new regulatory regime for the meat and poultry industry introduced in July 1996 was implemented through the Pathogen Reduction Hazard Analysis Critical Control Points (HACCP) System Final Rule, popularly known as 'Mega-Reg' because of its scale in seeking to replace all existing regulations governing the inspection of meat and poultry products. These regulations, applying to all food processors inspected by the FSIS and similar state agencies, require meat and poultry product processors to take preventative and corrective measures at each stage of the food production process where food safety hazards occur, using a variant of the HACCP system as defined by Codex. Each plant has the responsibility and flexibility to base its food safety controls on an approved HACCP plan. This plan must identify the critical control points (CCPs) detailed in the regulations and use the controls set out in the regulations in managing them. Sanitation Standard Operating Procedures (SSOPs) are also required. These must describe daily procedures sufficient to prevent direct contamination or adulteration of products. Additional requirements include mandatory *E.coli* O157 testing by slaughter operations, and compliance with performance standards for *Salmonella*.

Regular auditing of HACCP plans by independent experts is a common practice. However, ultimate responsibility for the acceptability of the HACCP plan rests with the FSIS. When recalls of product or sampling problems occur, the FSIS will usually require a re-evaluation of the HACCP plan. Facilities failing to implement 'proper HACCP programs' will face enforcement action that could mean withdrawal of the USDA's inspectors and plant shutdown. In these cases responsible management may be permanently barred from operating a food plant in the United States. Civil and criminal penalties, including fines and imprisonment, might also follow. The more severe penalties are reserved for fraudulent activity such as destroying or falsifying documentation, serious cases of negligence or the wilful contamination of the food supply.

The endorsement of the HACCP system by the United States had significant international implications. Meat and poultry inspection laws in the United States require that countries wishing to export meat and poultry products into the United States maintain an inspection system that is equivalent to that required by the FSIS for domestic production. This requirement meant effectively that the 40-odd countries approved to export meat and poultry products to the United States would have to produce and inspect products in accordance with HACCP principles. These countries discuss common food safety issues through the Codex Alimentarius, its committees, its staff and various meetings. Codex is a joint program of two United Nations agencies, the Food and Agriculture Organisation and the World Health Organisaton, designed to set common standards that facilitate international trade in food. The early adoption of HACCP principles by the European Union as well as the United States has meant that they have also been adopted by Codex Alimentarius as the starting point for food safety systems around the world.

3.3 HACCP implementation in practice

The nature of HACCP implementation in meat and poultry plants has been more traditional in the United States than in some other countries. As an example, 'Mega-Reg' requires continuous inspection of slaughter line operations and can thus be seen as layering HACCP onto existing inspection regimes rather than replacing the latter with the former. The key legislation lying behind the regulatory process predates HACCP as a concept and is based on a command and control approach requiring the constant presence of food inspectors. Short of this legislation being revised or replaced, there can be no full transfer of food safety from government inspectors to plant managers. Similarly, 'Mega-Reg' requires plant management to carry out microbiological sampling. This can be seen as antithetical to the concept of HACCP. Properly administered, HACCP obviates the need for routine microbiological sampling, replacing a reactive with a more proactive approach.

HACCP implementation under 'Mega-Reg' began initially in large meat and poultry operations, which had 18 months to comply, completing in early 1998. Small plants had 30 months to comply, completing in early 2000, and very small plants had 42 months. Preliminary results have been analysed by the Centers for Disease Control (CDC) and the USDA. Significant reductions in the levels of *Salmonella*, *Listeria* and *Campylobacter* contaminating raw meat have been documented. As an example, contamination rates for ground turkey fell by 45% from 1997–98, those for chicken by 45% and those for ground beef by 36%. Contamination rates for *Escherichia coli* O157:H7 have not been materially affected, but levels have not increased. However, overall contamination rates have remained high in some areas. In the case of *Salmonella*, 36% of ground turkey sampled was found to be contaminated, 11% of chicken and 4.8% of ground beef. In late 1998 there was a spate of product recalls caused by *Listeria monocytogenes* contamination and as many as 20 deaths caused by foodborne pathogens. The two largest recalls, at Bil Mar Foods in Michigan and at Thorn Apple Valley Foods in Arkansas, were reputed to involve 15 to 30 million pounds of product, making them some of the largest food product recalls in

HACCP in the United States: regulation and implementation   31

American history. Most experts have attributed these two recalls to, in one case, contamination as a result of poor GMP in the handling and storage of rework material, and, in the other, a failure to maintain the Standard Sanitary Operating Procedures (SSOPs) set out under 'Mega-Reg'. In May 1999 the FSIS responded by announcing a requirement for reassessed HACCP plans for ready-to-eat livestock and poultry products to be submitted, including *Listeria monocytogenes* as a specific hazard. These developments show that, while HACCP provides a systematic approach to food safety control, it relies on an effective understanding of key hazards and a systematic approach to implementation, including implementation of the relevant prerequisite programs (such as Good Manufacturing Practice (GMP) and Good Hygiene Practice (GHP)), to succeed.

There have been a number of other initiatives designed to remedy such problems as these and to complement and support HACCP systems, for example in developing more expertise in understanding foodborne pathogens. In January 1997 President Clinton announced a Food Safety Initiative (FSI) designed to improve the system for detecting outbreaks of food illness, promote research on emerging pathogens such as *E.coli* O157:H7 and *Cyclospora*, and educate consumers and the industry on safe food handling practices. Part of the FSI introduced in the autumn of 1997 is the Product Safety Initiative (PSI) designed to address safety along the entire food chain from farm to table, including the adoption of HACCP principles in agricultural production and in catering.

3.4 Beyond HACCP

Given continuing problems with outbreaks of foodborne disease, the food industry is continuing to look for new ways of managing risks. Two concepts under current discussion are kill steps and due diligence.

3.4.1 Kill steps
Kill steps are procedures that destroy residual bacteria in foods at the end of processing. It has been suggested that these can be used in conjunction with HACCP systems implemented within manufacturing operations. High temperature is the most frequently employed lethal agent, resulting in a straight-line inactivation curve. The level of inactivation is expressed in D values, which means decimal reductions at a given temperature. Two examples of kill steps are cooking of a product by the consumer and pasteurisation. Meat products are frequently subjected to post-processing pasteurisation, particularly ready-to-eat products that do not require further cooking prior to consumption. Post-processing pasteurisation is an established kill step for frankfurters (hot dogs), for example.

The effectiveness of kill steps depends on a number of factors, including the level of bacterial contamination of a product. Pasteurisation, for example, requires constant monitoring of bacterial loads in assessing product suitability and type of treatment. In general, heat activation has the disadvantage that it cooks or further cooks a product, altering its sensory and nutritional quality. Perhaps the ideal kill step is ionising radiation which is at least as effective as high temperature but does not affect product quality. Other methods include electron beam acceleration, which concentrates a stream of electrically generated electrons on to the surface of foodstuffs.

The application of some of the newer non-thermal kill steps is currently limited by the need for more research and effective commercial application. Irradiation has, on the other hand, been extensively researched. At present 41 countries, including the United States, allow the irradiation of about 100 different classes of food, on either an unconditional or a restricted basis. In 1997 a joint FAO/IAEA/WHO Study Group examined current toxicological, nutritional, microbiological and radiation chemical data, and concluded that there was no need for an upper dose limit to be imposed for food irradiation. The Study Group recommended that technological guidelines incorporating these findings be prepared and incorporated into Codex Alimentarius standards. The main obstacle has been consumer distrust of the technology. However, there are signs that attitudes are changing in the United States. In 1999 a joint survey by the Grocery Manufacturers of America and the US Food Marketing Institute showed that 80% of consumers would be likely to purchase an irradiated food product for themselves or their children if it carried the label 'irradiated to kill harmful bacteria'. Further outbreaks of foodborne disease may accelerate the implementation of kill steps such as irradiation as a complement to HACCP systems.

3.4.2 Due diligence

Due diligence is an ancient legal concept. It was developed as a way of establishing if an individual or organisation was guilty of negligence, by establishing a minimum standard of care against which a charge of negligence could be assessed. In the context of food production it addresses the question of whether the producer has done all that might reasonably be expected in the production of safe food. It assumes that, even if a product does cause illness, the producer is not at fault if he has exercised reasonable care in the way a product has been manufactured.

Due diligence can be seen as a radical concept in the area of food safety in that it implies that there can never be absolutely safe food, even with the implementation of HACCP systems. It focuses attention on producers accepting their special responsibilities in preparing food for others, and in meeting a commonly accepted industry standard for safe food production. The onus is then for stakeholders such as government and the food industry to establish common standards, such as GMP, quality or HACCP systems, and the framework for their implementation by individual producers. It also creates a responsibility for the appropriate agencies to monitor and improve those standards, the microbiological knowledge, technology and management structure which underpin them, and for individual food producers to keep abreast of those changes, in the constant battle with foodborne disease.

3.5 Bibliography

AHMED F E, *Seafood Safety*, Washington DC, National Academy Press, 1991.

ANON, *Hazard Analysis Critical Control Point System*, Geneva, World Health Organisation, 1995.

BLACK H C, *Black's Law Dictionary*, St Paul, MN, West Publishing Co., 1983.

CRAWFORD L M, 'The optimum microbiological food safety program', *Infectious Agents and Disease*, 1994 **3** 324–7.

CRAWFORD L M and RUFF E H, 'A review of the safety of cold pasteurization through irradiation', *Food Control*, 1996 **7**(2) 87–97.

NATIONAL ACADEMY OF SCIENCES, *An Evaluation of the Role of Microbiological Criteria for Foods and Food Ingredients*, Washington DC, National Academy Press, 1985.

NATIONAL ACADEMY OF SCIENCES, *Meat and Poultry Inspection*, Washington DC, National Academy Press, 1985.

PIERSON M D and CORLETT D A, *HACCP*, New York, Van Nostrand Reinhold, 1992.

SHAPTON D A and SHAPTON N F, *Principles and Practices for the Safe Processing of Foods*, Oxford, Butterworth Heinemann, 1991.

Part 2

HACCP on the farm and in primary processing

4

HACCP and farm production

A. M. Johnston, Royal Veterinary College, University of London

4.1 Introduction

Safe food produced on a farm, whether from animal or vegetable origin, must be free from pathogens which infect man and from contamination with poisons and residues. There is a general consensus that microbial agents constitute the major hazard to human health, but there must always be an awareness of possible hazard from residues or toxins. The production of meat, milk and eggs, regardless of new technology or changes in production methods, cannot be expected to achieve zero bacterial or chemical risk, but it may be easier to avoid residues. There is, however, the need to reduce the risk and, where possible, eliminate it at the farm level. The current use of the terms 'farm to table', 'stable to table' and 'plough to plate' clearly identifies the farm as one part of the production chain which must be considered in terms of food safety. Farming practices, in particular the apparent reliance in recent years on intensive farming systems, have been linked with the rise in foodborne illness in humans. The assumption of the on-farm risks that have to be considered, however, must be limited to those that might have an impact on human health. The difficulty frequently is to separate out the risks that influence only animal health on the farm from those which impact on human health, or both human and animal health, or may be perceived as being a risk by the consumer.

The majority of HACCP implementation to date has been within the production and manufacturing sections of the food industry. HACCP offers a risk assessment and management system that can be implemented prospectively, unlike other programmes such as animal herd health schemes on the farm that usually work retrospectively.

Beside the hazard identification and location of the exposure, there is the need to investigate along the food chain the critical steps where contamination occurs. In the end the decision must be taken where to situate inspection and what is the best way to ensure human health. The 'stable-to-table' concept relies on the evidence that the final product, which is consumed, results from subsequent steps of a longitudinal process. In order to further improve animal health and food safety for the consumer, HACCP is now being considered for use in the farmyard situation, because, after all, the farm animal either produces, or is itself, the product. With the need for food manufacturers to show due diligence throughout the food chain, as a defence the HACCP system has become the recognised standard and is increasingly being extended to encompass the entire farm-to-table continuum. In the United States, for example, the 1996 USDA food safety HACCP regulations which deal with slaughterhouses are seen to have an inevitable impact on farm production practices.[1]

Correct implementation of the HACCP system requires that scientifically documented steps and preventive measures exist that can be effectively applied at known critical control points (CCPs). Determination of critical control points for on-farm implementation for chemical, physical and certain biological hazards is currently possible but is considered lacking for microbiological hazards.[2] Much of the actions on farm are good manufacturing practice (GMP) or, in this case, good farming practice, and will never be CCPs.

4.2 Hazard analysis in animal rearing

Microorganisms are widely present in animals and in their environment. With animals disease is inevitable; perfectly healthy animals can also be carriers, and may be asymptotic excretors, of pathogens. The prevalence of pathogens on the farm, or a unit within a farm, depends on many factors, not least the type of husbandry, the environmental pressure on that farm and the standard of stockmanship. The human pathogen *Escherichia coli* O157 which is found on farms[3] and associated with ruminants demonstrates the problems of an organism that has a highly variable prevalence but is able to maintain itself in the herd, yet has a transient nature of shedding which appears to be influenced by feeding, transport and weather.

The diseases of animals which affect the safety of food are predominantly those that cause enteric disorder in the animals. In addition, the very environment in which animals are reared will always have a bacterial load with some level of pathogens. There will be organisms which are pathogenic to man but do not cause clinical illness in the animals, though they are present in the animal excreta and in the animal environment, such as *E. coli* O157. On the other hand, zoonoses such as *Chlamydia psittaci* or *Toxoplasma gondii* can cause significant losses on the farm but are most unlikely to affect food safety directly. Other pathogens may be excreted in large numbers before there is evidence of clinical illness in the animal or following apparent recovery from an illness.

In livestock production there are a number of points where controls can be applied. The first is at the birth of the animal, or at hatching in the case of poultry, and extends through all stages of animal production and includes the foodstuffs fed to the animals. The aim should be to have the young born fit and healthy with good levels of maternal immunity. In addition to their appropriate use in the neonate, vaccines can be given to the pregnant dam, such as the bovine combined rotavirus and K99 *E. coli* vaccine for calf scours, to help to protect the young in the first weeks of life. Animals and birds are usually kept in groups, either outside in fields or housed for all or part of the year. Access to the housed accommodation or to the pasture may be voluntary or controlled according to the farming system in place. Whichever system is used, the animals must be kept in the very best conditions with an overall aim to prevent disease in individual animals or in the whole herd or flock. The type of husbandry directly impacts on this. The most certain way to reduce or remove the risk of introducing disease organisms to animals is to use biosecure housing. This, of course, is contrary to the trend towards more extensive systems where there is the inevitable exposure to wildlife and vermin which are vectors of a number of important pathogens. The use of production systems which have biosecure housing does allow an 'all in all out' policy, followed by thorough cleaning and disinfection of the house before restocking. The original method was to apply this practice to each house on the site as it was emptied of animals or birds. More recently this practice has been extended to involve all animal accommodation on the site, every unit being emptied of livestock, then all cleaned and disinfected before any unit on that site is restocked.

In addition to keeping animals healthy, a critical part of husbandry is also to make sure they are kept visibly clean. This is of particular importance to reduce the possibility of contamination of milking animals and animals destined for slaughter so that they do not have dirty outer coats. A major influence on the cleanliness of the animals is the type of housing, the material used as bedding and, if the animals are kept outside, the underfoot conditions. There is a variety of housing systems used in practice, including straw bedded or deep litter yards, cow cubicles with straw, sand, rubber mats or even waterbeds as bedding, and sheds with slatted floors, or a combination of these. Straw bedding is a much-favoured system for comfort and cleanliness but is only satisfactory if the existing bedding is regularly replaced with clean straw. Failure to use good quality straw or empty out the yards regularly, as dung builds up, will lead to a problem with environmental organisms. This is of major concern for dairy cows housed in such a system, where failure to completely change the bed at regular intervals results in clinical mastitis caused by the environmental organisms. In some regions straw may not be available locally, which requires it to be transported from arable areas. A major factor in the effectiveness of any system in keeping the animals clean is the standard of management. Failure to attend to detail will lead to an increase in environmental organisms and inevitably also pathogens. The stockman therefore has a crucial role to play from both the animal health and public health perspectives.

Foodstuffs which are fed to animals must be free from both pathogens and undesirable residues. The role of animal feed in food safety has been highlighted in relation to both *Salmonella*, in particular *S. enteritidis* phage type 4 in poultry,[4,5] and bovine spongiform encephalopathy (BSE) in cattle[6] and more recently dioxins in animal feeds in Belgium.[7,8] Following the BSE epidemic, the long-established practice of using recycled animal protein has been questioned, with a ban on the use of ruminant or mammalian derived protein in animal feeds in some countries. Animal feeds are compounded from both home-grown and imported ingredients most frequently produced as a compounded, nutritionally balanced ration from commercial feed mills. The farmer may well prepare the feeds on the farm using either home-grown or purchased forage and cereals. It has been well documented that the ingredients for animal feeds may carry pathogens. The process of producing some forms of compounded feed, such as pelleted feed, requires a heat treatment stage which is effective against bacterial pathogens, but subsequent handling stages may allow recontamination. The farmer has a role to play in making sure the feed is stored in a manner that prevents contamination from external influences such as wildlife on the farm.

The bringing on to the farm of new animals, whether as replacement breeding stock or as animals to be fattened for slaughter, is frequently a way by which diseases are introduced. In most cases the major impact will be from diseases which affect animals but frequently such infections can include zoonotic organisms. It is of the utmost importance that incoming animals are kept separate from those already on the farm for the necessary period of quarantine and where possible that they come from a farm with a known health history.

With animals, whether farm animals or companion animals, disease is inevitable; perfectly healthy animals can also be carriers and may be asymptomatic excretors of pathogens. The prevalence of pathogens on a farm depends on many factors, not least the type of husbandry, the environmental pressure on that farm and the standard of stockmanship. It is also most important to recognise the difference between animal health, or disease control measures, and human health considerations when considering the legislation. There are, however, no specific statutory food safety controls applicable to on-farm production.

One of the easiest and perhaps more clearly defined parts of the farming operation to which the HACCP concept can be applied is the use of medications. This must include the decision-making process on whether to use and if so which medication as well as the mechanics of delivering the medications to the animals. While the treatment of bacterial disease in man and companion animals is invariably directed to the individual patient, the treatment of food-producing animals, especially pigs and poultry, is generally applied on the group or herd basis.[9] The three main reasons for antibiotic use in animals are therapy, prophylaxis, or strategic medication, and in farm animals performance enhancement. Therapy usually involves individual animals or a defined group of diseased animals for treatment of a previously identified disease. Prophylaxis or strategic medication is usually to contain the spread of infection and prevent illness in advance of clinical signs. Prophylactic treatment involves the medication of a herd or group of animals following the diagnosis of illness in one or more animals in the group, or on the basis of previous experience, usually when a proportion of animals are diseased during a defined period and the probability of most, or all, animals getting infected is high. The animal diseases requiring the most extensive use of antimicrobials for therapy or prophylaxis are respiratory and enteric diseases, especially of pigs and cattle, and mastitis in dairy cattle.

Large pig herds and poultry flocks, for example, can provide major logistical problems of antibiotic medication. Therapeutic or prophylactic antibiotics can be administered by in-feed or in-water medication. In-feed medication for a valid animal health reason must not be confused, as often happens, with the general term of feed additives.

The legal requirements covering the distribution of animal medicines differ according to the legal classification of the individual product, which also determines who may sell the product and under what restraint or control. Under *The Medicines (Restrictions on the Administration of Veterinary Medicinal Products) Regulations 1994*, no person is allowed to administer any veterinary medicinal product to an animal unless the product has been granted a marketing authorisation (product licence) for the treatment of a particular condition in the species being treated. Under the Regulations the veterinary surgeon is the primary prescriber of medicines, and in the UK it is usual practice for the veterinarians to prescribe and dispense medicines for animals. This applies to both food-producing and companion animals. For food-producing animals the veterinarian or person acting under his or her direction may only administer a product that contains substances found in products licensed for use in food-producing animals and must keep records.

There has been pressure on the industry to use production methods that will deliver the animal to slaughter at a predetermined weight with the required carcass conformation in the shortest time and at lowest possible cost. This has led to the use of growth promotion techniques, including sub-therapeutic levels of antibiotics in the feed and steroid hormones during the growing phase. The use of substances having a thyrostatic, oestrogenic or gestagenic effect for growth promotion purposes has been prohibited within the European Union (EU), or for products to be imported into the EU, since January 1989. The counter-argument to justify the use of steroid hormones is that they are naturally occurring substances and if the withdrawal periods are followed there is no risk to human health. This issue was reviewed by the Scientific Committee on Veterinary Measures Relating to Public Health in 1999. They considered that the scientific evidence necessary to make a balanced scientific judgement is lacking but it is known that one, 17beta oestradiol, is a complete carcinogen and as such is able to initiate and promote cancer. The committee considered that there was sufficient uncertainty in terms of consumer public health that the ban on their use in the EU should continue.[10]

The use of antibiotics, without veterinary prescription, for the purposes of increasing growth in food animal production started in the early 1950s. Following an outbreak of food poisoning due to multi-drug resistant salmonella, an expert committee chaired by Professor Swann reviewed the use of antibiotics in agriculture. Their report in 1969[11] resulted in significant changes in the use of antibiotics, including their use for growth promotion purposes. More recently there has again been considerable concern about the use of antibiotics, especially for growth promotion purposes, in animals and specifically about food being a vector of antibiotic resistance from animals to humans. This has led to a number reports from groups of experts, nationally and internationally, considering the use of antibiotics in animals, in man and for plant protection purposes.[12,13] There is agreement that there should be prudent use of antibiotics in veterinary and human medicine with little justification for the uncontrolled use of antibiotics at sub-therapeutic levels to promote growth. The major concern is if there is evidence of medical equivalence for the antibiotic, either where the same drug is used in man and in animals or if there is known antibiotic resistance. This is particularly relevant if there is a possible impact on the effectiveness of important antibiotics used in human medicine, especially when the antibiotic is one of last choice for life-threatening infections. Debate on the growth promotion debate will undoubtedly continue, but already there is evidence of sectors of the industry stopping the use of antibiotic growth promoters as part of their production system. It is easy to say that there should be no use of these products just to sustain cheap food production systems and make animals grow faster. However, use of some of the very same 'antibiotic growth promoters' appear to reduce disease in the animals, and stopping their use would require a greater use of therapeutic antibiotics. There is a balance, which can be achieved between the two schools of thought, which requires the husbandry systems to be changed to reduce the need for use of antibiotics in any form. The issue of consumption of residues in food of animal origin is perhaps of less concern, as there is mandatory testing for residues and a requirement only to use drugs which are licensed for use in food-producing species within EU Member States.

One option would be to eradicate specific agents which cause disease if they are identified on the farm. This, however, depends first on the agent being identified in the herd or flock or in individual animals harbouring the agent. In addition to there being an accurate 'test' available, there is the need to decide whether eradication is really necessary for both animal health and human health reasons. The aim must be to prevent entry of the agent into individual animals, not just into the herd or flock.

The biological way forward for disease control using vaccines promises to be an important alternative to the need for use of antibiotics. While it has always been important to use available vaccines in the appropriate manner, with the increasing efficacy, and at the same time specificity, of modern vaccines precise diagnosis becomes a must. There is therefore a future for the veterinary clinician on the farm in improving the health status of the food-producing animals, following proper assessment of all relevant factors, including the provision of a farm veterinary health plan. The success of any scheme for any farm or unit requires, as a minimum:

- surveillance of possible diseases or risks;
- establishment of a management structure to reduce the need to react but with action plans in place so that it is possible to react promptly and appropriately if necessary;
- active supervision at all levels;
- investigation of all possible, or actual, problems or variations from the normal, which requires accurate records and monitoring.

4.3 Setting up the HACCP system

Before a HACCP programme can be implemented in any system it is essential that all personnel be committed to the same goals. Farm resources must be sufficient to achieve the correct monitoring steps. The hazards will depend on the individual farm production system which will vary between farms and within one farm, in both the species kept and the production system used. It is therefore impossible to design one HACCP plan that can be applied to all farms.

The HACCP system derived from Codex Alimentarius 1991 consists of seven principles. The sequence of applying HACCP as described by Noordhuizen and Welpelo[14] comprises 12 steps. Use of these 12 steps in relation to farming can be seen in Table 4.1.

With the on-farm situation there are some obvious differences compared with the traditional food industries that have used HACCP. The people involved with Step 1 are the farm staff, usually consisting of a farmer or farm manager and in most cases only one to three members of staff, if any, who are often members of the farmer's family. In addition there are external advisers who need to be consulted, such as the farm's veterinarian and the animal feed specialist. In Step 2 the product is the slaughter animal (more specifically the meat that will be derived from that animal), milk or eggs. The intended use (Step 3) of the HACCP process is to ensure good health for the herd or flock and refers to disease agents or other hazards that the individual animal should be free of to ensure that carcase meat and offal, eggs or milk do not pose a threat to consumer health. The consumers of meat and offal can include, in addition to the healthy, people who are at greater risk, e.g. the immuno-compromised, children, the elderly, pregnant women, and people with allergies to pharmaceutical compounds such as penicillin. The construction of a flow diagram in Step 4 is important as it helps to identify all the aspects of the farm production process that influence product quality as well as animal and human health. In Step 5, while farmers are often unfamiliar with many concepts of food safety and hygiene, it is critical that they are consulted to make adjustments to the flow diagram as they have a fundamental understanding of their farm and how it operates.

Table 4.1 Steps in applying HACCP

Steps		Examples and specification
Step 1	Identification of persons involved	Farmer and employees. External experts.
Step 2	Description of products	Animals, meat, eggs, milk, wool.
Step 3	Identification of intended use	Disease agents the herd should be free of.
Step 4	Construction of flow diagram	Description of animal production process as communication tool.
Step 5	On site verification of flow diagram	Allows specific adjustments and first review of potential hazards.
Step 6	Listing of hazards at each process element [Principle 1]	Check hazards for severity and probability risk quantification needed.
Step 7	Application of a HACCP decision tree [Principle 2]	Selection of CCP for each hazard.
Step 8	Establish target levels and tolerance for each CCP [Principle 3]	Animal replacement: free of specific disease agents. Diagnostic tests: antigen-testing vs. serology.
Step 9	Establish a monitoring system [Principle 4]	CCPs are linked to a monitoring system. Monitoring aims at detecting loss of control at an early stage, and at providing information for correction action.
Step 10	Establish corrective actions [Principle 5]	Needed for each CCP selection. Correction also needed when monitoring indicates trend towards loss of control.
Step 11	Verification of the application [Principle 6]	Check correct functioning with respect to steps 6–10 necessary for introducing and maintaining system.
Step 12	Documentation [Principle 7]	Relevant processes, demonstrable control, certification and insurance.

The next step is to implement these principles. A good method of doing this is to construct a flowchart for each species showing where all potential hazards may occur. It is essential that the whole plan is practical and readily usable on the farm, to maximise compliance. The flowchart should identify all biological, chemical and physical risks that can occur, and assess them in terms of Critical Control Points (CCP). An action can only be classified as a CCP if it is possible to eliminate or significantly reduce it and this is likely to be a major stumbling block in a farm situation. The problem with following the HACCP principles on farms is that they are usually controls which are good working practices and not CCPs. The three factors that are required before a hazard can become a CCP are identification, measurement, and control measures. Each farm system will have different CCPs as a result of specific managerial and environmental considerations.

As identified by Noordhuizen and Welpelo,[14] herd or flock health management requires the identification of specific disease hazards and their related preventative measures concerning the occurrence and spread of undesired disease agents. Risk assessment and risk management achieves this. It is important to understand the limitations of the HACCP system on farms. Mitchell[15] highlights the major reasons for failure which in relation to the farm would be as follows:

1. Failure to establish relevant monitoring systems (Principle 4).
2. Failure to establish proper corrective actions (Principle 5), despite monitoring systems highlighting the need for correction.
3. Failure to consider all hazards appropriate to the farm.
4. Difficulty in implementing theoretical aims practically in the farm environment.
5. Over-complication of HACCP plan leading to failure of compliance.
6. The farm system is not yet ready for the HACCP system.

4.4 HACCP plans for cattle

The various production stages for cattle are summarised in Table 4.2 which should be considered with Table 4.3. The issues on the beef farm are very similar to those for the meat production systems of other species of animals but very different from the issues for lactating animals.

4.4.1 Beef cattle

The beef farm may raise the animals on farm as a sucker herd followed by the fattening stage. Animals may be sold on for the final stages of the fattening process. The farmer may have no breeding animals and rear through to fat animals bought in as baby or weaned calves. Whichever system is used, it is crucial that each animal is identified, and full records of any movements between farms, auction markets and the abattoir must be kept. Cattle going for slaughter are graded by conformation criteria. In addition to deciding that the cattle are ready for slaughter, they must be inspected to ensure they do not have any condition making them unfit for human consumption. To avoid contamination of the carcase during the slaughter process the animals should be unsoiled on leaving the farm and not become soiled during transport or at auction.

4.4.2 Dairy unit

The dairy industry has had many years of experience of working to high standards of milk quality and safety. This has been helped, at least for milk from cows, by a combination of financial inducement for high standards or financial penalty for failure(s) along with legislative control. Although there is the single

Table 4.2 Summary of production stages for cattle

Procedure	Problem	Prevention
Replacement breeding animals or purchased for fattening	Buying in disease, e.g. salmonella, tuberculosis, pneumonia	Purchase from known disease-free source – check identification. Do not introduce to herd until certain they are not carriers or excretors
Vaccination	Viral diseases, pneumonia, possibly clostridial rotavirus/*E. coli*	Vaccination of breeding stock to ensure maximum passive immunity transfer to calves and before risk period
Feed	Contamination of incoming feed and when in store with enteric bacteria and moulds Transmissible spongiform encephalopathy	Vermin-proof stores; good quality hay and silage No mammalian derived protein in feed
Environment	Spread of disease by direct contact between cattle discharges, aerosol or by handler	Use good quality straw for bedding. Clean pens using 'all in all out' principle. Good ventilation if housed
Use of medicines	Injection site abscess Residues in meat Antibiotic resistance	Sterile needles and good technique Withdrawal periods adhered to Avoid need for antibiotics by good husbandry, clean environment and good colostrum intake by neonate
Pasture contamination	Waterlogged pasture encourage coccidia and fluke Nematode infestation Hydatid, *C. bovis*, infestation	Adequate drainage or fence off and use of coccidiostat and flukicide Pasture management and use of anthelmintic Regular worming of dogs and appropriate exclusion period if sludge applied
Foot care	Welfare Arthritis possible	Early recognition and treatment Routine foot trimming and dipping
Housing during fattening	Build up of faeces on hide	Good housing and husbandry to avoid soiling. May be necessary to wash or clip before dispatch for slaughter
Housing before slaughter	Cattle coming off wet fields or fodder crops can be very soiled	Put out deep, clean, dry straw bedding for a few days or until suitable to go for slaughter

Table 4.3 Farm HACCP

Process step	Risk: H, M, L	Control	Criteria	Control measures	Monitoring	Corrective action	Records
1. Breeding female	In poor health and/or carriers of disease/parasites. Susceptible to *Salmonella* infection (H)	GMP	All animals in good health. Free from signs of clinical infection	Therapeutic treatment of animals suffering from infections. If suspect *Salmonella* isolate animal(s) from other livestock and seek veterinary advice	Daily inspection of all animals by specified person	Veterinary advice with clinical infections or unknown causes of ill health	Medicines book. Diary of illness in animals entered into database weekly
		GMP	Improve herd/flock resistance to clinical and sub-clinical disease	Minimise risk of disease by optimum husbandry including, e.g., control of parasitic infestations	Daily inspection. Feedback of meat inspection data from abattoir	Seek veterinary advice if prophylaxis appears to be ineffective, e.g. parasite infestation detected during PM meat inspection	Medicines book. Diary of illness. Keep record of all meat inspection results. Enter into database daily
		GMP	Good body condition	Maintain ideal condition score (CS)	Daily inspection of all animals in flock by specified person	If condition score incorrect adjust diet appropriately	Keep record of CS and diet on database

Table 4.3 Continued

Process step	Risk: H, M, L	Control	Criteria	Control measures	Monitoring	Corrective action	Records
2. Breeding male, in addition to 1 above	Can introduce disease onto farm (H)	GMP	If rented or bought in should be disease free and in good health	Do not rent sires. Quarantine new sires after purchase, appropriate vaccination and prophylaxis	Specified person to inspect	If bought in sires show signs of ill health isolate immediately and seek veterinary advice	Keep record of all movements on database
3. General	Animals in poor health and/or carriers of disease/parasites. Susceptible to *Salmonella* infection (H)	GMP	All animals in flock in good health. Free from signs of clinical infection	Therapeutic treatment of sick animals. Isolation of animals which are ill or abort. Cull barren animals or those with history of mastitis	Daily inspection by specified person. Pregnancy diagnosis	Veterinary advice with clinical infections or unknown causes of ill health	Movement records. Diary of illness and results of pregnancy diagnosis. Enter into database weekly. Medicines book up to date
		GMP	Animals kept in good conditions	Provide high standard of husbandry	Monitoring of staff performance	Train staff before start job and update as necessary	Document training
4. Parturition							
Cleaning and disinfection of pens	Environmental build up of *Salmonella* (M)	GMP	No environmental contamination with *Salmonella*	Pens cleaned between groups on an all-in/all-out basis	Weekly visual inspection of pen cleanliness by management	If cleaning insufficient, repeat cleaning process	Keep record of pen disinfection and cleaning

Stage	Hazard	Type	Target	Control	Monitoring	Corrective action	Records
New-born	Poor passive immunity. Risk of infection with enterobacteriacea (H)	GMP	Ensure sufficient quantity and quality of colostrum within first 6 hours	Help to suckle if having difficulty. Store colostrum to feed if extra colostrum not available from dam	Designated person to check whether neonate has fed within first 5 hours after birth	Feed with mother's, bought in or stored colostrum using stomach tube	Keep record of when colostrum given
	Hypothermia (H)	GMP	Ensure adequate colostrum received within first 8 hours	Help to suckle if having difficulty. Store colostrum to feed if extra colostrum required	Designated person to check whether neonate has fed within first 5 hours after birth	Feed with mother's or stored colostrum using stomach tube	Keep record of when lamb receives colostrum
		GMP	Temperature between 39–40°C	Temp. 37–39°C: ensure fed, place below warming lamp. Temp. below 37°C: place in warming box, give intraperitoneal injection of glucose solution	Designated person to check and take temperature if suspect hypothermia	If neonate has hypothermia carry out control measures	Keep record of animals treated for hypothermia
		GMP	Sufficient teats and milk	Check udder and number of teats	Check before parturition	Foster extra piglets, lambs	Record reason for fostering

Table 4.3 Continued

Process step	Risk: H, M, L	Control	Criteria	Control measures	Monitoring	Corrective action	Records
Bedding in pens	Build up of infective material on surface layer of bedding	GMP	Clean dry bedding (straw) in pen	Place large quantities of fresh good-quality straw bedding to all pens every day, twice per day when weather wet. Individual mothering pens: add fresh straw before every new ewe and lamb(s)	Designated person to check cleanliness of bedding in pens daily	If bedding in pen not clean, add sufficient straw to cover pen surface	Keep record of number of straw bales used per day
Place mother and newborn in mothering pen	Poor bond between neonate and mother leading to poor performance/health in lamb due to rejection	GMP	Good bond between mother and neonate	Place in mothering pen for 48 hours if mother does not accept progeny	Designated person to check for rejected newborn	If rejected, place mother and progeny into foster pen or feed artificially	Record all rejected neonates and success of fostering
Identification	Difficult to determine which newborn belongs to which mother	GMP	Mother and progeny should be clearly identifiable	Apply visible ID such as marker spray soon after birth. Use ear tags for individual animal identification	Daily management observation to ensure that staff identify lambs correctly	Apply identification to lambs or ewes that are unmarked or incorrectly marked. Replace identification equipment if necessary	Record all lamb and ewe identification marks, ear tags, etc.

Castration, disbudding of calves and tailing of lambs	Stress reduces ability to resist infection (M)	GMP	Disbud, castrate and tail with minimum of pain and suffering	To be carried out by competently trained individual fully conversant with legal requirements	Management (or designated person) to check daily whether castration and tailing done correctly	Advise person carrying out tailing/castration if incorrect procedure being used	Record date of birth and time of disbud, tailing or castration
5. Put out into field							
Grazing	Contamination with pathogens (M)	CCP_2	Do not allow pasture to be grazed when untreated faecal material has been applied	No grazing on land which has had sewage sludge, slurry or manure applied unless within the guidelines for application	Check records weekly to ensure sheep or cattle are not grazing grassland or forage that has not been sufficiently rested	If animals are grazing land which has not been sufficiently rested move them to a different field	Date of sludge or manure application on all fields. Sludge treatment method
	Contamination of pasture by geese	CCP_2	Bird scare device to deter geese from grazing pasture		Weekly observation by management for signs of geese. Weekly inspection of bird scare device by designated member of staff	If geese present use additional bird scare or use shooting as a control measure. Repair or replace faulty bird scare	Keep record of bird scare inspection. Keep record of sightings of geese

Table 4.3 Continued

Process step	Risk: H, M, L	Control	Criteria	Control measures	Monitoring	Corrective action	Records
Drinking water	Contamination with *Salmonella*, *Campylobacter*, cryptosporidia (M)	CCP_2	Drinking water free from pathogens	Use mains water only. Clean drinking troughs thoroughly annually	Sample water troughs annually and test for *Salmonella*	If trough is positive for *Salmonella* clean and disinfect immediately. Retest and if still positive, re-clean, disinfect and test water supply. Identify source of contamination	Record results of all water samples
			Clean drinking water	Ensure drinking troughs are cleaned out regularly	Daily visual inspection by designated member of staff of all drinking troughs in use	Removal of visible contamination. Empty and clean if contaminated with faeces, birds, etc.	Record findings of daily visual inspection
6. Prior to housing or parturition	Pneumonia, clostridial infection (L)	GMP	Animals free from pneumonia or clostridial infection	Vaccination prior to housing or parturition. Check ventilation of buildings	Specified person to ensure that vaccination is at correct time	If not vaccinated do so at next opportunity	Diary and medicines book

7. Control of parasites	Infestation with helminths (H) and ectoparasites (L)	GMP	Free from clinical and subclinical helminth and ectoparasitic infestation	Administration of appropriate anthelmintic, depending on helminth species of concern, e.g. if wet grazing need to use flukicide	Daily inspection of all animals by specified person. Look for signs of helminth infestation such as diarrhoea and ectoparasites, e.g. hair loss. Post-mortem results of meat inspection	Immediate treatment if symptoms of infestation. Seek veterinary advice if treatment appears ineffective. Helminth infestation: modify prophylaxis if necessary	Keep livestock medicines book up to date. Keep record of meat inspection results for lambs, if possible to obtain
8. Weaning	Post-weaning stress	GMP	—	Careful handling to minimise stress	Implement further inspection the week following weaning	Prompt treatment of animals showing signs of ill health	Records of any ill health or treatment
9. Over wintering outside	Animals develop poor condition, ill health due to adverse conditions	GMP	Animals have dry area to shelter from wind, rain and snow	Provide shelter such as straw bales by hedge or other temporary windproof structure. Provide dry lying areas within shelter with tarpaulin, tin or other suitable cover to protect from rain or snow	Designated person to inspect shelter daily. Designated person to observe to determine if some animals are not gaining shelter	If shelter is insufficient or damaged provide additional and/or replacement shelter	Record of daily inspections of shelter

Table 4.3 Continued

Process step	Risk: H, M, L	Control	Criteria	Control measures	Monitoring	Corrective action	Records
		GMP	Provision of clean water	Ensure water troughs are clean and that water is not frozen	Designated person to inspect water supply in fields daily	If supply frozen break ice and clad pipes if necessary. If water in danger of freezing increase checks to 3 times daily. If water contaminated clean trough	Record of daily inspections of water troughs
			Correct nutrition for weather conditions	Ensure that sufficient roughage is available and that nutrition is at desired level	Designated person to inspect daily	If in poor condition provide supplementary feed or house	Record of daily inspections
10. General precautions							
Drinking water	Contamination with enteric pathogens (H)	CCP_2	Drinking water free from enteric pathogens	Use mains water whenever possible. Clean the drinking bowls and buckets once every month	Sample water bowls prior to housing and test for *Salmonella*	If water bowl is positive for *Salmonella* clean and disinfect immediately. Retest if still positive, re-clean, disinfect and test water supply	Record results of all water samples

Hazard	CCP		Control measure	Monitoring	Corrective action	Records
		Clean drinking water	Clean and disinfect all drinking bowls and buckets before and after housing of sheep	Daily visual inspection by designated member of staff of all drinking bowls and buckets in use	Removal of visible contamination. Empty and clean if contaminated with faeces	Record findings of daily visual inspection
Contamination of feed with *Salmonella* (H)	CCP$_2$	Clean feed	Ensure feed is stored under clean and dry conditions	Store feed in closed bins that are dry and vermin proof. Bagged feed cover with bird-proof sheeting. Use blower so that loose feed does not come into contact with ground. Ensure feed store is dry and clean	Specified person to check integrity of feed bins/feed store once per week	If feed bins are damaged move any feed to a new bin and repair or replace damaged feed bin	Record findings of weekly feed bin/feed store checks
Infection with, e.g., *Salmonella*, *Leptospira* (H)	CCP$_2$	Rat/mice population	Control rat and mouse population	Poison baits around buildings. Seek advice of specialist pest-control contractor	Weekly inspection of baits by specialist contractor	Replacement of baits and poison if necessary by contractor	Keep records of all dead rats and mice found
			Have 3 metres of open ground surrounding livestock building and feed storage area	Keep whole farm tidy. Do not stack pallets or leave farm machinery by livestock buildings/feed storage	Weekly visual inspection by management	Removal of rubbish and proper storage of equipment, farm materials and machinery	Record of rubbish or equipment requiring removal

Table 4.3 Continued

Process step	Risk: H, M, L	Control	Criteria	Control measures	Monitoring	Corrective action	Records
Staff	Spread of *Salmonella* from other livestock (H)	CCP$_2$	Clean clothes and boots	Staff must change protective clothing and use disinfectant foot dips before and after entering areas	Managerial observation	Enforcement of measure by management	Record of occasions when hygiene measure requires enforcement
Visitors	Introduction of *Salmonella* (H)	GMP	Minimise presence of visitors	Vehicles parked away from buildings	Managerial/staff checking of enforcement of these measures	Ask unauthorised visitors to immediately leave farm area. Remove vehicles from vicinity of buildings	Visitors book. Visitors should sign in and out
		GMP	All visitors to wear clean protective clothing	Changing facilities near housing	Managerial/staff checking of enforcement of these measures	Ask visitors to immediately go and change into protective clothing	Record use of protective clothing. Sign clothing in and out

	CCP$_2$	Ensure visitors/staff do not tread infectious agents into farm buildings	Obligation for staff and visitors to use disinfectant foot dip before entering into livestock buildings. Designated person to change foot dips weekly	Weekly managerial inspection of foot dips	Replenishment of foot dips when necessary	Record use of foot dip solution. Record inspection of foot dips
Wild birds						
Infection with *Salmonella*, *Campylobacter* (H)	CCP$_2$	Minimise birds roosting in building roof	Use bird scare such as bird of prey silhouette or sonic bird scare	Daily visual inspection of building by specified person	Shoot pigeons	Keep records of all dead birds found
	GMP		Remove spilt, waste feed	Daily visual inspection of building by specified person	Clean up any spilt, waste feed	Keep record of spilt feed and disposal

raw product, milk, which is consumed or will go for processing, it can be from a number of species of animal. The main milk-producing species is cattle, with sheep and goats also milked commercially. Milk can also be harvested from less common animals such as camels, buffalo and horses and may be done so commercially in the future.

The hazards in milk are mainly from faecal and environmental contamination of the teat and udder, but both chemical and microbiological hazards can be present in the milk within the udder. The chemical contaminants may be due to feeding practices (aflatoxins, dioxins, nitrates), from husbandry practices (pesticides), from veterinary medicines, and from pollution (heavy metals, radioactive elements). Microbiological hazards include the zoonotic organisms present in the milk as contaminants of the milking process, and organisms which are excreted in the milk from the udder.[16-19] The form of the milk-producing animal does not help, with the udder at the rear of the animal and under the anus. The major microbiological risk is from faeces, in particular when the faeces are soft or very liquid. Sheep and goat faeces are typically voided as pellets which reduces to some degree the faecal soiling of the animal and the hands of the milker. The relevant aspects of milk production relevant to control of the hazards include routine milking schedules, the importance of an efficient, well maintained milking machine, management of the housing, and mastitis control. Bacteria reach the milk from contamination of the udder surface, from within the mammary gland and from the inner surfaces of the milking equipment, including the bulk storage tank. Milk from a cow with clinical mastitis can easily have 10^6 organisms per millilitre which if it were allowed to pass in to the bulk tank could have serious consequences. Subclinical mastitis is a problem for the farmer, not only for the health of the udder but also as the presence of an increase in the somatic cell count of the milk lowers the quality of the product, particularly for manufacturing. Mastitis is considered to be of three types – contagious, environmental and 'summer' mastitis. The commonest organisms involved in the different forms of mastitis are shown in Table 4.4.

Prevention of mastitis starts with the selection of the replacement animal which must have good udder conformation. The teats and the teat end in particular must be maintained in good condition. Keratin and fatty acids on the skin have antibacterial properties. Most infections, however, are through the teat canal which in effect is the first line of defence against infection of the udder. Teat end damage is a sequel of poor milking equipment, possibly badly serviced, or incorrect handling of the milking cluster during the milking process. Coating of the teat end in disinfectant, by spray or dipping, after each milking has the effect of leaving some disinfectant in the bottom of the teat canal which protects the udder while the canal is still open after milking and reduces the pathogens on the skin of the teat.

The infusion of a long-acting intramammary antibiotic into each quarter of the udder at the end of the lactation will eliminate bacteria present in the udder and protect the udder in the early dry period. It has been shown that use of dry cow therapy is effective against *E. coli* and in the subsequent lactation the

Table 4.4 Common microorganisms involved in bovine mastitis

Contagious mastitis organisms	Infection occurs during milking
Streptococcus agalactiae	Obligate bacteria of the udder which may also colonise the teat canal and is extremely contagious
Staphylococcus aureus	Frequently a chronic infection of the udder but may also be involved in peracute or gangrenous mastitis. It is a normal inhabitant of teat skin and can be difficult to eliminate from the udder
Streptococcus dysgalactiae	Commonly found in teat lesions and in the tonsil and is sometimes considered midway between a contagious and an environmental organism
Mycoplasma bovis, M. californicum	Highly contagious and a big problem in North America
Environmental mastitis	**Infection occurs either at milking or between milking**
Escherichia coli	Present in large numbers in faeces; infection is often associated with faulty milking machines
Klebsiella pneumoniae	A soil commensal and commonly found on damp and poorly stored sawdust which may be used as bedding for the milking animals
Pseudomonas aeruginosa	Less common and is associated with water stored at low heat and used to wash teats before milking
Summer mastitis organisms	
Actinomyces pyogenes with the anaerobe *Peptococcus indolicus* or with *Strep. dysgalactiae*, or both	Problem of animals during summer months, linked with flies and damage to the teats

Table 4.5 Actions taken during the milking process

Milking procedure	Action
Identify cow on entry to milking parlour	GMP
Ensure correct ration given	GMP
Remove cows with signs of mastitis	CCP
Dry wipe or wash teats (pre-milk disinfection possible)	CP
Apply cluster immediately after preparation	GMP
Remove cluster at end of milking or use automatic cluster removal to avoid over-milking	GMP
Teat end disinfection	GMP
Milking staff should wear rubber/vinyl gloves	CP
Clean and disinfect plant after each milking session	CCP
Prompt cooling of milk	CCP
On-farm pasteurisation	CCP
Test for Somatic Cell Counts and Total Bacterial Counts	CP
Test for residues	CCP

number of mastitis episodes is significantly reduced. For cows in the summer mastitis season it is worth taking them through the milking parlour each day and spraying their udders with fly repellent. In addition, if straw yards are used for housing the cows it must be changed regularly to avoid build-up of environmental organisms.[20]

The process of milking animals can be divided into a number of defined actions which impact on the product and the well-being of the animal. There are few control points, with most actions being good practice (Table 4.5). The factors which impact on milk hygiene are shown in Fig. 4.1.

Milk is very sensitive to taint which can occur from three sources: chemicals, bacteria and the use of certain feeds. Chemical taint is the most important and a common cause is use of the wrong chemicals, such as phenols which are banned from use in the dairy or milking parlour, or not following the product data sheet instructions for use. The commonest feed ingredients which cause taint are brassica plants, fishmeal and weeds such as wild garlic. The use of home

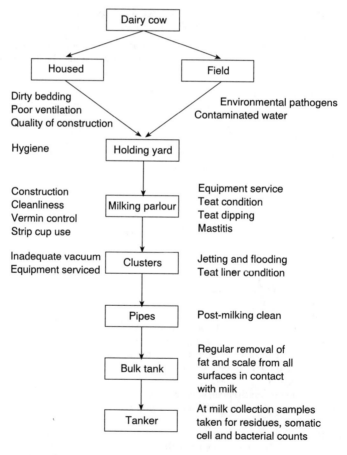

Fig. 4.1 Factors which impact on clean milk production.

remedies or alternative remedies may cause a problem if, for example, they contain aromatic substances such as aniseed or linseed oil.

4.5 HACCP plans for sheep and goats

The primary output from a sheep farm is the slaughter animals, but sheep may be milked to produce drinking milk or further processing, whereas it is the converse in goat herds. The comments on milking cows are equally applicable to sheep, although the scale of the operation is usually smaller. There may be a sector of the industry in which hand-milking is carried out, but mechanical milking equipment is readily available. To ensure that a HACCP plan applied to a sheep flock remains simple and workable it must be separated from any flock health scheme. There are many diseases that pertain to the overall health and productivity of the flock and, possibly, to the quality of the meat but that will not directly affect human health. The production of sheep as summarised in Table 4.6 should be considered along with the general production information in Table 4.3.

In the sheep flock there is the additional concern for the health of the stockman from zoonoses which cause abortion, such as *Chlamydia psittaci*, *Toxoplasma gondii*, *Campylobacter* spp., *Salmonella* spp., and the use of chemicals, e.g. organophosphates, as dips. In terms of the diseases mentioned in Table 4.6, evidence of parasitic infestation will be the most likely indication of action being required at post-mortem meat inspection.

Table 4.6 Summary of production stages for sheep

Procedure	Problem	Prevention
Replacement animals	Buying in disease, e.g. *Maedi visna* virus, *Caseous lymphadenitis*	Purchase from known disease-free source; do not introduce to flock until certain they are not carriers or excretors
Vaccination	Clostridial diseases, pneumonia and abortion agents – cause mortality, morbidity and condemnation at meat inspection	Vaccination of all sheep with booster of clostridial vaccine before lambing to ensure maximum passive immunity transfer to lambs
Feed	Contamination of incoming feed and in stores with enteric bacteria and *Toxoplasma gondii* and moulds Lamb feed with coccidiostat Transmissible spongiform encephalopathy agent	Vermin-proof stores; avoid contamination of hay by cat faeces; good quality hay and silage Apply withdrawal period No mammalian derived protein in feed and genotype breeding males for susceptibility

Table 4.6 Summary of production stages for sheep

Procedure	Problem	Prevention
Environment	Spread of disease by direct contact between sheep, from discharges (e.g. uterine fluids and placenta), aerosol or by handler	Use good quality straw and remove placentas from lambing yards. Clean and disinfect pens. Good ventilation if housed
Antibiotic use	Injection site abscess Residues in meat Resistance, e.g. following prophylaxis for watery mouth (*E. coli*) infection	Sterile needles and good technique Withdrawal periods adhered to Avoid need by clean environment and good colostrum intake
Pasture contamination	Waterlogged pasture encourage coccidia and fluke Nematode infestation Hydatid, *T. ovis, T. hydatigena* infestation	Adequate drainage or fence off and use of coccidiostat and flukicide Pasture management and use of anthelmintic Regular worming of dogs
Foot care	Welfare Arthritis and pyaemia possible	Early recognition and treatment Routine foot trimming and dipping
Dipping	Ectoparasites – fleece damage and possible emaciation Post-dipping lameness (*Erysipelas rhusiopathiae*)	Routine dipping or injectable product Keep dip solution clean with possible use of antbacterials in solution. Use spray
Crutching	Ewes	Reduce faecal contamination at lambing or at milking; avoid flystrike
	Lambs	Reduce faecal contamination at slaughter and flystrike in summer
Housing before slaughter	Lambs coming off wet fields or fodder crops can be very soiled	Put out deep, clean, dry straw bedding for a few days or until suitable to go for slaughter

4.6 HACCP plans for a poultry unit

Poultry meat and eggs and their products are recognised sources of human salmonellosis. One of the big food scares followed the announcement in the UK by Edwina Currie in 1989 about *Salmonella* in eggs. There then followed a dramatic drop in egg sales and the subsequent raft of measures put in place by government to control *Salmonella enteritidis* and *S. typhimurium*, which included slaughter arrangements for infected flocks. *Salmonella* infections in poultry often remain undetected due to the lack of clinical symptoms in the flock.

The poultry industry has evolved over the years with the result that there is a breeding pyramid with very high health status, elite, grandparent and parent breeding birds at the top (Fig. 4.2). Recognising the problem of vertical transmission of *S. typhimurium* and *S. enteritidis*, great care is taken to ensure these birds are free from *Salmonella* spp.

Control of infections in these breeding flocks can be by slaughter of the breeding flock, an option which is not reasonable for layer flocks and broiler birds. Systematic approach at each link in the poultry production chain is

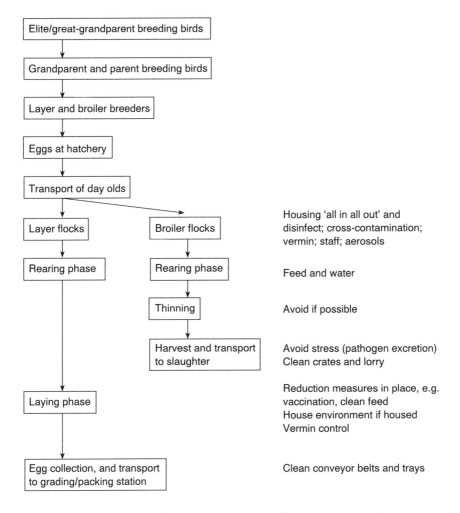

Fig. 4.2 Factors which impact on clean poultry and egg production.

necessary if the flock prevalence, and the meat and eggs produced, of pathogens is to be markedly reduced. The recommendations contained in the Richmond Committee Report[4] in 1990 dealt specifically with the poultry industry and made a number of recommendations relating to housing, husbandry and feed which are valid today. The major concern still remains with the likely contamination of a percentage of the meat due to colonisation of the birds at the farm production with cross-contamination during transport. Withdrawal of food from the slaughter birds before transport to reduce the excretion of *Salmonella* by the birds on arrival at the slaughter plant has been common practice. However, Corrier and his fellow workers identified the crop as being a significant source of *Salmonella* contamination during slaughter and feed withdrawal before transport increased the likelihood of contamination.[21]

As eggs may be consumed raw or used as raw ingredients in uncooked dishes it is essential that they are free from pathogens. The egg industry has moved from battery houses with the layer birds in cages to floor-based and outdoor welfare-friendly systems. The results of investigations have suggested that other systems have a higher proportion of *Salmonella* isolations than battery systems.[22] A code of practice was published by MAFF in 1996 which provides guidance on good hygiene principles and practice on the production site, at the grading and packing station and during distribution and storage. The British Egg Industry Council operates a Code of Practice for Lion Quality eggs. This has been revised to include a requirement by the flocks to use vaccination for *Salmonella enteritidis*. These measures collectively appear to be reducing the incidence of *S. enteritidis* in the laying flocks. In contrast some countries do not permit the use of this vaccine and rely on the effectiveness of control measures on the farm.

A relatively large number of vaccines are used in the poultry industry for major viral diseases in both breeding and commercial layers. In addition to the vaccine which is available for *S. enteritidis* there is also a vaccine against coccidiosis. The *Salmonella* vaccine is not used at the very top of the breeding pyramid where control is by very high biosecurity on the site along with rigorous control by careful monitoring of feed, staff and water supply, etc., supported by microbiological monitoring.

Organisms identified as potentially hazardous to food production are *S. enteritidis, Campylobacter jejuni* and *Listeria monocytogenes*. The hazards and CCPs that have been identified in the on-farm production of broiler meat are well described by the 1990 Report on the Microbiological Safety of Food.[4] The aim of the HACCP system in broiler production is the reduction of contamination of birds leaving the farm, with the previously described organisms. Clinically affected birds should be identified before the flock leaves the farm. The problem will remain with asymptomatic carrier birds which are shedding or become shedders following catching and transport prior to arrival at the processing plant. In order to prevent the passage of pathogens from old birds to day-old chicks, the best policy is to implement an all-in, all-out practice, that is, the cleaning and disinfection of all houses and equipment prior to arrival of a new flock. This will prevent residual contamination with dust, dirt and pathogens. Cleaning should include drinkers and feeders, with removal of litter. Bacteriological samples can be taken following cleaning to assess the efficacy of the disinfection, and records of all cleaning procedures should be kept. Additionally, adequate cleaning and disinfection of the farm and equipment between each crop will limit the spread of infection between flocks. It has been the practice to empty each house, clean and disinfect, then restock. More recently the emptying of all houses on a site with a complete clean and disinfection of the whole site along with composting of the litter before restocking has been recommended.

Vehicles bringing birds on to the farm, visiting personnel, and employees should all be suitably disinfected and recorded, and all livestock movements should be noted. Vermin should additionally be controlled, and in particular bird/rodent proof feed bins should be provided to prevent access by these animals resulting in potential contamination of food sources. This will be difficult, if not impracticable, with free-range systems. Additional CCPs that should be considered are those involving the use of pharmaceutical (medicines and food additive) agents. All drugs used should be clearly recorded, and withdrawal times should be closely adhered to.

4.7 HACCP plans for a pig unit

Diseases are present within swine populations that could pose a potential health risk to consumers. These include agents such as *Salmonella* spp., *Streptococcus suis, Yersinia enterocolitica, Trichinella spiralis* and *Toxoplasma gondii*. In addition, there must be constant awareness of emerging agents, such as Nipah virus in the pig population of SE Asia in 1999.

The use of good husbandry practices on the farm is critical in the production of pigs intended for slaughter. The pig industry is a good example of the difficulty in applying HACCP on farms with the tremendous variation in the type of pig production systems in use. Over many years the pig industry in some countries has developed an integrated production chain involving the farm and slaughter plant and using data from the findings at post-mortem meat inspection. This allows a study of the effects of the production system on disease occurrence in the groups of pigs. The interrelations between the farm-level circumstances and the health status of animals have been described by many workers. Examples include a comprehensive review by Stark in 1999 of the environmental risk factors on respiratory disease in swine,[23] while Bandick *et al.* reported that slatted floor systems without straw bedding were associated with abscesses on the exterior surface of the animals.[24] In addition, a number of adverse factors have been identified in pig production:

- multiple sources of piglets
- overcrowding of pigs

- deficient isolation of groups
- continuous fattening (holding back smaller pigs)
- large farms with large herds
- big barns
- adverse climates
- dust and handling liquid manure without precautions against ammonia.

There has been considerable pressure on the farming industry from consumer groups to move from the intensive production system to outdoor, extensive systems. This move to outdoor, extensive production exposes the pigs to a much greater challenge from the environment and increases the risk of zoonotic infections and infestations. The intensive pig production units which have buildings with a high level of biosecurity are best suited to a HACCP plan. While the aim must be to apply HACCP-like principles on the farm, there is a practical limit to the number of critical control points which can be applied even in the intensive units. Bacterial contamination of pigs entering the slaughterhouse must be carefully controlled to prevent contamination of the carcase and subsequent introductions of bacterial pathogens into the human food chain. This can be achieved by introducing HACCP onto the farm to limit the spread of potentially pathogenic organisms between individuals, and subsequent populations.

Similar requirements to those previously described for poultry can be implemented in commercial, intensive pig production units, with similar goals. There is then the requirement to identify the responsibility at each stage of the production chain. Quality Assurance schemes, including major producer or retailer schemes, are starting to address these concerns by implementing regular veterinary inspections of all accredited farms, with minimum requirements of hygiene and welfare, thus further raising standards. In slaughter pig production, preharvest food safety has been emphasised with integration of Total Quality Management and HACCP so that significant improvements can be expected.[25,26] This must include a surveillance and control programme for disease and presence of pathogens. This may be, for example, as the current requirement to sample pig carcases within Member States of the European Union for evidence of *Trichinella spiralis*, or the Danish national scheme described by Mousing and his co-workers in 1997 based on the quantification of the within-herd prevalence of *Salmonella enterica*.[27] This was the most extensive attempt on a national level to control *S. enterica* in pork, and the epidemio-surveillance of *S. enterica* carriage by pigs in Denmark was reviewed by Christensen *et al.* in 1999.[28] Using a central database, the results were used to assign the herds to one of three levels which determined the acceptability of the herd or the action necessary. In addition, a clear relationship was shown between antibody levels in serum and that in muscle tissue fluid, allowing detection of *Salmonella* infection at the herd level through meat juice analyses in the slaughter plant.

Critical to the success of any future system must be the exchange of information from the findings at meat inspection in the abattoir. In fact this is not just a trace-back from the abattoir system but also a trace-forward of information about individual pigs or groups of pigs. This would depend on there being an accurate recording system on the pig farm which must contain information on a number of items and events, including health and performance indicators. The minimum information to be recorded on the farm and available for inspection is:

- performance data of pigs, e.g. growth rate, feed conversion
- dates of visits by veterinary services
- health problems
- morbidity and mortality data
- results of veterinary examinations
- results of laboratory examinations and tests, e.g. serology for *Mycoplasma*, *Salmonella*
- treatments, e.g. vaccination, medication
- use of feed additives
- results of preselection.

For pig production the provision of feed which is microbiologically clean and free from residues is as important as in other species. This is easier to achieve if the pigs are housed with the feed delivered in closed vehicles and delivered to the feeding trough via an auger from the storage silo. This is not possible for pigs which are outside where the feed may be on the ground and also provides an attraction to birds and vermin. Feeders for outdoor pigs which restrict access by birds have been developed but will never be totally bird and vermin free. Pigs by nature are rooting animals and will thus be exposed to hazards while rooting and when wallowing in water holes.

With housed pigs, major stress factors are the moving of groups of pigs along with mixing of litters to form larger groups. This can lead to the onset of enteric and respiratory illness. Leaving the piglets in the pen in which they were born following weaning until they reach slaughter weight is considered to reduce the possibility of disease and reduce antibiotic use.

4.8 Summary: the effectiveness of HACCP on the farm

The potential benefits of on-farm HACCP for improving the health status of livestock, for reducing or controlling foodborne pathogens and for quality assurance has been commented on by several authors.[4,29–34] With regard to cattle and sheep, most attention has been focused on dairy cattle, particularly with regard to antibiotic residues in milk.[31,35] None of the papers give examples of actual on-farm HACCP plans for cattle or sheep farms or consider the practicality of implementing such a system in a non-intensive farming environment. The poultry industry has applied HACCP-like principles as part of the *Salmonella* reduction programme. Noordhuizen and Frankena[33] give an example of a generic HACCP-like approach to the control and prevention of salmonellosis on pig farms but again not for a specific farm. Furthermore Pierson[29] stated in 1995 that the animal and feed production HACCP plans that he had come across were essentially GMP plans in a HACCP format without any true CCPs in place. Therefore the provision of actual examples of farm-level HACCP plans and discussion of the practicalities, strengths and weaknesses of such a system are required.

Farm-level factors and their impact on health of animals intended for slaughter must have flexibility to take into account regional risks but must be a vertically integrated production and slaughter chain. Critical to food safety is the recording of the following:

- data on the herd as a whole, including information about administration of medicinal products and immunisation programmes
- data on the health status, including information from the disease records and the general body condition of the animals going to transport
- data on the performance of each group, and the herd as a whole, e.g. daily liveweight gain, mortality and morbidity figures
- knowledge about farm environmental factors, which are crucial for a good result of fattening, including data on the buildings
- feed quality control at the farm level, including feed supplier quality assurance
- traceability of individual animals and groups of animals at all times, including movements on to the unit.

To this information must be added the 'feedback information' from the slaughter plant or the processing plant. This would include findings at post-mortem meat inspection from the slaughterhouse, including any effects of transport such as the presence of Pale Soft Exudative (PSE) or Dark Firm Dry (DFD) meat or other defects such as injuries, filthiness, fatigue or stress. The monitoring of pathogens and residues, identified and agreed as being appropriate to the production system and to the geographical region in which the animals are produced, is part of the necessary epidemiological surveillance.

It is perhaps easier for the farm to apply the HACCP concept when considering residues. This will include residues from the use of medications and from other sources. A summary is provided in Table 4.7. The legislative requirements for medications to be used on food producing animals is a major factor on the safety of that food from the animal(s) but care has to be exercised to ensure residues from other sources do not enter the food chain. The assessment of risk must therefore consider all obvious sources of residues and also recognise the risk from unintentional access to source of a potential residue or to residues following an illegal act or operation.

One major problem is that when HACCP is said to be used it is usually as part of the farm's Quality Assurance scheme when in reality it is Safety = Welfare with little consideration of true food safety issues. There is a real need for the whole area to be properly established so that any HACCP or Risk Assessment or Risk Management approach is set up to *manage* and not to *react*.

Table 4.7 Dairy, beef and sheep HACCP (residues)

Process step	Risk: H, M, L	Control	Criteria	Control measures	Monitoring	Corrective action	Records
All animals	Animals in poor health, possible sign of toxic substance (L)	GMP	All animals in flock/herd in good health, free from signs of clinical infection	Veterinary treatment of animals and identification of source of toxic chemical	Daily inspection of all animals in herd by speciific person	Veterinary advice with apparent toxic infections or unknown causes of ill health in livestock	Keep livestock medicines book up to date. Keep record of illness in animals. Enter into computor database weekly
Grazing land	Contamination of pasture with toxic plants leading to intake by livestock	CCP_1	Only allow access to pasture that is free of toxic plants	No grazing on land where toxic plants are present	Daily inspection of fields with animals by designated person. Inspection of new fields prior to moving animals in	If cattle are grazing land that is contaminated with toxic plants move them to a different field	Record findings of field inspection on database

Table 4.7 continued

Process step	Risk: H, M, L	Control	Criteria	Control measures	Monitoring	Corrective action	Records
Contamination of grazing land with heavy metals and other toxic elements		CCP$_1$	Only graze land that is free from toxic elements in the soil such as heavy metals	Take soil samples from all grazing land every five years. Have soil tested for toxic elements. Find out the history of land use for the fields and surrounding land (e.g. adjacent to old lead mine)	Designated person to check results of soil tests. Investigate and corroborate any positive results	Do not allow cattle to graze contaminated field. If cattle are already present in field remove from field and seek veterinary advice regarding potential animal and public health consequences. Investigate source of contamination	Record results of all soil tests
Contamination of grazing land with heavy metals and other toxic elements by application of sewage sludge (H)		CCP$_1$	Any sewage sludge applied to the land should be treated and free from heavy metal and other toxic contaminants	Sewage sludge should be certified as free from heavy metals and other toxic contaminants.	Test sample of every batch of sewage sludge applied to land. Designated person to check	If heavy metals or other contaminants found in sludge do not graze animals in field	Record date of sewage sludge application on all fields. Have record of sewage sludge tests

Medication for therapeutic or prophylactic reasons	Antibiotic or drug residues in milk or meat (H)	CCP$_1$	Adhere to withdrawal times	Record all medication use in farm medicines book. Should include who administered the medication and the final withdrawal date. Ensure that staff correctly administer medication and identify animals. *Only use licensed veterinary products*	Management to check farm medicine book to ensure all required data is correctly entered. Designated person to check that sufficient withdrawal period for medication has been met for *all* animals before they are sent for slaughter or milked for human consumption	If withdrawal times for medication has not been met, *do not* send animals to slaughter or milk for human consumption	Keep farm medication book up-to-date and correctly filled in
Clean feed	Contamination of feed with toxic substance such as mycotoxin (M)	CCP$_2$	Ensure feed is stored under clean and dry conditions	Store feed in closed bins. Keep feed dry and out of contact with ground. *Do not* store with anything else	Specified person to check integrity of feed bins once per week. Dairy feeds tested for aflatoxin	If feed bins are damaged move any feed to a new bin and repair or replace damaged feed bin	Record findings of weekly feed bin checks

Table 4.7 continued

Process step	Risk: H, M, L	Control	Criteria	Control measures	Monitoring	Corrective action	Records
Contamination of feed with toxic substance such as pesticide (H)		CCP_1	Safe storage of all farm chemicals, such as insecticides, separate from feed store, livestock, etc.	Safe and secure storage of chemicals and other hazardous substances in biulding/container approved for such purposes. Full COSHH assessment of farm chemicals and environment. Do not harvest within prescribed withdrawal period	Weekly inspection by specified person of chemical storage and farm buildings and livestock areas to ensure safe storage and use of substances covered by COSHH	If chemical spill or contamination found *immediately follow COSHH action plan and secure area.* All feed, bedding or other contaminated substances to be disposed of appropriately according to COSHH report/ chemical manufacturers recommendations. If animal contaminated seek appropriate veterinary advice	COSHH report and action plan to be kept up to date. Record weekly inspection findings in report book, signed and dated by specified person responsible for inspection

| Clean water | Toxic chemicals in water leading to contamination of animals, resulting in residues in livestock products and animal public health risk, e.g. milk (H) | CCP$_1$ | Water free from toxic impurities, whether heavy metals, manufactured chemicals or other contaminants | Use mains water only. Clean the drinking water bowls once every week. Water buckets to be emptied and rinsed every day before adding fresh water. Do not use lead pipes | Sample drinking water tanks, bowls and troughs once every 6 months and test for purity and toxins | If water is found to be contaminated, stop use immediately and supply animals with water from an uncontaminated source. Find out source of contamination, if due to on-farm substance covered by COSHH audit *immediately follow COSHH action plan* | Record results of all water samples in report book, signed and dated by specified person responsible for inspection |

To implement a HACCP system successfully the farm should already be observing all GMPs. There must also be a real commitment from the management to develop a HACCP system with effective communication with and training of the farm staff and others involved in any way with the farm operation. An effective monitoring system will provide information from accurate records for future use. It will also enable management to take timely decisions before a process gets out of control.

The success of any scheme for any farm or unit requires, as a minimum:

- surveillance of possible diseases or risks;
- appropriate measures for necessary actions put in place;
- active supervision at all levels;
- investigation of all possible, or actual, problems or variations from the normal.

The success of animal production practices cannot be based solely on the reduction of foodborne disease. The data from the slaughter plant and the further processing stages must all be linked with the data from the farm. The slaughter plant and further processing can provide valuable data from routine testing programmes for zoonotic pathogens such as *Salmonella* and *Trichinella spiralis* as well as for residues. For the link to be effective there must be baseline data from the live animal stage.

Under such circumstances food safety is critical for any farmer, farm group or national industry to maintain or increase market share. Therefore farms need not only to improve food safety but also to provide documentation that verifies what steps are taken and what controls are in place. For such purposes the implementation of the HACCP system on the farm has real potential to improve public health and animal health and welfare. The international use of the HACCP system by food manufacturers and producers is a logical progression enabling harmonisation of international food safety regulations and the removal of non-tariff barriers to trade arising from food safety.[36] The considerations must also apply to foods and animals imported from countries where the controls, for example, on antimicrobial use will not always be as rigorous as in the UK. Increasingly agricultural produce, the raw material on which the rest of the food industry relies, is a commodity on a worldwide market. There is increasing competition for producers of beef and lamb as well as pork/pigmeat and poultry meat.

A cornerstone of future assurance to consumers, the EU and the rest of the world will be that proper supervision and checks are being carried out on the farm with adequate records being maintained. To provide this assurance the minimum aim must be 100% compliance with current legislation with evidence available that this level of compliance is being maintained. There has been in recent years an increase in the number of farm-assured schemes and the direct influence by the major retailers on agricultural practices through their producer schemes. These farm quality assurance programmes stress the importance of a strong working relationship between producers and their veterinarians and emphasise that efficient management practices on the farm are a way of improving the safety of the food supply.

The reputation of the stakeholders – the farming industry and the professions – must not be compromised in any way for whatever reason. However, the consumer must recognise that there is a cost to all the improvements to the on-farm situation. Often the concerns about the whole food chain are associated with food scares and presented as a perceived worry about food-related issues that has little if anything to do with the reason for the food scare. On the other hand, if the controls placed on the industry are too stringent, there will be such an increase in the cost of production that the result will be increased imports of produce from countries where the standards of husbandry and slaughter are lower than in, for example, the UK.

The role for food from non-traditional species must also be considered in the future. World supplies of animal-derived protein are limited and in some parts of the world under considerable pressure. It is possible to harvest more from the wild provided care is taken while drawing on wildlife reserves. Already game farming and fish farming in particular have changed the availability of different types of meat.

It is also most important when considering the need for legislation to recognise the differences between disease in respect of animal health or human health. At present there are no specific statutory food safety controls applicable to on-farm production. It is very easy to say that more control is necessary on the farm and even to increase the legislative controls on farming, but legislation is not always the answer, especially if there is no audit of compliance. Equally in the EU and the worldwide marketplace, there is little point in disadvantaging a country's agriculture such that it is priced out of the market and the food is imported from farming systems of lower standards, in terms of both animal welfare and food safety, but at lower cost. An example could be the banning of sow stalls on welfare grounds in the UK with a significant extra cost to the UK pig producer, which has not been applied to any other country. Equally of concern at this time are the increasing reports of animal medicines available illegally, even by mail order, with suggestions that they are 'on the Internet'. They must be very tempting to farmers at this time of economic crisis in farming, not least when they are at less cost than the veterinary surgeon can purchase the same drug.

The industry has increasingly become a target for consumers campaigning for changes in animal welfare and husbandry systems as well as expressing concern for the environment. These concerns about the food animal production systems and the methods by which the product is harvested, including how animals and birds are slaughtered to produce meat, are very relevant to the whole subject of veterinary public health.

In conclusion the application of HACCP 'behind the farm gate' is still in its early stages; however, as consumer demand for good quality, disease-free products increases the need for the implementation of such control systems will be higher. The aim is to produce animals in a manner concurrent with these aims, with the minimum of medical/pharmaceutical intervention. This will include improvements to husbandry, appropriate use of vaccines if available, even changes to the management of the farm. While on-farm HACCP is not a panacea that will remove foodborne pathogens and other health risks from food of animal origin, it is a system with widely understood principles for identifying significant risks and their control. HACCP allows implementation of an effective documented system that will eliminate or reduce the likely occurrence of foodborne hazards. In addition HACCP is an internationally recognised system for quality assurance that is understood and accepted by the rest of the food industry, including livestock producers and customers.

Disease control in animals is multi-faceted and the more traditional 'fire-brigade' responses without consideration of preventive measures are no longer acceptable. In professional hands with diligent attention to Good Veterinary Practice they are valuable, versatile and safe components with a vital and specific role to play in control of bacterial disease in animals.

The success of animal production practices cannot be based only on an increase or a reduction of human foodborne disease. There must be a gathering of information relating to animal production, including the influence of changes in management practices that may play a role in pathogen prevalence. Epidemiological surveillance will enable the prediction or projection of risk factors and of emerging issues so that perception can be replaced by reality based on scientifically reliable data.

Production and health information from poultry units has been used for a number of years to target the level of post-mortem meat inspection necessary for each batch of broilers delivered to the slaughter plant. There is a strong possibility that all inspection systems will change to one based on an analysis of risk. An important part of any new system will be the monitoring of salmonella on the farms of origin. Studies of the type by Edwards et al.[37] and Fries et al.[38] are required to provide the basis for any alternative system of integrated meat inspection. Such studies might give background for designing a truly targeted organoleptic post-mortem inspection system that yields a net benefit to consumer health protection.

4.9 References

1 BUNTAIN B, The role of the food animal veterinarian in the HACCP era, *J Am Vet Med Assoc*, 1997 **210** 492–5.
2 CULLOW J S, HACCP (Hazard Analysis Critical Control Point): is it coming to the dairy? *J Dairy Sci*, 1997 **80** 3449–52.
3 CHAPMAN P A, SIDDONS C A, WRIGHT D J, et al. Cattle as a source of Verocytotoxin-producing *Escherichia coli* O157 infections in man, *Epidemiology and Infection,* 1993 **111** 439–47.
4 The Microbiological Safety of Food. Part I. Report of the Committee on the Safety of Food (Chairman Sir Mark Richmond), London, HMSO,

1990, pp 45–58.
5   Ministry of Agriculture Fisheries and Food, The Report of the Expert Group on Animal Feedingstuffs (The Lamming Report), London, HMSO, 1992, p 2.
6   JOHNSTON A M, Bovine Spongiform Encephalopathy, *CPD Veterinary Medicine*, Rila Publications, London, 1998 **1** 26–9.
7   ASHRAF H, European dioxin – contaminated food crisis grows and grows, *Lancet*, 1999 **353** 2049.
8   ERICKSON B E, Dioxin food crisis in Belgium, *Anal Chem*, 1999 **71**(15), 541–3.
9   JOHNSTON A M, Use of antibiotic drugs in veterinary practice, *BMJ*, 1998 **317** 665–8.
10 Opinion of the Scientific Committee on Veterinary Measures Relating to Public Health. Assessment of potential risks to human health from hormone residues in bovine meat and meat products. European Commission, Brussels, 30 April 1999, also on www.europa.eu.int/comm/dg24/health/sc/scv/outcome_en.html.
11 Report of the Joint Committee on the Use of Antibiotics in Animal Husbandry and Veterinary Medicine, Chairman Professor M Swann. London, HMSO, 1969.
12 Advisory Committee on the Microbiological Safety of Food. Report on Microbial Antibiotic Resistance in Relation to Food Safety. London, The Stationery Office, 1999.
13 Opinion of the Scientific Steering Committee on Antimicrobial Resistance. European Commission Directorate-General XXIV, Brussels, 28 May 1999. Also on www.europa.eu.int/comm/dg24/health/sc/ssc/outcome_en.html.
14 NOORDHUIZEN J P T M and WELPO H J, Sustainable improvement of animal health care by systematic quality risk manaagement according to HACCP concept, *Veterinary Quarterly*, 1996 **18** 121–6.
15 MITCHELL R T, Why HACCP fails, *Food Control*, 1998 **9** 2–3 101.
16 SHARP M W, Bovine mastitis and *Listeria monocytogenes*, *Vet Rec*, 1989 **125** 512–13.
17 EDWARDS A T, ROULSON M and IRONSIDE M J, A milkborne outbreak of serious infection due to *Streptococcus zooepidemicus* (Lancfield group C), *Epidemiol Infect*, 1988 **101** 43–51.
18 SHARP M W and RAWSON B C, Persistent *Salmonella typhimurium* PT 104 infection in a dairy herd, *Vet Rec*, 1992 **131** 375–6.
19 WOOD J, CHALMERS G and FENTON R et al., *Salmonella enteritidis* from the udder of a cow, *Can Vet J*, 1989 **30** 833.
20 BERRY E A, Mastitis incidence in straw yards and cubicles, *Vet Rec*, 1998 **142** 517–18.
21 CORRIER D E, BYRD B M, HARGIS M E et al., Presence of *Salmonella* in crop and caeca of broiler chickens before and after preslaughter feed withdrawal, *Poultry Science*, 1999 **78** 45–9.
22 GEUE L and SCHLUTER H, A salmonella monitoring in egg production farms

in Germany, *J Vet Med*, 1998 **45** 95–103.
23 STARK K D C, Epidemiological investigation of the influence of environmental risk factors on respiratory disease in swine – a literature review, *Vet J*, 1999 **159** 37–56.
24 BANDICK N, DAHMS S, KOBE A, TITTMANN A, WEISS H and FRIES R, Zusammenhänge zwischen Mastbedingungen und Ergebnissen der amtlichen Fleischuntersuchung, *Proc. Deutsche Veterinärmedizinische Gesellschaft, 38. Arbeitstagung des Arbeitsgebietes Lebensmittelhygiene in Garmisch-Partenkirchen*, 29.9–2.10.1999, Vol I, pp 279–85.
25 BARENDSZ A W, Food management and total quality management, *Food Control*, 1998 **9** 163–75.
26 DECLAN J, BOLTON A H, OSER G J, COCOMA S, PALUMBO S A and MILLER A J, Integrating HACCP and TQM reduces pork contamination, *Food Technology*, 1999 **53** 40–3.
27 MOUSING J, KRYVAL J, JENSEN T K, AALBAEK B, BUTTENSCHON B and WILLEBERG P, Meat safety consequences of implementing visual inspection procedures in Danish slaughter pigs, *Vet Rec*, 1997 **140** 472–7.
28 CHRISTENSEN J, BAGGENSEN D L, SORENSEN V and SVENSMARK B, Salmonella level of Danish swine herds based on serological examination of meat juice samples and *Salmonella* occurrence measured by bacteriological follow-up, *Prev Vet Med*, 1999 **40** 277–92.
29 PIERSON M, An overview of the Hazard Analysis Critical Control Points (HACCP) and its application to the animal production food safety, *Proc Hazard Analysis and Critical Control Point Symposium*, 1995 1–10. The 75th Annual Meeting of the Conference of Research Workers in Animal Diseases, 12 November 1995.
30 TROUTT H F, GILLESPIE J and OSBURN B I, Implementation of HACCP program on farms and ranches, in *HACCP in Meat, Poultry and Fish Processing*, eds PEARSON A M and DUTSON T R, pp 36–57, London, Blackie, 1995.
31 SISCHO W M, KIERNAN N E, BURNS C M and BYLER L L, Implementing a quality assurance program using a risk assessment tool and operations, *J Dairy Sci*, 1997 **80** 777–87.
32 HANCOCK D and DARGATZ D, Implementation of HACCP on the farm, *Proc Hazard Analysis and Critical Control Point Symposium*, 1995 1–10. The 75th Annual Meeting of the Conference of Research Workers in Animal Diseases, 12 November 1995.
33 NOORDHUIZEN J P T M and FRANKENA K, Epidemiology and quality assurance applications at farm level, *Prev Vet Med*, 1999, 93–110.
34 The Microbiological Safety of Food, Part II. Report of the Committee on the Safety of Food (Chairman Sir Mark Richmond), London, HMSO, 1991.
35 CULLOR J S, An HACCP learning module for graduate veterinarians, *J Am Vet Med Assoc*, 1996 **209** 2049–50.
36 CASWELL J A and HOOKER N H, HACCP as an international trade standard,

Am J Agr Econ, 1996 **78**, 775–9.
37 EDWARDS D S, CHRISTIANSEN K H, JOHNSTON A M and MEAD G C, Determination of farm-level risk factors for abnormalities observed during post mortem meat inspection of lambs: a feasibility study, *Epidemiology and Infection*, 1999 **123** 109–19.
38 FRIES R, BANDICK A, DAHMS S, KOBE A, SOMMERER M and WEISS H, Field experiments with a meat inspection system for fattening pigs in Germany. World Congress on Food Hygiene, 29th August 1997, Den Haag, The Netherlands, 9–14.

5

HACCP in primary processing: red meat

C. O. Gill, Agriculture and Agri-Food, Canada

5.1 Introduction

From the turn of the twentieth century, developed countries have operated systems of meat inspection for the purposes of ensuring the safety and integrity of the meat supplied for human consumption. The principal concern was to exclude from the meat supply animals and carcasses with overt symptoms of systemic disease, and overtly diseased tissues from otherwise healthy carcasses.[1] Somewhat secondary concerns were to discourage grossly unhygienic practices at meat packing plants, and to prevent meat from animal species regarded locally as inedible from entering the meat supply under false description.[2] Meat inspection activities were therefore focused on the live animal and the carcass undergoing dressing, for detection of any disease condition at the earliest practicable time.[3]

Since the middle of the twentieth century, the incidence of disease in meat animals has been greatly reduced by improvements in animal husbandry practices such as the processing of feeds to assure their freedom from specific parasites and pathogens; the identification and culling of animals infected with specific pathogens, to obtain herds free of the targeted organisms; or vaccination of animals against diseases. The reduction of the incidence of a variety of diseases in many regions has progressed to the extent that the value of continuing with the inspection procedures designed to detect them is highly questionable.[4] Thus, for many years now, the greatest risk to public health from red meats has been their contamination with enteric pathogens, which are often carried by symptomless animals.[5,6] Obviously, no amount of inspection can identify symptomless carriers or the presence of pathogenic bacteria on apparently wholesome meat. Nonetheless it was believed by regulatory authorities that the matter could be addressed by traditional inspection procedures; that is, by increasingly detailed definitions of unacceptable and desired practices at meat plants, and increasingly detailed specifications for the construction and design of plant and equipment, with all requirements being enforced by continuous surveillance at each plant.[7]

Some thought otherwise. Given the variability of the material being processed, the variety of processes, the increasing speed of processing, and the dubious ability of inspection to detect microbiologically compromised product, it was suggested that assurance of the microbiological safety of meat would require the implementation of HACCP systems.[8] However, until recently, regulatory authorities persisted with the traditional approach, during which time there was no indication that the microbiological condition of meat was improving in any way.[9]

Finally, in response to some well-publicized outbreaks of meat-borne disease, the United States Department of Agriculture promulgated a policy of replacing traditional meat inspection practices, in respect of matters other than the detection of overt disease in animals and carcasses, with HACCP systems for assuring the hygienic adequacy of meat plant processes.[10] Other national meat inspection agencies soon followed this long-considered course. Only then did it become apparent that procedures for implementing HACCP systems of assured efficacy at meat packing plants were wholly lacking.

There are seven stages in the classic construction of a HACCP system (Table 5.1). The first stage is a hazard analysis, in which a team of individuals with, collective, general expertise in the type of process that is being considered and specific familiarity with the process for which the HACCP system is being constructed, identify all possible hazardous conditions of the product that might develop in the course of the process.[11] Then, the operations in the process where each hazard may be wholly prevented or minimized, or wholly eliminated or minimized, are identified as the Critical Control Points for the process. A CCP where a hazard is wholly prevented or eliminated is type 1; when the hazard is only minimized the CCP is type 2.[12,13] Clearly, if the hazard analysis is faulty, the CCPs will be unrecognized or misidentified, and the control system will not operate to assure the safety of the product.

The problem was and is that for raw meats pathogenic bacteria may be added to or may grow on the product during almost any stage of processing, while the

Table 5.1 The actions classically required for constructing a HACCP system

1. Conduct a hazard analysis
2. Identify the CCPs[a]
3. Establish performance criteria for each CCP
4. Establish monitoring for each CCP
5. Identify corrective actions for failure at each CCP
6. Document the system
7. Establish a verification procedure

[a] CCP = Critical Control Point

product cannot receive a treatment that will assuredly remove all pathogenic bacteria from it. Thus, by the usual definition, almost every operation in a meat packing plant can be considered a CCP. Moreover, every CCP would be type 2, where microbiological contamination might be minimized but not entirely prevented or removed.

Faced with the obvious impossibility of constructing HACCP systems for processes in which all operations can formally be regarded as CCPs, and in the usual absence of any data which identifies the microbiological effects of any operation in any process, plant managements have often attempted to derive HACCP systems from existing Quality Assurance procedures which are uncertainly related, or sometimes obviously not related, to product safety. Similarly, regulatory authorities have elaborated procedures for HACCP system implementation which are based largely on traditional inspection practices; that is, on the assumption that the microbiological performance of a process can be decided by inspection of product and equipment for visible contamination, with the provision that the process is documented and performed in accordance with all other requirements of the regulatory authority.[14–16] As a result, the systems for supposedly assuring meat safety that are now being implemented at meat packing plants are not HACCP systems at all. Instead, they are Quality Management Systems for assuring the quality of compliance with regulatory requirements. Such systems will function to assure meat safety only if microbiological safety is assured as a consequence of meeting the regulatory requirements. That has certainly not been the case in the past, and there is no reason to suppose that it is the case now.

This chapter therefore offers for consideration an alternative approach to HACCP implementation at meat packing plants. A complete HACCP system should, of course, include procedures for controlling physical and chemical as well as microbiological hazards and the same people may well be responsible for assuring the control of the three types of hazard. However, physical and chemical hazards will not be considered as they are of generally lesser concern, and can be controlled by more usual procedures than the microbiological hazards that arise in meat plant processes.[17]

5.2 Microbiological data: collection and analysis

A HACCP system which is designed to control microbiological contamination must be based on microbiological data.

The efforts and costs required for the collection of microbiological data are relatively large, so the amount of microbiological data that it is practicable to collect from any process is very limited. Also, there is inevitably a more or less lengthy lag between the collection of samples from a process and their analysis for bacteria of interest. Consequently, microbiological data cannot be used for the routine, on-line control of processes. That must be achieved by the maintenance of appropriate Standard Operating Procedures (SOPs) at the CCPs.

Microbiological data must be used instead to characterize the hygienic performance of a process, with identification of the CCPs; to determine the microbiological effects of any changes to previously existing operations, or introduction of novel operations, particularly when the changes to the process are intended to improve the hygienic performance; and to verify the hygienic performance of a process.

Although it would be desirable to examine samples for the pathogenic bacteria of concern, those are generally too infrequent and few on meat and associated equipment to be useful in the construction and operation of a HACCP system.[18] Instead, samples must be examined for more numerous organisms that are indicative of the possible presence of pathogenic types. The indicator organisms that can be suggested for process characterization, and validation and verification of HACCP systems[19] are total aerobic counts, coliforms, generic *Escherichia coli*, generic *Aeromonas* and generic *Listeria*.[20] Total aerobic counts are an indicator for the general microbiological condition of product and equipment. The coliform group includes both psychrotrophic organisms which will grow at chiller temperatures and mesophilic types such as *E. coli*. *E. coli* is the accepted indicator for contamination with faecal material.[21] When recovered coliforms are largely *E. coli*, faecal material and/or ungulate ingesta are the likely sources. When *E. coli* are a small fraction of the coliform population, the immediate source is likely to be environmental, or in the case of pork packing processes possibly the mouths of animals.[22] *Aeromonas* isolated from meat or equipment may include *Aeromonas hydrophila*, and *Listeria* may include *Listeria monocytogenes*, both the named species being cold-tolerant pathogens.[23] Both *Aeromonas* and *Listeria* on meat usually derive from in-plant rather than animal sources,[24] with *Aeromonas* occurring in high numbers in pooled water and moist detritus at packing plants, while *Listeria* is recovered commonly from the drier detritus that persists in some equipment and from heavily polluted areas such as drains.[25]

To evaluate the hygienic performance of an operation or process it is necessary to enumerate rather than simply detect the presence of indicator organisms. As bacteria grow and die exponentially, it is proper to compare logarithms rather than untransformed numbers of bacteria. Thus, it has been a usual practice when comparing the numbers of bacteria recovered from groups of samples to transform the number recovered from each sample to a log value and to compare the means of the log values for each set of counts.[26] That procedure may lead to an erroneous assessment of the hygienic performance of a process when data from successive stages of a process are compared, because decreasing variance between bacterial numbers on the product as a process proceeds will alone give increasing mean log values.[27]

The microbiological condition of meat should properly be decided by reference to the log of the mean numbers present on the product. The distribution of bacteria on raw meat tends to approximate the log normal.[28] The log mean for a set of n counts is then related to the mean log by the formula

$$\log A = \bar{x} + \log_n 10 \left(\frac{s^2}{2}\right)$$

where log A is the log mean, $\bar{x}$ is the mean log and s is the standard deviation of the log counts.[29]

It follows that the mean log will always be less than the log mean unless the standard deviation is zero, and processing that results in a reduction of variance between counts will produce an increase in the mean log even when no bacteria are added to the product during the process (Fig. 5.1). Consideration of mean log values for the numbers of bacteria on the product before and after processing could then suggest that bacteria had been added to the product during the process, and precipitate a fruitless search for the non-existent source of the supposedly additional bacteria.

In practice, the log mean numbers of bacteria on a raw meat product at any stage of a process can be reasonably estimated from 25 samples collected at random from the product passing through the process, provided that bacteria are recovered from 20 or more of the 25 samples.[30] A total aerobic count is likely to be recovered from most if not all samples in a set even when the sample size is small. However, other indicator organisms may be few, so it is desirable to collect large samples so as to obtain a high incidence of positive samples for all indicator organisms.

Until meat is comminuted, bacteria are present only on the surface. Surfaces may be sampled by excision of tissue or swabbing. Excision is commonly believed to recover more bacteria than swabbing.[31] However, excision is not a practical procedure for recovering bacteria from carcasses or cuts that are moving on high speed processing lines. Moreover, it appears that there is in fact little if any difference between the numbers recovered at packing plants by excision or swabbing meat surfaces with mildly abrasive materials such as cellulose acetate sponge or medical gauze (Table 5.2).[32] As the interest is in log numbers, there is no need for delimitation of an exact surface area to sample by swabbing, which is just as well, because applying a

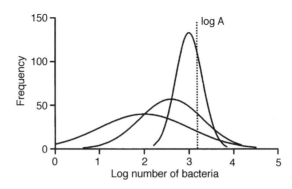

Fig. 5.1 Three different distributions of log numbers of bacteria for the same log mean numbers (log A) on the product.

Table 5.2 Log mean numbers of total aerobic bacteria recovered from beef or pig carcasses sampled by excision or by swabbing with cellulose acetate sponge, medical gauze or cotton wool buds[32]

Sampling method	Log mean numbers (log cfu/cm^2)	
	Beef	Pig
Excision	2.90	2.37
Sponge	3.02	2.35
Gauze	3.21	2.38
Cotton wool	2.61	2.12

template when attempting the swabbing of product which is moving on a high speed line is as impractical as sampling by excision. Even so, the area that can be sampled during a few seconds is limited. Thus, it has been found convenient for product undergoing processing at meat packing plants to sample by swabbing with a moistened 5 cm × 5 cm gauze swab an area of approximately 10 cm × 10 cm.[33]

All the bacteria of a type of interest recovered by such swabbing of a 100 cm^2 area can be enumerated, if necessary, by pummelling the swab with diluent, using a Stomacher® (Seward Medical Ltd, London, UK), and passing the whole of the fluid obtained through a hydrophobic grid membrane filter.[34] The filter can then be incubated on an appropriate selective medium for enumeration of an indicator organism. Usually total aerobes are far more numerous than the other groups of organisms. A single sample can therefore be used to enumerate both total aerobic counts, using about 10% of the fluid from the sample, and an indicator organism, using the rest of the fluid. In the case of coliforms and *E. coli*, both can be enumerated on a single filter.[35]

Even with enumeration at the level of 1 cfu/100 cm^2, an indicator organism may be recovered from fewer than 20 samples in some sets. Estimation of the numbers on the product is then still possible, by reference to the logs of the total numbers recovered from each set of 25 samples.[30] That value is, of course, equivalent to a crude estimate of the log mean. It has been observed in practice that the differences between the values for log total number recovered for sets of bacterial counts tend to be similar to the differences between the log mean values for the sets.

An alternative approach to enumerating relatively rare indicator organisms would be, in some situations, to increase the area sampled. To be useful, the increase has to be large, as the incidence of positive samples can be expected to only double for each ten-fold increase in the surface area sampled (Table 5.3).[32] Sampling of areas of 1000 or 10 000 cm^2 will usually be practicable only when product is stationary or removable from a processing line. However, for most purposes, sampling for 100 cm^2 areas seems sufficient for discernment of changes in the numbers of bacteria on product as a result of processing. In general, changes in log mean numbers or log total number recovered of one log

Table 5.3 Numbers of samples positive for *Escherichia coli* (+ve) in, and the log total numbers of *E. coli* recovered from, sets of 25 samples (log *N*) obtained by swabbing areas of 10, 100 or 1000 cm^2 on beef or pig carcasses[29]

Sampled area (cm^2)	Carcass type			
	Beef		Pig	
	+ve	Log *N*	+ve	Log *N*
10	6	1.15	3	0.90
100	13	2.24	5	1.95
1000	20	3.93	15	2.76

unit or more would be required to distinguish unambiguously an increase or decrease in bacterial numbers as a result of an operation or process. Values which differ by no more than 0.5 log units must be considered similar.[36]

5.3 HACCP implementation: the general approach

The classic approach to HACCP implementation is ineffective for controlling microbiological hazards in processes for raw meat production because knowledge of the microbiological effects of the individual operations in any process is generally lacking. Indeed, there is often little or no knowledge of the microbiological effects of any of the processes performed at a packing plant. The microbiological methods which have been described in the previous section can be used to remedy that lack of knowledge. To do that, the stages of HACCP system construction must be expanded from seven to some 12 stages (Table 5.4).

The activities which occur at any but small meat packing plants are too numerous to comprehend in detail if they are viewed as all being elements of a single production process. Therefore, it is necessary to divide the activities into discrete processes which can be investigated sequentially. Activities are divided into processes as seems convenient with regard to plant layout, procedures, products and management practices. The only provisions are:[37]

1. every activity that occurs in the plant must be placed in a process, and
2. no activity may be placed in more than one process.

With such limited requirements there is no reason why the list of processes should be the same for all plants. For example, the skinning, eviscerating, and trimming, washing and otherwise cleaning of beef carcass could be viewed as three processes of 1. skinning, 2. eviscerating, and 3. carcass cleaning, or as a single carcass dressing process. Despite that, processes are likely to be similarly defined at many plants because of broadly similar arrangements for processing and management of activities at most plants (Table 5.5). When deciding the list of processes, the HACCP team should identify the initial and final operations of

Table 5.4 The actions required for constructing an effective HACCP system for controlling the microbiological contamination of meat during a meat packing plant process

1. Describe the process
2. Establish consistent procedures for performance of the process
3. Identify the microbiological characteristics of the process
4. Establish the CCPs[a]
5. Implement actions to improve hygienic performance at each CCP
6. If appropriate, implement novel decontaminating operations
7. Establish SOPs[b] for each operation
8. Identify corrective actions for failure to maintain any SOP at a CCP
9. Identify the microbiological characteristics of the improved process
10. Establish microbiological criteria for process performance
11. Establish a verification procedure
12. Document the system

[a] CCP = Critical Control Point
[b] SOP = Standard Operating Procedure

each, and their relationships to one another, to ensure that no operation is overlooked, and that none is duplicated in different processes. Each process must then be examined separately, to determine its microbiological effects upon the product and to control the microbiological contamination of the product occurring during the process.

In principle, it would be desirable to proceed by first determining the microbiological condition of the product emerging from each process. Initial controlling and improving activities could then be focused on the process with the greatest deleterious effect on the microbiological safety of the product, as improvement of other processes would be largely irrelevant to product safety while the microbiological condition of the product was essentially determined

Table 5.5 Some processes likely to be performed at most meat packing plants

1. Reception of stock
2. Slaughter, and pre-dressing treatment of carcasses
3. Carcass dressing
4. Collection of offals
5. Collection of head meats
6. Carcass cooling
7. Carcass grading
8. Storage and shipment of carcasses
9. Carcass breaking
10. Storage and shipment of boxed meat
11. Storage and shipment of bulk meat
12. Cleaning of stock holding areas
13. Cleaning of slaughter and pre-dressing treatment facilities and equipment
14. Cleaning of carcass dressing facilities and equipment
15. Cleaning of carcass breaking facilities and equipment
16. Cleaning of personal equipment

by the process with the poorest hygienic performance. In practice, however, it is probably expedient in most cases to consider processes as they occur sequentially in a plant.

When considering a process it must first be described by discussion with those responsible for and involved in it. The description should commence with a plan of the facility or facilities within which the process occurs, with the identification of all items of equipment used in the process, the way in which the process is manned and the forms of the incoming and outgoing product. If the equipment used and the manning are variable, then the equipment used in and the manning of each recognized, alternative form of the operation must be identified. The persons responsible for decisions about the form of a variable process and for assuring that the selected form is in fact adopted must be identified.

For example, a beef carcass dressing process which is designed to deal primarily with carcasses from feedlot animals may have to be changed from its usual form to accommodate carcasses with hides bearing much hardened tag that can impede skinning operations, the carcasses of culled dairy cows, or unusually large carcasses from bulls or uncommon, large breeds. The changes may involve slowing of the line and implementing of additional operations, such as ones for removing tag from areas where the skin must be cut, removing the udder, or raising the forepart of the carcass to prevent its contacting the floor. Lots of cattle which will require some changes to the dressing process should be identified before slaughter, by a designated person, in accordance with established criteria. The information about which changes are required for which carcasses should be conveyed, by an established procedure, to the person responsible for overseeing the dressing process.

That person should implement standard procedures to ensure that the appropriate, additional operations are in place and that the line is operating at the appropriate speed before any carcasses of an unusual form are processed. For such changes to proceed without error it is obviously essential that all relevant information be conveyed directly, unambiguously and in good time to those who must act on it, and that the appropriate, prescribed actions are taken in response to the information received.

Each form of the process should be divided into a series of operations. An operation will usually be defined on the basis of the actions of individual workers and/or the use of individual, large items of equipment. Of course, an operation may be performed by more than one worker, and one worker may engage in more than one operation.

The description of a process provided by those involved with it may contain some accounts of ideal rather than the actual practice; mention of some practices which have been abandoned and failure to mention some practices which have been recently introduced; uncertainly about the actions performed by individual workers at some workstations; lack of detail about some operations; and little or no note or varying practices by individual workers performing the same operation. Therefore the initial description must be checked against the actual, routine performance of the process.

Each operation in the process should be checked against the description on at least three occasions, to develop a detailed description of how each operation is performed in practice. When operations are performed consistently, the detailed descriptions will become the provisional operating procedures for those operations. When any operation is found to be performed inconsistently, it will be necessary to develop a set of provisional operating procedures for it. The provisional operating procedures for an inconsistent operation must be developed with the assistance and agreement of both those who are performing and those who are supervising the operation, as workers are likely to rapidly abandon any arbitrarily imposed operating procedures that can be considered awkward or uncomfortable to implement. Procedures for maintaining the documented operating procedures should be agreed and implemented, and the process should be operated in the agreed manner for 2 or 3 weeks, with appropriate checking, to ensure that it is being routinely performed in a consistent manner.

It is essential that consistent procedures are established for a process before identification of its microbiological characteristics is attempted, as an inconsistent process is by definition uncontrolled, and it is not possible to establish from limited amounts of data the characteristics of an uncontrolled process.

The condition of product entering and leaving the consistent process with respect to the numbers of appropriate indicator organisms on it should be determined by random sampling of product entering and leaving the process. If the data indicate that the numbers of bacteria on the product are not increased during the process, then the process can be considered under control with respect to microbiological contamination and devoid of CCPs, provided that the documented operating procedures are maintained.

If the data indicate that the numbers of any of the indicator organisms increase or decrease as a result of the process, then control of the process to minimize increases or maximize decreases should be sought. Such optimum control of the microbiological effects of the process will be possible if the CCPs are properly identified.

The process should be inspected to identify those operations where contamination or decontamination of the product seem to be occurring. Then, randomly selected items of product entering and leaving the operation should be sampled at a site or sites which is or are affected by the operation.[38]

Operations which are found to be depositing on or removing from the product relatively large numbers of the indicator organism are the CCPs. Numbers are assessed as large relative to the numbers added or removed by the operation which has the greatest microbiological effect, as only that operation and perhaps one or two others with quantitatively similar microbiological effects will determine the microbiological condition of the product leaving the process.

The critical operations can then be examined to decide whether or not their hygienic performances might be improved by practicable changes to equipment, manning, or operating procedures. The microbiological effects of any supposedly improving change to an operation which is implemented should be determined by appropriate sampling of product entering and leaving the operation, as before. Changes which are demonstrated to be microbiologically effective may be incorporated as standard for the operation, and the process description should be amended accordingly. Changes which are microbiologically ineffective should be discontinued, and other approaches to improving the hygienic performance of the operation should be sought.

The improvements to the hygienic performance of a process which can be obtained by physically and economically practicable modifications of existing operations may be limited. Then, consideration should be given to the implementation of novel decontaminating operations in a process. Any such operation which is implemented should be examined by appropriate microbiological sampling of product entering and leaving the operation, first to determine whether or not the operation is in fact microbiologically effective, and secondly to identify the operating procedures for assuring its efficacy. Any novel and effective decontaminating operation must be a CCP.

When the process is in its final form, the documented operating procedures can be designated the Standard Operating Procedures (SOPs) for the process. To ensure the acceptable hygienic performance of the process, the SOPs at the CCPs must be maintained. Thus, procedures for adequately frequent, routine checking of the operating procedures at the critical operations must be implemented. Checks on operating procedures may involve observation of workers' actions, inspection of product or equipment for filth or detritus, and/or confirmation of the operating conditions set for equipment. In addition, the process must be regularly if less frequently assessed against the standard description, to ensure that operating procedures generally do not drift from those which are standard, and that any intended changes to the plant, equipment or operating practices are fully documented and tested for their microbiological effects.

When a process is operating under SOPs, procedures for dealing with failure to maintain the SOPs at each of the CCPs must be developed. The procedures for dealing with SOP failure should be invariant and preferably implementable by the line workers involved with the critical operations. The responses to failure must be actions to correct misprocessing and to separate for appropriate corrective treatment or disposal any product that has been misprocessed. Merely reporting misprocessing to supervising staff for subjective assessment of or decision on misprocessed product is not a proper response to failure at a CCP.

When a process is operating under SOPs with established procedures for dealing with processing failures, it can then be regarded as under control. The microbiological characteristics of the controlled process can then be identified by determining the microbiological effects of each critical operation, and the microbiological condition of the product leaving the process. That microbiological information should be incorporated in the process description, and the condition of the product leaving the process should be used to define microbiological criteria for the accepted performance of the process. Verification of the process would then involve assessment of the process against the process description, and sampling of the final product to determine whether or not it complies with the microbiological criteria for product from the process.

Finally, the documentation of the HACCP system should be organized in a form that is useful to those who must operate the system to assure the microbiological safety of the product.

The investigative approach to implementing HACCP systems at meat packing plants is necessary because it is impossible to be certain of the microbiological effects of any operation or process in the absence of appropriate microbiological data. Indeed, apparently similar operations or processes at different plants can have very different microbiological effects upon the product. Despite that, some general account of the microbiological effects of meat plant processes is necessary and useful for directing the initial activities for HACCP implementation at any plant. Such general accounts of processes are therefore provided in the following sections. However, readers should bear in mind that in particular instances a general description may not describe at all the hygienic characteristics of a particular process.

5.4 Stock reception

Animals presented for slaughter must be inspected by a member of the meat inspection authority for symptoms of overt disease or injury, which would necessitate rejection of the animal for use as human food and/or its emergency slaughter. Such inspection with appropriate responses to symptoms of disease or injury should be part of the HACCP system.

The other factor which is considered at some plants is the general cleanliness of the stock. For stock like cattle and sheep which give carcasses that are skinned, it seems self evident that more contaminants are likely to be transferred to the meat from the hides during skinning of dirty than of clean stock.[39] In addition, intensively reared stock in particular may become so coated with mud and manure on bellies, flanks and legs that cutting of the hide is mechanically impeded. Consequently, stock may be washed before slaughter, as is usual with sheep in Australasia;[40] or when animals are judged to have hides carrying excessive tag the dressing process for carcasses from them may be slowed or otherwise modified to accommodate difficulties with removing the hides, as is usual with cattle in some regions.[41] Thus, there are two aspects to consider when deciding how to deal with dirty stock within a HACCP system. These are: the extent to which dirty hides increase the contamination on carcasses when the carcasses are dressed at a normal rate and with the maintenance of all SOPs for the dressing process; and what operations must be introduced to allow the controlled processing of some types of dirty stock.

With regard to the first of those aspects, research has generally failed to confirm the intuitively expected relationship between visible filth on hides and the microbiological condition of carcasses (Table 5.6).[42] Instead, it has been

Table 5.6 The mean log numbers of total aerobic bacteria recovered from carcasses from groups of cattle assessed as having different levels of visible filth on the hides[42]

Hide condition score[a]	Mean log numbers (log cfu/cm^2)
0	3.03
1	3.39
2	3.49
3	3.58
4	3.39
5	3.29
6	3.13

[a] Hide condition scale: 0 = clean, 9 = hide very dirty and covered with tag

observed that when hides are wet the microbiological contamination of carcasses tends to be greater irrespective of the apparent cleanliness of the hide (Table 5.7).[43] Thus, washing of animals before they are slaughtered cannot be expected to improve the microbiological condition of carcasses, but washing may result in the microbiological condition of carcasses being degraded if the hides are not dry by the time skinning commences.

Any treatment other than washing to remove tag from the live animal, for the purpose of permitting unmodified operation of the dressing process, would seem undesirable, if only because the animals would likely be stressed, which could well have adverse effects upon the quality of the meat. The only actions in respect to a HACCP system to be taken in response to animals being excessively dirty would then be to assess all groups of animals according to a formal scoring system as acceptably clean or excessively dirty, with segregation of the one type from the other and operation of an appropriately modified slaughter and predressing treatment process or dressing process for processing of the excessively dirty carcasses.

Table 5.7 The effect of washing animals on the mean log numbers of total aerobic bacteria recovered from dressed sheep carcasses[43]

Condition of the hide	Washed or unwashed	Mean log numbers (log cfu/cm^2)
Short wool, clean	Washed	4.33
	Unwashed	4.00
Short wool, dirty	Washed	3.94
	Unwashed	4.05
Long wool, clean	Washed	4.47
	Unwashed	3.94
Long wool, dirty	Washed	4.74
	Unwashed	4.30

5.5 Slaughter and predressing

Slaughter and predressing treatment processes involve the stunning, sticking and bleeding out of animals, and the shackling and raising of carcasses to a rail. It may also involve electrical stimulation of cattle or sheep carcasses after bleeding; operations to clear tag from the hides of cattle or sheep carcasses; or the operations of scalding, dehairing, singeing and polishing (blackscraping) of pig carcasses.

Electrical stunning is usually employed for sheep and pigs or, less commonly and for pigs only, stunning by CO_2. However, cattle are usually stunned by means of a captive bolt applied to the forehead. The captive bolt will obviously tend to drive contaminants from the hide into the brain, but the other forms of stunning may not be wholly aseptic. For example, to ensure the necessary good electrical contact, electrodes which penetrate the skin are used with some automatic electrical stunning equipment for pigs. During the stunning operation, the animal may fall to one side against the electrode, which penetrates deeply and tears the flesh when the animal is moved forward.

After stunning, sheep and pigs are usually dropped to a conveyor where each is arranged with the head over one side. The animals are then stuck and allowed to bleed out before each carcass is shackled by one hind leg and raised to a rail. Stunned cattle are usually dropped from the knocking box to the floor, and are shackled and raised to the rail before they are stuck. Sticking wounds in the throats of cattle and pigs are usually relatively small, even when the blood is collected by means of a hollow handled (vampire) knife, but sticking of sheep may involve a gash cut to sever the neck from the ventral aspect through to the backbone. Irrespective of the size of the wound, flesh is obviously likely to be contaminated during the sticking operation.

Despite the contamination of tissues around any wounds made during slaughter, contaminants apparently do not spread to other parts of the carcass.[44] Thus, while SOPs for slaughtering operations should aim to prevent unnecessary wounding of animals, the minimizing of necessary wounding and the adequate cleaning of instruments used during slaughter, none of the slaughtering operations can likely be considered a CCP. Indeed, given the relationship between wet hides and increased contamination of meat during skinning, wetting of the hides of cattle and sheep while the stunned animals are on the floor or sticking conveyor might ultimately have a greater deleterious effect than any wounding, and should be avoided.

If excessively dirty carcasses of cattle or sheep must be treated to allow the normal operation of the carcass dressing process, then any such treatment would probably be a CCP. In that case, either excessively dirty carcasses would have to be reliably identified, or the treatment would have to be applied to all carcasses irrespective of the conditions of the hides. A treatment could conceivably involve the removal of tag by shearing of the embedded hair from the hide, possibly only along the lines where cuts would be made in the hide during dressing, or possibly with the breaking up of hardened layers of tag on some parts of the hide. However, such operations do not seem to have been described in the literature.

The only operation described in the literature that might be relevant to the matter of cleaning excessively dirty carcasses was an experimental one for dehairing beef carcasses before they were dressed.[45] The treatment did not result in any reduction in the bacterial load on carcasses, but it did appear to reduce visible contamination and, if applied to all carcasses, could avoid any perturbation of the dressing process that might arise from a need to clean excessively dirty carcasses.

Although dehairing of beef carcasses is only an experimental treatment, the cleaning and dehairing of pig carcasses which are then dressed without being skinned is usual at most pork packing plants. That cleaning and dehairing of pig carcasses involves the four sequential treatments of scalding, dehairing, singeing and polishing. Carcasses are usually scalded by being drawn by one shackled leg through a tank of water, at a temperature of about 60 °C, for about 8 minutes. The treatment tends to remove dirt adhering to the skin, and destroys most of the bacteria on the carcass surface.[46] The carcass is then withdrawn from the scalding tank, unshackled, and passed through a dehairing machine in which it is scraped and rotated by broad rubber flails attached to revolving drums while warm water is circulated over the carcass from a tank beneath the equipment. The condition of the recirculated water is well suited for the growth of bacteria, while the treatment forces or washes faeces and saliva from the carcasses. Thus, the carcasses that emerge from the dehairing equipment are usually heavily contaminated with bacteria.[47]

The dehaired carcass is suspended from a trolley on a processing rail by means of a gambrel passed between the tendon and the bone of each rear hock. The gambrelled carcass is passed through some arrangement of gas flames to burn any residual hair on the skin. The singeing of the skin causes a more or less large reduction of the numbers of bacteria on it.[48] However, even heating of the skin during singeing is essentially impossible, so large numbers of bacteria persist on some parts of the carcass. The singed carcass is passed to the polisher where it is flailed with thin cords and scrubbed by stiff brushes to remove carbonized scurf and hair. During that treatment the bacteria which survived the singeing treatment are spread over the carcass and are likely augmented by bacteria which persist in the polishing equipment. Therefore, despite appearing clean the polished carcasses will usually carry substantial numbers of both spoilage and pathogenic bacteria which will persist on the dressed carcasses unless some decontaminating treatment is employed.

Recontamination of scalded carcasses during dehairing by flailing would seem unavoidable, and uniform heating of carcass surfaces during singeing to largely destroy bacteria at all points on the surface would seem impractical. However, singeing operations can be adjusted to maximize the destruction of bacteria during that operation, while polishing apparatus could be designed and operated to minimize any persisting bacterial flora in the equipment. If that were done singeing and polishing operations might be regardable as CCPs, but at present there would be few pork packaging plants where the microbiological effects of the singeing and polishing operations are known, and very few if any where those operations are purposely used to enhance the microbiological conditions of carcasses.

5.6 Carcass dressing

Carcass dressing processes for most species can be considered as proceeding in three, sequential, broad phases of skinning, eviscerating and cleaning the carcasses. In the first phase, the skin is cut open and progressively stripped from the carcass, usually commencing with the hindquarters on carcasses which are suspended by the rear legs but, with sheep at least, sometimes commencing with the brisket and shoulders on carcasses which are suspended by the front legs.[49] In the second phase the head, the viscera and sometimes the tail are removed, and the carcass may be split along the backbone. In the third phase the carcass is trimmed to remove excess fat, bruised tissue and visible contamination, and is washed. Vacuuming treatments to remove visible contamination from carcasses may also be applied during any of the three phases. Thus, a beef carcass dressing process will typically involve over 30 distinct operations (Table 5.8).

There is wide variation in the microbiological condition of the carcasses from different beef carcass dressing processes, and no consistent relationship between the numbers of different indicator organisms on carcasses from different processes. Thus carcasses from some processes can carry relatively high numbers of total aerobic counts but few *E. coli* while both groups of bacteria are relatively few on carcasses from other processes, and on carcasses from some processes the coliforms are largely *E. coli* while on carcasses from other processes *E. coli* are only a small fraction of the coliforms present (Table 5.9).[30] However, the microbiological condition of the carcasses leaving any process tends to be consistent irrespective of whether the carcasses come from feedlot beef animals or culled dairy cows (Table 5.10)[30] or of the season of the year, in regions where there are large seasonal differences (Table 5.11).[30] Those observations indicate that the manner in which the dressing process is performed is generally of far greater importance for determining the microbiological condition of dressed carcasses than are the conditions of the hides of the incoming stock.

Because of the lack of consistent relationships between the numbers of various types of bacteria on product from different plants, the microbiological performance of dressing processes with respect to safety must be assessed by reference to the numbers of *E. coli* or other organisms indicative of possibly hazardous contamination, rather than by reference to total aerobic or coliform counts, which may be unrelated to health hazards.

Some bacteria will inevitably be deposited on the meat during the skinning of carcasses. Contamination during skinning must therefore be minimized by

Table 5.8 Operations in a high line speed beef carcass dressing process

1. Stun
2. Shackle
3. Bleed
4. Skin right rear hock
5. Skin right butt
6. Remove right, rear hoof; hook right leg
7. Skin left, rear hock
8. Skin left butt
9. Remove left, rear hoof; hook left leg
10. Open brisket skin
11. Open tail skin
12. Skin rump
13. Vacuum rear hocks
14. Skin tail
15. Vacuum butts
16. Remove horns, ears and front hooves
17. Skin brisket
18. Skin back
19. Remove hide
20. Trim head
21. Split sternum
22. Trim forelegs
23. Free and wrap bung
24. Remove head; tie esophagus
25. Remove viscera
26. Split carcass
27. Change from dressing to main chain hook
28. Trim butt
29. Trim rump
30. Trim brisket
31. Remove tail
32. Remove hanging tender
33. Remove mesenteric fat
34. Remove diaphragm remnants
35. Trim neck
36. Weigh
37. Wash

adopting practices which limit direct and indirect contacts between the outer surface of the hide and the meat.[50] It is likely that the majority of the *E. coli* on carcasses are deposited on the meat during only a few skinning operations. Those CCPs should become apparent during the examination of the microbiological effects of the operations in the dressing process (Table 5.12).[51] Actions to improve the performance of the critical skinning operations are likely to involve only relatively minor changes to working practices, manning, or the arrangement of equipment (Table 5.13).[52] Upgrading of the facilities without proper consideration of working practices cannot be expected to improve the microbiological performance of a process.[53]

Table 5.9 Log mean numbers of total aerobic bacteria, coliforms and *Escherichia coli* on beef carcass sides leaving the carcass dressing processes at 10 beef packing plants[30]

Plant	Log mean numbers		
	Aerobes (log cfu/cm^2)	Coliforms (log cfu/100 cm^2)	E. coli (log cfu/100 cm^2)
A	3.42	1.96	2.06
B	3.12	2.03	2.01
C	4.28	3.05	1.98
D	3.62	2.51	1.74
E	4.89	2.94	1.28
F	3.70	1.89	0.79
G	2.78	1.39	0.75
H	2.20	0.77	0.70
I	3.01	1.56	0.58
J	2.04	—[a]	—

[a] — Numbers recovered too few for calculation of the statistic

Enteric organisms will be present in faecal material in the rectum or around the anus, and in the mouths of cattle. Operations involving the bung or head during the eviscerating phase of dressing can then result in hazardous contamination of the carcass. However, contamination from those sources can apparently be largely avoided if the freed bung is enclosed in a plastic bag during an operation in which a worker handling the bung does not have to contact any other part of the carcass, and the head is removed by a worker who does not handle other parts of the carcass.[54,55] Rupture of the gut during evisceration must be considered a failure of control and should automatically precipitate actions to remove the affected carcass and offal from routine processing, with cleaning or replacement of any affected equipment before the resumption of normal processing. Affected product should be discarded or subjected to cleaning and decontaminating treatments which will

Table 5.10 Log mean numbers of total aerobic bacteria, coliforms and *Escherichia coli* on sides of beef cattle or cow carcasses leaving the carcass dressing processes at three packing plants[30]

Plant	Carcass type	Log mean numbers		
		Aerobes (log cfu/cm^2)	Coliforms (log cfu/100 cm^2)	E. coli (log cfu/100 cm^2)
A	Beef	3.62	2.51	1.74
	Cow	3.82	2.60	1.73
B	Beef	3.70	1.89	0.79
	Cow	3.50	1.75	0.73
C	Beef	2.87	1.39	0.75
	Cow	3.21	1.90	0.64

Table 5.11 Log mean numbers of total aerobic bacteria, coliforms and *Escherichia coli* recovered at different times from carcass sides leaving a beef carcass dressing process[30]

Sampling period	Log mean numbers		
	Aerobes (log cfu/cm^2)	Coliforms (log cfu/100 cm^2)	*E. coli* (log cfu/100 cm^2)
June 1995	3.83	2.33	2.30
June 1995	2.18	1.88	1.84
May 1996	3.12	2.03	2.01
June 1996	3.35	3.39	2.57
July 1996	4.01	2.38	2.03
July 1996	3.21	2.54	2.26
Jan. 1997	3.50	2.34	2.16
Feb. 1997	3.25	1.44	1.13
Apr. 1997	3.19	1.60	1.50
May 1997	3.43	1.96	1.87

ensure that their microbiological condition is comparable with that of normally processed product.

Cleaning and decontaminating treatments for beef carcasses have traditionally involved trimming and washing. In North America at least it has also been a usual practice to vacuum clean areas such as the hocks where it is difficult to avoid contamination of the meat with hair and other material from the pelt. More recently, apparatus for applying hot water and/or steam to a carcass surface that is being vacuum cleaned, and apparatus for treating washed sides of beef with solutions of organic acids or other antimicrobials, hot water or steam have been installed at beef packing plants. Thus, cleaning and decontaminating treatments are of two distinct types: those that are applied to limited areas where visible contamination is apparent, and those that are applied to the whole carcass side irrespective of the presence or otherwise of visible contamination.

Application of treatments of the first type is guided by the visible contamination which is observed on the carcass. Unfortunately there is no

Table 5.12 Log total numbers of *Escherichia coli* recovered from 25 samples obtained from sites on carcasses related to specific hindquarters skinning operations or to hindquarters skinning as a whole in three beef carcass dressing processes[51]

Operation	Log total numbers (log cfu/2500 cm^2)		
	Process A	Process B	Process C
Cut crotch	5.47	5.03	3.52
Skin hock	4.82	4.62	3.97
Skin butt	3.99	0.95	2.75
Skin rump	4.04	2.49	2.61
Skin hindquarters	4.72	3.42	2.44

Table 5.13 Log total numbers of *Escherichia coli* recovered from 25 samples obtained from sites on carcasses related to specific hindquarters skinning operations or to hindquarters skinning as a whole before or after improvement of the hindquarter skinning operations in a beef carcass dressing process[52]

Operation	Log total numbers (log cfu/2500 cm^2)	
	Unimproved	Improved
Cut crotch	5.47	4.05
Skin back	4.82	4.69
Skin butt	3.99	3.21
Skin rump	4.04	1.46
Skin hindquarters	4.72	3.44

relationship between the extents of visible and microbiological contamination.[56] Consequently, the treatments for removing visible contamination from limited areas are largely without effect on the microbiological condition of the carcass even when, as with trimming, some bacteria will be removed along with a portion of the surface (Table 5.14).[57] Despite that, trimming might sometimes be effective if it is applied to all sides, irrespective of the appearance, to a site of appropriate size which is known to be often relatively heavily contaminated with enteric organisms (Table 5.15).[58] Treatments such as vacuuming while treating with water or steam of pasteurizing temperatures will be ineffective even then, unless the whole of the treated surface is raised to more than 80 °C for at least 10 seconds. That is essentially impossible with current treatments on high speed lines in which a cleaning head with an orifice of area about 50 cm^2 is applied to an area of over 1 m^2 during a period of less than 20 seconds. Treatments of limited areas will therefore usually only clean but not decontaminate carcasses, and so should not be regarded as CCPs.

Table 5.14 Log mean numbers of total aerobic bacteria, coliforms and *Escherichia coli* recovered from lamb hindquarters before or after trimming, vacuum cleaning or hot water vacuum cleaning[57]

Operation	Stage of the operation	Log mean numbers		
		Aerobes (log cfu/cm^2)	Coliforms (log cfu/100 cm^2)	E. coli (log cfu/100 cm^2)
Trimming	Before	3.97	4.02	3.66
	After	3.96	4.06	3.90
Vacuum cleaning	Before	3.67	3.47	3.41
	After	3.56	3.35	3.03
Hot water vacuum cleaning	Before	3.32	4.39	4.32
	After	3.27	4.33	4.21

Table 5.15 Log total numbers of total aerobic bacteria, coliforms and *Escherichia coli* recovered from 25 samples from an anal site on cooled beef carcasses before or after routine trimming of the site during a carcass breaking process[58]

Stage of the operation	Log total numbers		
	Aerobes (log cfu/25 cm^2)	Coliforms (log cfu/2500 cm^2)	E. coli (log cfu/2500 cm^2)
Before	4.10	3.50	3.18
After	3.88	2.32	2.16

Among treatments of the whole carcass side, washing, like the limited areas treatments, is generally effective for removing visible contamination, but ineffective for removing bacteria. That need not be the case, as it is apparently possible to wash sides in a manner which removes substantial numbers of bacteria (Table 5.16).[59] However, such an effect would at present likely be serendipitous as the microbiological effects of the washing operations are unknown at most meat plants.

The primary purpose of the carcass washing operation is currently the removal of visible contamination and blood. For that purpose, large quantities of water must be used. It is therefore generally uneconomical to include antimicrobiological agents in wash waters. Instead, if they are used, they are applied in a separate operation after washing. Unfortunately, it is impossible to achieve a consistent coverage of all parts of the carcass surface without using large volumes of solution, and bacteria on a meat surface tend to be protected against the lethal effects of antimicrobial agents. Thus, commercially impractical treatments with large volumes of hot and relatively concentrated solutions are required to achieve consistent and large reductions in total bacterial numbers by treating carcasses with organic acids.[60] Moreover, bacteria are highly varied in their susceptibilities to antimicrobial agents. For example,

Table 5.16 Log total numbers of total aerobic bacteria, coliforms and *Escherichia coli* recovered from 25 samples from beef carcasses before or after washing in three carcass dressing processes[59]

Plant	Stage of the operation	Log total numbers		
		Aerobes (log cfu/25 cm^2)	Coliforms (log cfu/2500 cm^2)	E. coli (log cfu/2500 cm^2)
A	Before	3.95	2.38	2.01
	After	4.46	2.29	2.81
B	Before	4.33	3.74	2.12
	After	4.99	2.89	2.74
C	Before	4.00	3.22	3.12
	After	3.35	2.21	1.45

enteric organisms such as *E. coli* and *Salmonella* are notably resistant to the lethal effects of the organic acids which are used in commercial decontaminating treatments for beef carcasses.[61] Treatments of carcasses with solutions of antimicrobial agents are then likely to be ineffective for reliably destroying substantial numbers of pathogenic organisms, and so are unlikely to be CCPs.

Pasteurizing beef sides with hot water or steam can be effective for reducing numbers of *E. coli* by more than two orders of magnitude (Table 5.17).[62,63] Treatment with steam must take place within a sealed chamber, with steam applied at greater than atmospheric pressure or after evacuation of air, so that steam condenses evenly onto the whole of the carcass surface.[64] For effective heating of the surface with steam the surface must be clean, as any surface underlying debris will be protected from heating, and dry, as otherwise the film of water on the carcass surface rather than the surface itself will be heated by the steam. Scrupulous cleaning and drying are not necessary for hot water pasteurization to be effective, but water must be delivered onto the carcass as sheets rather than as sprays because the large surface area presented by droplets from a spray head results in rapid cooling of the water with ineffective heating of the carcass surface.[65]

Pasteurizing treatments for carcasses need have no large or persisting effects on the appearances of skin, fat, cut bone or membrane-covered surfaces. However, cut muscle surfaces will be dulled and darkened or bleached by effective pasteurizing treatments. As such discoloration is undesirable, there can be a tendency to reduce the treatment temperature and/or time to minimize the effect on the appearances of carcasses. That should be avoided, as the treatment may then become ineffective.[66] Instead, the operating parameters for a pasteurizing treatment must be established by reference to microbiological data. Then, the carcass pasteurizing treatment will be a CCP.

Sheep carcass dressing processes are similar to those for beef in that bacteria are deposited on the carcass mainly during the skinning operations. Mechanization of some skinning operations and inverted dressing, where the forequarters are skinned first then the hide is pulled from the rump and hind legs while the carcass is suspended by the forelegs, may somewhat reduce

Table 5.17 Log total numbers of total aerobic bacteria, coliforms and *Escherichia coli* recovered from 25 samples from beef carcass sides before or after a pasteurizing treatment with hot water or steam[62,63]

Treatment	Stage of the treatment	Log total numbers		
		Aerobes (log cfu/25 cm^2)	Coliforms (log cfu/2500 cm^2)	E. coli (log cfu/2500 cm^2)
Hot water	Before	5.21	3.84	3.79
	After	3.09	0.90	0.00
Steam	Before	5.23	4.06	3.84
	After	4.19	1.69	1.11

Table 5.18 Log mean numbers of total aerobic bacteria, coliforms and *Escherichia coli* on pig carcasses after polishing and after dressing at two packing plants

Plant	Stage of processing	Log mean numbers		
		Aerobes (log cfu/cm^2)	Coliforms (log cfu/100 cm^2)	E. coli (log cfu/100 cm^2)
A	After polishing	3.54	1.72	1.53
	After dressing	3.75	1.58	1.41
B	After polishing	3.60	1.38	1.02
	After dressing	3.78	3.04	2.52

bacterial contamination. However, any reduction would seem to be small as the microbiological condition of dressed sheep carcasses is apparently similar at many plants and has apparently remained unchanged over many years.[49] It seems likely that the unavoidably extensive handling of the small sheep carcasses severely limits control over the deposition of bacteria on carcasses.[67] Thus, with sheep carcasses, an effective decontamination treatment, such as carcass pasteurizing, may be a necessary treatment and CCP for assuring the microbiological condition of the meat.

With pig carcasses which are not skinned, the microbiological condition of the carcasses at the end of a dressing process is often little different from that of the carcasses leaving the polisher (Table 5.18). In other operations, large numbers of bacteria may be deposited on carcasses from the mouth and/or in faecal material from the intestine.[38] In most processes, the mouth is probably the major source of the enteric organisms deposited on carcasses during dressing.[22] Improvement of the microbiological condition of carcasses will then usually require the implementation of a pasteurizing treatment. Such a treatment can be applied to the polished carcasses, preferably after work on the head has been completed, to give carcasses which carry few *E. coli* at the

Table 5.19 Mean scores for the overall appearances and for the appearances of individual types of tissue on sides of pork which were not pasteurized or were pasteurized before or after the sides were split. Appearances were scored on a scale where 1 = very undesirable and 7 = very desirable[69]

Tissue evaluated	Mean scores			
	Treated split		Treated unsplit	
	Untreated	Treated	Untreated	Treated
Overall	6.49	4.57	5.38	5.67
Fat	6.45	5.40	5.87	5.53
Cut bone	6.45	5.88	5.95	6.10
Membrane	6.37	5.56	5.98	5.75
Cut muscle	6.44	4.14	5.39	5.32

104   HACCP in the meat industry

end of the dressing process, and without any damage to the appearance of the dressed carcass.[68] However, treatment of the carcass after evisceration but before it is split seems to be optimal, as by then all operations where carcass contamination might not be wholly avoidable have been completed, while little cut muscle is exposed to be discoloured by the decontaminating treatment (Table 5.19).[69]

5.7   Collection and cooling of offals

Offals include a range of diverse tissues which can be considered as falling into three broad groups with respect to their collection and cooling. Those groups are the mainly muscle tissues of head meats, weasand meat and tongues, and appendages such as tails, ears and trotters; visceral organs such as the heart, liver, kidneys, spleen and thymus; and portions of the gut, such as parts of cattle and sheep stomachs which are sold as tripes, the small intestines of pigs which are sold as chitterlings, and the tubes of connective tissue stripped from the outsides of intestines which are sold as natural casings.[70]

The tissues of the first group will be more or less heavily contaminated with bacteria at the time of their removal from the carcass whatever collection methods are adopted. The critical operations in the collection processes for those tissues are then the cleaning treatments and any decontaminating treatments to which they are subjected. Washing alone, when performed for a sufficient time with large volumes of water, can reduce the numbers of bacteria on head meats, tongues and cattle tails (Table 5.20),[71] and probably on similar product such as weasand meat. Dehairing, scrubbing or other vigorous cleaning treatment may be required for substantial reduction of the numbers of bacteria on products largely covered by skin, like ears and trotters.

However vigorous the cleaning, the numbers of bacteria remaining on offals of the first group are likely to remain high. A pasteurizing treatment of such

Table 5.20 Log mean numbers of total aerobic bacteria, coliforms and *Escherichia coli* on beef tongues and tails before and after washing, and beef cheeks and lips after washing at a packing plant[67]

Product	Stage of processing	Log mean numbers		
		Aerobes (log cfu/cm^2)	Coliforms (log cfu/100 cm^2)	E. coli (log cfu/100 cm^2)
Tongue	Before washing	4.84	4.34	4.27
	After washing	2.13	<1.00	<1.00
Cheeks	After washing	3.35	2.48	2.17
Lips	After washing	2.42	1.77	1.40
Tails	Before washing	3.73	4.86	4.66
	After washing	2.60	2.89	2.58

products would then seem desirable, although such a treatment is at present not usual for any of those products.

The visceral organs which compose the second group of offals can be removed without being much contaminated with bacteria.[72] The organs must be inspected for symptoms of overt disease, and are usually placed on trays along with, if in separate compartments from, the intestines of the animal. Unfortunately, the requirements for inspection always override any consideration of preventing microbiological contamination of the organs. Consequently, they may be heavily contaminated during operations for their removal from the carcass and inspection, but changes to the process to improve their microbiological condition will be difficult to implement if the proposed changes are seen as conflicting in any way with the inspection procedures. Organ offals are usually washed before they are packed, but the extent to which microbial loads on the products are reduced by washing in commercial processes does not seem to have been reported.

Portions of the gut which are used for food will always be heavily contaminated with bacteria associated with faeces and ingesta even after the extensive washing that must be applied to remove most of the visible contamination. Portions of gut used as food for humans may be sold raw,[73] but much of those tissues are cooked or otherwise processed before they are sold to consumers. Thus, chitterlings are usually subjected to prolonged boiling and are pressed in moulds to form a compact mass suitable for slicing; beef tripes are usually scalded then soaked in an alkaline peroxide solution which bleaches and swells the tissues; and casings are usually preserved by dry salting or immersion in strong brine solutions. All of those treatments can destroy most of the bacteria present on the product.[74] The final treatments of the gut portions are therefore CCPs in the collection processes for those products.

For the offals which are not processed, like those in the third group, the growth of bacteria must be controlled or prevented by chilling or freezing the products. While temperatures remain above 7 °C, mesophilic, enteric pathogens present on the product will be able to grow, while cold-tolerant pathogens are capable of growth at temperatures down to 0 °C or below.[75] The rates at which bacteria grow tend to increase rapidly with temperature (Fig. 5.2),[76] so the rate at which the temperature of an offal is reduced, from body temperature to at least the chiller temperature range, is as important for product safety as the final temperature attained by the product.

The small sizes and the extensive washing with cold water of the individual pieces of tissue amongst the offals of the first group will generally ensure that they are at a temperature of about 25 °C by the time that they are packed (Table 5.21).[77] If they are then placed in boxes or other containers of moderate size and the containers are each exposed to an adequately rapid flow of cold air, then they will cool sufficiently rapidly to preclude any extensive growth of mesophilic pathogens. In contrast, large organs such as beef livers or hearts may cool little between the times they are removed from carcasses and the times of their arrival at a packing station. Those organs can then be at a temperature of over 35 °C

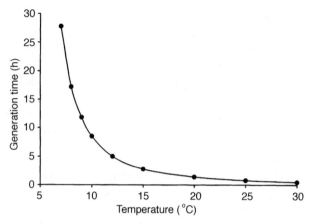

Fig. 5.2 Effect of temperature on the generation time of *Escherichia coli*.[72]

when packed, and if placed in large boxes or containers will cool only slowly at the centres of such containers even when those are exposed to high flows of cold air. Under such circumstances extensive growth of pathogenic bacteria on the product is possible.

The worst practice, which obtains at some plants, is to collect warm offals into bulk containers which are held for lengthy periods at the collection point before the containers are filled and moved to a packing station. Under those circumstances, a flora predominantly of *E. coli* can develop.[78] The best practice with offals is their suspension on hooks along a rack which is placed in a chiller or freezer in an area of high airflow so that the offals cool rapidly with drying of the surfaces.

The adequacy of the control over offal cooling can be assessed by the collection of temperature histories from randomly selected units moving through a process, and integration of the temperature histories with respect to models

Table 5.21 Temperatures at the centres of newly packed boxes of offals at a beef packing plant[77]

Offal	Temperature (°C)	
	Maximum	Average
Liver	38	36
Heart	39	36
Hanging tender	36	34
Tongue	30	26
Cheek	29	25
Lip	29	25
Weasand meat	27	24
Tail	27	24

Table 5.22 Values for the proliferation of *Escherichia coli* on cooling offals determined by the enumeration of bacteria or calculated from product temperature history data[76]

Offal	*E. coli* proliferation (generations)	
	By enumeration	By calculation
Liver	13.6	13.3
Heart	10.7	10.9
Kidney	10.9	9.5
Brain	7.0	7.3
Sweetbread	2.7	2.8

which describe the dependency on temperature of the growth of indicator organisms, such as *E. coli*.[79] Packing and cooling procedures can then be adjusted to ensure that the maximum proliferation of pathogens at any point within any product unit is maintained within tolerable limits. The growth predicted for *E. coli* at a monitored point within a mass of organ offals is likely to be close to the growth that would occur amongst any such organisms that were there (Table 5.22).[76] Bacterial growth values estimated from product temperature histories collected from randomly selected sites in randomly selected product units may then be substituted with some confidence for directly determined bacterial numbers when assessing the microbiological effects of offal cooling processes. Similar procedures can also be applied for assessing the adequacy of procedures for cooling meat which is cut from warm carcasses in a hot boning process.[80]

5.8 Carcass cooling

Because of the metabolic activity in pre-rigor muscle, the temperatures of carcasses tend to increase in the immediate post-slaughter period, from about 37 °C to about 40 °C. Dressed carcasses are usually cooled before they are further processed, with regulatory authorities requiring that the meat be at no more than 10 °C or, for the European Union, 7 °C before it is moved from the slaughtering plant or further processed. In principle, that should mean that the temperature at the centre of the thickest portion of the carcass, that is in the deep leg at the proximal part of the thigh, is below 10 °C or 7 °C for every carcass leaving the carcass cooling process.[81] Such temperatures are commonly obtained for sheep and pig carcasses cooled overnight, but are often not obtained for the larger beef carcasses in the same time.[82] Thus, in practice, rather warm beef carcasses may be regularly processed provided that the mean deep temperature of carcasses leaving a process is considered to meet the specified temperature.

The concern about the temperatures of cooling carcasses arises from the possibility of pathogenic bacteria growing rapidly on the surfaces of carcasses while they remain warm.[83] That such growth can occur is certain, but the extent to which it occurs in commercial cooling processes must be considered when seeking control of those processes.

Carcasses are cooled in batches. That is, the cooler is filled over a period that may extend over a whole working day. The cooler is often operated while it is being filled, and is operated overnight after it is filled. At most plants, each batch of carcasses is unloaded from the cooler during the day after it was loaded, but at some beef plants carcass cooling is extended to a second day to ensure adequate cooling of all the carcasses. The unloading of a cooler is often more rapid than the loading, so carcasses may be subjected to cooling for different times in the same process (Table 5.23).[84,85]

The rate at which a carcass cools will be determined by the size of the carcass, the air temperature and rate of flow of air over its surfaces. In practice, the air conditions within a cooler are far more variable than the sizes of the carcasses undergoing a cooling process, so air conditions rather than carcass size dictate the rates of cooling.[86] The air conditions experienced by a carcass will depend upon the pattern of flow of refrigerated air around the cooler, the arrangement of each individual carcass and the surrounding carcasses in relation to that air flow, and the spacing between carcasses. As a result it is not unusual to find that some carcasses in recently filled coolers are exposed to air at about 0 °C flowing at a rate of several metres per second, while around others the air can be still and at a temperature of about 10 °C.

The performance of a carcass cooling process can be assessed by the collection of temperature histories from the warmest region of the surface on each of 25 randomly selected carcasses passing through the process, with calculation of the growth of *E. coli* permitted by each temperature history, as in the assessment of offal cooling processes.[87] However, such an evaluation will allow consideration of only a worst-case situation, as it cannot take account of the extent to which other parts of the carcass surface are at any time cooler than the warmest area, or of factors other than temperature that may limit bacterial growth. Thus while a temperature function integration procedure can be used to set minimum acceptable performance criteria, to compare processes or in routine monitoring, it is desirable to assess the performance of a cooling process from

Table 5.23 Residence times of carcasses in chillers during the carcass cooling processes at four plants[84,85]

Plant	Carcass type	Residence time (h)		
		Max.	Min.	Average
A	Beef	28.0	15.8	21.7
B	Beef	24.0	20.0	22.6
C	Sheep	27.3	17.5	21.5
D	Pig	24.5	14.8	20.5

microbiological data also, to relate the worst case to the overall performance and to identify any factors other than temperature that affect bacterial growth.

In the classic carcass cooling process, the carcass is exposed to a flow of cool, unsaturated air. In those conditions water will evaporate from the warm carcass surface, which will therefore dry as well as cool. Such surface drying has long been regarded as essential for ensuring the microbiological stability of cooling carcasses,[83] and therefore cooling processes in which surface drying is prevented are not allowed by some regulatory authorities.

The desirability of surface drying is supported by various studies which showed that the numbers of total aerobic bacteria on carcasses can be prevented from increasing, or can even be reduced when effective drying of the carcass surface accompanies cooling. However, such findings are somewhat misleading in that bacteria differ greatly in their susceptibilities to the inhibitory and lethal effects of drying. As it happens, Gram negative organisms are generally more sensitive to drying effects than are Gram positive species.[88] There should then be little surprise in the recent finding that drying results in larger reductions in the numbers of *E. coli* than in the numbers of the meat microflora as a whole.[85]

Despite the desirability of surface drying from the microbiological point of view, it is economically undesirable because evaporation of water from the carcass surface equals loss of saleable weight.[89] Consequently, the spraying of carcasses with water at regular intervals of a few minutes during the first few hours of the cooling process has become the usual practice in North America. Despite the carcass surfaces being wet when the surface of the carcass is warmest, increases in bacterial numbers can apparently be avoided (Table 5.24).[84,85]

The lack of increase in bacterial numbers during spray chilling is probably because bacteria are washed from the carcass during spraying. Certainly, the control of bacterial numbers during spray chilling does not necessarily involve any better control of *E. coli* than of the microflora as a whole. However, spray-

Table 5.24 Log mean numbers of total aerobic bacteria and *Escherichia coli* on carcass before and after four carcass cooling processes[84,85]

Carcass type	Carcass process	Log mean numbers			
		Aerobes (log cfu/cm^2)		*E. coli* (log cfu/100 cm^2)	
		Before	After	Before	After
Beef	Spray chilling	4.03	3.58	<1	<1
Beef	Spray chilling then surface freezing	3.12	2.45	2.01	<0
Sheep	Air chilling	3.33	2.86	3.57	1.49
Pig	Blast freezing then spray chilling	1.83	2.57	<1	<1

chilling processes can apparently be operated to give a differential destruction of *E. coli* similar to that observed with surface drying.[84] That destruction is probably due to freezing of the film of water on the carcass surface at the end of the spraying period, but further investigation is required for the mechanism of *E. coli* destruction to be properly established.

Freezing of the carcass surface need not always be such as to produce a reduction in bacterial numbers. The cooling of beef or sheep carcasses too rapidly after slaughter can cause muscle fibres to contract and produce toughing of the meat.[90] The carcasses are therefore cooled so that much if not all of the muscle tissue remains above chiller temperatures until rigor has developed. With pig carcasses, however, the onset of rigor is more rapid than in the carcasses of cattle or sheep, so rapid cooling soon after dressing is possible. It has then become a common practice to pass pig carcasses through a tunnel where they are exposed for about an hour to air at about $-20\,°C$ moving at several metres per second. That treatment causes crust freezing of some carcass surfaces, and aids in the achievement of a relatively uniform temperature for carcasses after overnight cooling. However, rapid freezing as in a blast of freezing air is less injurious to microorganisms than is slower freezing,[91] so it is unsurprising that blast chilling does not produce any substantial reduction in the number of bacteria on carcasses which are sprayed during subsequent cooling.[85]

Although it appears that carcass cooling processes can be operated to reduce the numbers of enteric organisms on carcasses, it is unlikely that any commercial cooling processes are intentionally operated in such a manner. Instead, they are often operated to minimize loss of carcass weight during the cooling process as far as that is compatible with achieving acceptable final temperatures for the product. Any decontaminating effect of the cooling process is then fortuitous, while in the absence of appropriate microbiological data growth of hazardous organisms may proceede wholly unregarded. Thus proper control of a carcass cooling process requires the definition and subsequent maintenance of operating parameters that ensure at least the containment and preferably the reduction of the numbers of *E. coli* on carcasses.

5.9 Carcass breaking; equipment cleaning

Carcass breaking processes at large plants are often complex. With beef carcass sides, a typical process involves the progressive removal of portions from the hanging side, with each portion being passed to a separate collection or processing line which involves one or more conveyors and often band, circular or reciprocating saws, and powered knives. Workers on the processing lines are equipped with knives, sharpening steels, scabbards and meat hooks, and usually wear gloves, aprons and other articles formed from steel mesh to protect against wounding. All such items are regarded as personal equipment, and are retained by individual workers between work shifts.

Regulations usually require that workers entering a carcass breaking facility be suitably attired in clean clothing, pass through a boot wash, and wash their hands. The air temperature of the facility is maintained at about 10 °C. Those measures are considered sufficient for controlling the ingress of hazardous contamination with workers, and preventing the substantial growth of any pathogenic bacteria that enter on the meat.[92] Cleaning of the equipment used for carcass breaking is not regarded as part of the process control system. Instead, it is designated a 'pre-HACCP requirement', and is judged to be adequate if the cleaning of large items of equipment is conducted in accordance with written procedures, the equipment appears to be free of visible contamination on inspection after cleaning, and occasional sampling of cleaned, meat-contacting surfaces does not recover excessive numbers of aerobic bacteria.[14] Cleaning of personal equipment is almost invariably left to the discretion of the individual worker.

That approach to controlling carcass breaking processes embodies fundamental misunderstanding of the risks that may arise in such processes. The major risk to consumers from a carcass breaking process would be the persistence in improperly cleaned equipment of a population of pathogenic organisms which spread from the source during use of the equipment, to contaminate all product that passes through the process. Thus, the assured, adequate cleaning of equipment is essential for ensuring the safety of the meat.

The microbiological performance of a carcass breaking process can be determined by enumerating indicator organisms on carcasses entering and on product leaving the process. For example, when such a procedure was applied at four beef packing plants, increased numbers of *E. coli* were recovered from product at the end of three of the carcass breaking processes (Table 5.25).[93] Large populations of bacteria that included *E. coli* can be recovered from both large items of equipment and personal equipment, particularly steel mesh gloves, used in carcass breaking processes after the equipment is supposedly cleaned (Table 5.26). Increases in the numbers of aeromonads or listerias can

Table 5.25 Log total numbers of *Escherichia coli* recovered from 25 samples from carcass sides entering or loin primal cuts leaving the carcass breaking processes at four beef packing plants[93]

Plant	Log total numbers (log cfu/2500 cm^2)	
	Carcass	Loins
A	—^{a}	4.97
B	2.05	4.38
C	1.57	2.74
D	1.59	1.86

^{a}— None detected

Table 5.26 Log total numbers of total aerobic bacteria, coliforms and *Escherichia coli* recovered from cleaned equipment used in a beef carcass breaking process

Sample type	Number of samples	Log total numbers		
		Aerobes	Coliforms	E. coli
Water on conveyors	12	5.20	3.45	—[a]
Conveyor belt guards	12	7.21	6.85	5.02
Motor housings	20	7.62	5.85	4.41
Belt drive guards	20	8.01	7.33	6.04
Conveyor belt rollers	25	8.29	7.87	5.10
Steel mesh gloves	25	8.90	5.51	4.30
Rubber gloves	25	2.43	1.11	0.90

[a]— None detected

also be used to identify contamination from improperly cleaned equipment, particularly when relatively large numbers of *E. coli* on the incoming product render uncertain the discernment of any increases in the numbers of that indicator organism (Table 5.27).[94] It follows that to ensure that no contamination from improperly cleaned equipment was occurring, it would be desirable to demonstrate that there was no increase in the numbers of any of several indicator organisms as a result of a carcass breaking process.

The difficulty with cleaning carcass breaking equipment arises at least in part because there are no established design standards for equipment used in meat plants.[95] Instead, the acceptability of equipment is judged by regulatory authorities, item by item, on the basis of data provided by manufacturers and the in-plant performance as assessed by supervising veterinarians. As often no-one involved in an assessment of meat plant equipment has any training relating to the hygienic design of equipment, it is unsurprising that basic design faults are overlooked; and even if the basic design is hygienically satisfactory, it may be

Table 5.27 Log total numbers of *Escherichia coli* and aeromonads recovered from the shoulders, loins and legs of carcasses entering or the corresponding primal cuts leaving a sheep carcass breaking process[94]

Carcass portion	Stage of the process	Log total numbers (log cfu/2500 cm^2)	
		E. coli	Aeromonads
Shoulder	Before breaking	3.49	–
	After breaking	2.45	3.25
Loin	Before breaking	1.99	1.08
	After breaking	3.14	2.81
Leg	Before breaking	1.69	1.40
	After breaking	2.84	2.27

compromised by in-plant modifications which are carried out without consideration of the possible hygienic effects.

The questionable value of a system which essentially ignores the microbiological condition of the personal equipment that is used with meat hardly requires comment.

Obviously, the cleanliness of all equipment used in the carcass breaking process, and indeed in all other processes where raw meats may be contaminated at meat packing plants, should be the subject of cleaning processes which are performed, monitored and verified in full accordance with HACCP procedures. In addition, the development of design criteria and industrial standards for meat plant equipment would be highly desirable, as that could greatly assist in avoiding the hygienic deficiencies of much current equipment.

The categorizing of equipment cleaning as a pre-HACCP requirement outside the HACCP system proper is evidently inappropriate. This suggests that the whole concept of pre-HACCP requirements should be re-examined, as it seems that all matters that are now so designated should be covered by either design or performance criteria, rather than be specified individually at each plant, or be part of a proper HACCP system when the matter is an essential element for ensuring product safety (Table 5.28).[15] As it is, the pre-HACCP requirement approach appears largely counterproductive to ensuring product safety, as it seems often to draw attention and controlling activities away from the condition of the product to matters which are peripheral to product safety but which can be easily if inappropriately assessed by inspection.

5.10 Smaller plants

The investigative approach to implementing HACCP systems to ensure the microbiological safety of meat is obviously economically impractical for small plants. For such plants, the HACCP systems would have to be erected on the basis of Good Manufacturing Practices (GMPs) identified from microbiological data obtained at larger plants but tested for utility at selected, small plants. A programme of GMP identification and verification that involved both industry and regulatory authorities would be necessary to progressively develop understanding of the hygienically critical practices at smaller plants.

The GMP approach would seem feasible at smaller plants because with relatively simple processes employing little equipment, control of or changes to working practices would usually involve instruction of only a few workers. However, the difficulties in arriving at appropriate GMPs at each plant may still be considerable. For example, only one species is usually slaughtered at large plants, but several may be slaughtered with different frequencies at smaller plants. The microbiological conditions of carcasses from different species are likely to differ even when they are processed in a similar manner (Table 5.29).[96] Therefore, appropriate GMPs for at least the reception, slaughter and dressing of each species that is processed at a plant would have to be identified and implemented.

Table 5.28 Matters covered in pre-HACCP requirements of the Canadian Food Inspection Agency

A. Premises
    A.1 Building exterior and environment
    A.2 Building interior
        2.1 Design, construction and maintenance
        2.2 Lighting
        2.3 Ventilation
        2.4 Waste disposal
    A.3 Sanitary Facilities
        3.1 Employee facilities
        3.2 Equipment cleaning
    A.4 Water, steam and ice supplies
        4.1 Water and ice
        4.2 Steam
        4.3 Records

B. Transportation and storage
    B.1 Transportation
        1.1 Food carriers
        1.2 Temperature control
    B.2 Storage
        2.1 Incoming material storage
        2.2 Non-food chemicals receiving and storage
        2.3 Finished product storage

C. Equipment
    C.1 General equipment
        1.1 Design and installation
        1.2 Food contact surfaces
        1.3 Maintenance and calibration
        1.4 Maintenance records
        1.5 Calibration records

D. Personnel
    D.1 Training
        1.1 General food hygiene
        1.2 Technical training
    D.2 Hygiene and health
        2.1 Cleanliness and conduct
        2.2 Communicable diseases/injuries

E. Sanitation and pest control
    E.1 Sanitation
        1.1 Sanitation program
        1.2 Sanitation records
    E.2 Pest control
        2.1 Pest control program
        2.2 Pest control records

F. Recalls
    F.1 Recall system
        1.1 Procedures
        1.2 Product code identification
        1.3 Recall capability
    F.2 Distribution records

G. Records

Table 5.29 Log mean numbers of total aerobic bacteria, coliforms and *Escherichia coli* on the carcasses of six species dressed at a small packing plant[96]

Species	Log mean numbers		
	Aerobes (log cfu/cm^2)	Coliforms (log cfu/100 cm^2)	E. coli (log cfu/100 cm^2)
Cattle	2.37	1.93	1.56
Pigs	3.16	4.27	2.64
Deer	2.15	<1.00	<1.00
Buffalo	3.39	<1.00	<1.00
Ostrich	2.98	<1.00	<1.00
Emu	3.22	<1.00	<1.00

Or again, to arrive at appropriate GMPs for a process it will be necessary to take account of the available facilities, the use of the product, or other factors that constrain or modify plant practices. A general compendium of GMPs cannot then be a single, prescriptive list, but must identify options and alternatives that should be adopted in various, identified circumstances. Certainly, GMPs for small packing plants should not be dictated in vague terms from general principles and current assumptions, but should be identified, verified and refined on the basis of case studies as an ongoing part of regulatory activities.

5.11 Microbiological criteria

Microbiological criteria for the acceptability of raw meat has for long presented a problem, because of the wide variability in the numbers recovered in different samples from the same product, the likely growth of bacteria during any storage of chilled meat, and the expected presence of enteric pathogens in some fraction of all raw meat. Nonetheless, it is patently impossible to operate an effective system for control of microbiological contamination without reference to numbers of bacteria, and without specification of microbiological criteria for decision as to whether or not the control system is working.

Recognition of the need for microbiological criteria in relation to HACCP implementation has prompted the USDA to promulgate criteria for the numbers of *E. coli* recovered from the three sites assumed to be the most heavily contaminated on cooled carcasses, and for the frequency of recovery of *Salmonella* from those sites (Table 5.30).[14] The criteria are based on data obtained from a survey of the conditions of cooled carcasses at US plants, and so do no more than questionably define current, general commercial performance. As plants are required to meet those criteria if they are to continue to operate, the derivation of criteria from the current condition of meat was unavoidable. However, if the safety of meat is to improve there is need for microbiological

Table 5.30 USDA microbiological criteria for generic *Escherichia coli* and *Salmonella* on beef and pig carcasses[14]

E. coli

	Beef	Pig
Testing frequency	1 test/300 ccs	1 test/1000 ccs
Samples considered (n)	13	13
Lower limit of marginal range (m)	5 cfu/cm^2	<10 cfu/cm^2
Upper limit of marginal range (M)	100 cfu/cm^2	10 000 cfu/cm^2
Number permitted in marginal range	3	3

A process fails to meet the criteria if in the 13 samples most recently collected four or more samples yield *E. coli* at numbers $> m$ but $\leq M$, or one or more samples yield *E. coli* at numbers $>M$.

Salmonella

Carcass type	Number of samples	Permitted positive samples
Beef cattle	82	1
Cow	58	2
Pig	55	6

A process fails to meet the criterion if the number of *Salmonella* positive samples exceeds the permitted number in any set of the stipulated number of samples

criteria which define a performance that is attainable in commercial practice with optimum operation of a HACCP system.

From the various studies of meat plant processes it appears that, when the assessment of process performance is based on the random sampling of product by swabbing 25 areas each of approximately 100 cm^2, and with appropriate prevention of contamination and application of an effective decontaminating treatment or treatments, it is possible to produce cooled carcasses of any species of red meat animal on which the log mean numbers of total aerobic bacteria are <2 per cm^2, and on which *E. coli*, aeromonads and listerias are each recovered at log total numbers <1 per 2500 cm^2. As contamination during the breaking of carcasses is wholly avoidable, those numbers should be if anything less on cuts and manufacturing beef when product is finally packed at meat packing plants. However, such superior performance will become general only if there is movement towards assessing HACCP systems against the microbiological condition of the product produced rather than against subjective judgements and adherence to hygienically irrelevant procedural details.

5.12 References

1. THORNTON H and GRACEY J F, *Textbook of Meat Hygiene*, 6th edn, London, Balliere Tindall, 1974.
2. BLAMIRE R V, 'Meat hygiene: the changing years', *State Vet J*, 1984 **39** (114) 3–13.
3. WALLEY T, *A Practical Guide to Meat Inspection*, 3rd edn, Edinburgh, Pentland, 1896.
4. HATHAWAY S C and McKENZIE A I, 'Postmortem meat inspection in New Zealand: the dismantling of traditional systems', *Surveillance*, 1990 **17** (1) 6–8.
5. TAUXE R V, '*Salmonella*: a postmortem pathogen', *J Food Protect*, 1991 **54** 563–8.
6. SMITH J L and FRATAMICO P M, 'Factor involved in the emergence and persistence of food-borne diseases', *J Food Protect*, 1995 **58** 696–708.
7. HATHAWAY S C and McKENZIE A I, 'Postmortem meat inspection programs; separating science and tradition', *J Food Protect*, 1991 **54** 471–5.
8. AMERICAN PUBLIC HEALTH ASSOCIATION, *Proceedings of the 1971 National Conference on Food Protection*, Washington DC, US Government Printing Office, 1972.
9. MACKEY B M and ROBERTS T A, 'Improving slaughtering hygiene using HACCP and monitoring', *Fleischwirtsch Int*, 1993 (2) 40–5.
10. US DEPARTMENT OF AGRICULTURE, 'Pathogen reduction: hazard analysis and critical control point systems; proposed rule', *Fed Regist*, 1995 **60** 6774–889.
11. US DEPARTMENT OF AGRICULTURE, National Advisory Committee on Microbiological Criteria for Foods, 'Hazard analysis and critical control point system', *Int J Food Microbiol*, 1992 **16** 1–23.
12. INTERNATIONAL COMMISSION ON MICROBIOLOGICAL SPECIFICATION FOR FOODS, 'The hazard analysis critical control point approach to control of food safety and quality', in *HACCP in Microbiological Safety and Quality*, ICMSF, London, Blackwell, 1988.
13. TOMPKIN R B, 'The use of HACCP in the production of meat and poultry products', *J Food Protect,* 1990 **53** 795–803.
14. US DEPARTMENT OF AGRICULTURE, 'Pathogen reduction: hazard analysis and critical control point systems; final rule', *Fed Regist*, 1996 **61** 38805–989.
15. CANADIAN FOOD INSPECTION AGENCY, *Food Safety Enhancement Program*, Ottawa, Health Canada.
16. SOUL P, *The UK Hygiene Assessment System*, York, UK Meat Hygiene Service.
17. BRYAN F L, 'Application of HACCP to ready-to-eat chilled foods', *Food Technol*, 1990 **44** 70–7.
18. FLOWERS R S and SILLIKER J H, 'The role of microbiological testing in food safety programs: appropriate or inappropriate uses of microbiological

testing', in *The Role of Microbiological Testing in Beef Food Safety Programs*, Kansas City, American Meat Science Association, 1999.
19. MOSSEL D A A, WEENK G H, MORRIS G P and STRUIK C B, 'Identification, assessment and management of food-related microbiological hazards: historical, fundamental and psycho-social essentials', *Int J Food Microbiol*, 1998 **39** 19–51.
20. GILL C O, 'Microbiological contamination of meat during slaughter and butchering of cattle, sheep and pigs', in *The Microbiology of Meat and Poultry*, eds DAVIES A R and BOARD R G, London, Blackie Academic, 1998.
21. GILL C O, 'Current and emerging approaches to assuring the hygienic condition of red meats', *Can J Anim Sci*, 1995 **75** 1–13.
22. GILL C O and JONES T, 'Control of the contamination of pig carcasses by *Escherichia coli* from their mouths', *Int J Food Microbiol*, 1998 **44** 43–8.
23. PALUMBO S A, 'Is refrigeration enough to restrain foodborne pathogens?' *J Food Protect*, 1986 **49** 1003–9.
24. GILL C O and JONES T, 'The presence of *Aeromonas, Listeria* and *Yersinia* in carcass processing equipment at two pig slaughtering plants', *Food Microbiol*, 1995 **12** 135–41.
25. JOHNSON J L, DOYLE M P and CASSENS R G, '*Listeria monocytogenes* and other *Listeria* spp. in meat and meat products: a review', *J Food Protect*, 1990 **53** 81–91.
26. KILSBY D C, 'Sampling schemes and limits', in *Meat Microbiology*, ed BROWN M H, London, Applied Science Publishers, 1982.
27. BROWN M H and BAIRD-PARKER A C, 'The microbiological examination of meat', in *Meat Microbiology*, ed BROWN M H, London, Applied Science Publishers, 1982.
28. HILDERBRANDT G and WEISS H, 'Sampling plans in microbiological quality control 2. Review and future prospects', *Fleischwirtsch Int*, 1994 (3) 49–52.
29. KILSBY D C and PUGH M E, 'The relevance of the distribution of microorganisms within batches of food to the control of microbiological hazards from foods', *J Appl Bacteriol*, 1981 **51** 345–54.
30. GILL C O, DELANDES B, RAHN K, HOUDE A and BRYANT J, 'Evaluation of the hygienic performances of the processes for beef carcass dressing at 10 packing plants', *J Appl Microbiol*, 1998 **84** 1050–8.
31. DORSA W J, CUTTER C N and SIRAGUSA G R, 'Evaluation of six sampling methods for recovery of bacteria from beef carcass surfaces', *Lett Appl Microbiol*, 1996 **22** 39–41.
32. GILL C O and JONES T, 'Microbiological sampling of carcasses by excision or swabbing', *J Food Protect*, 2000 **63** 167–73.
33. GILL C O, BADONI M and JONES T, 'Hygienic effects of trimming and washing operations in a beef-carcass-dressing process', *J Food Protect*, 1996 **59** 66–9.
34. JERICHO K W F, KOZUB G C, LOEWEN K G and HO J, 'Comparison of methods to microbiologically evaluate surfaces of beef carcasses by hydrophobic

35. ENTIS P and BOLESZCZUK P, 'Direct enumeration of coliforms and *Escherichia coli* by hydrophobic grid membrane filter in 24 h using MUG', *J Food Protect*, 1990 **53** 948–52.
36. JARVIS B, *Statistical Aspects of the Microbiological Analysis of Foods*, Amsterdam, Elsevier, 1989.
37. GILL C O, McGINNIS J C and JERICHO K W F, *Implementation of HACCP Systems for Beef Carcass Dressing Processes*; HACCP Manual No 1, Lacombe Research Centre, Agriculture and Agri-Food Canada, 1996.
38. GILL C O and JONES T, 'Assessment of the hygienic characteristics of a process for dressing pasteurized pig carcasses', *Food Microbiol*, 1997 **14** 81–91.
39. HADLEY P J, HOLDER J S and HINTON M H, 'Effects of fleece soiling and skinning methods on the microbiology of sheep carcasses', *Vet Rec*, 1997 **140** 570–4.
40. PETERSEN G V, 'The effect of swimming lambs and subsequent resting periods on the ultimate pH of meat', *Meat Sci*, 1983 **9** 237–46.
41. RIDELL J and KORKEALA H, 'Special treatment during slaughtering in Finland of cattle carrying an excessive load of dung; meat hygienic aspects', *Meat Sci*, 1993 **35** 223–8.
42. VAN DONKERSGOED J, JERICHO K W F, GROGAN H and THORLAKSON B, 'Preslaughter hide status of cattle and the microbiology of carcasses', *J Food Protect*, 1997 **60** 1502–8.
43. BISS M E and HATHAWAY S C, 'Microbiological and visible contamination of lamb carcasses according to preslaugher presentation status: implications for HACCP', *J Food Protect*, 1995 **58** 776–83.
44. MACKEY B M and DERRICK C M, 'Contamination of the deep tissues of carcasses by bacteria present on the slaughtering instruments or in the gut', *J Appl Bacteriol*, 1979 **46** 355–66.
45. SCHNELL T D, SOFOS J N, LITTLEFIELD V G, MORGAN J B, GORMAN B M, CLAYTON R P and SMITH G C, 'Effects of postexsanguination dehairing on the microbial load and visual cleanliness of beef carcasses', *J Food Protect*, 1995 **58** 1279–302.
46. NICKELS C, SVENSSON I, TERNSTROM A and WICKBERY L, 'Hygiene and economy of scalding with condensed water vapour and in tank', *Proc 22nd Meeting of Meat Research Workers,* Malmo, 1976.
47. GILL C O and BRYANT J, 'The presence of *Escherichia coli*, *Salmonella* and *Campylobacter* in pig carcass dehairing equipment', *Food Microbiol*, 1993 **10** 337–44.
48. GILL C O and BRYANT J, 'The contamination of pork with spoilage bacteria during commercial dressing, chilling and cutting of pig carcasses', *Int J Food Microbiol*, 1992 **16** 51–62.
49. BELL R G and HATHAWAY S C, 'The hygienic efficiency of conventional and inverted lamb dressing systems', *J Appl Bacteriol*, 1996 **81** 225–34.

50. TROEGER K, 'Evaluating hygienic risks during slaughtering', *Fleischwirtsch*, 1994 **74** 624–6.
51. GILL C O, McGINNIS J C and BRYANT J, 'Microbial contamination of meat during the skinning of beef carcass hindquarters at three slaughtering plants', *Int J Food Microbiol*, 1998 **42** 175–84.
52. GILL C O and McGINNIS J C, 'Improvement of the hygienic performance of the hindquarters skinning operations at a beef packing plant', *Int J Food Microbiol*, 1999 (in press).
53. HUDSON W R, ROBERTS T A and WHELEHAN O P, 'Bacteriological status of beef carcasses at a commercial abattoir before and after slaughtering improvement', *Epidem Infect*, 1987 **91** 81–6.
54. SCHUTZ F, 'On-the-rail slaughter of cattle as related to hygiene requirements', *Fleischwirtsch*, 1991 **71** 306–8.
55. NESBAKKEN T, NERBRINK E, ROTTERUD O-J and BORCH E, 'Reduction of *Yersinia enterocolitica* and *Listeria* spp. on pig carcasses by enclosure of the rectum during slaughter', *Int J Food Microbiol*, 1994 **23** 197–208.
56. JERICHO K W F, BRADLEY J A, GANNON V P J and KOZUB G C, 'Visual demerit and microbiological evaluation of beef carcasses: methodology', *J Food Protect*, 1993 **56** 114–19.
57. GILL C O and BAKER L M, 'Trimming, vacuum cleaning or hot water-vacuum cleaning effects on the hindsaddles of dressed lamb carcasses', *J Muscle Foods*, 1998 **9** 391–401.
58. GILL C O and JONES T, 'The microbiological effects of breaking operations on hanging beef carcass sides', *Food Res Int*, 1999 **51** 123–32.
59. GILL C O, BRYANT J and McGINNIS J C, 'Microbial effects of the carcass washing operations at three beef packing plants', *Fleischwirtsch Int*, 2000(2) (in press).
60. BELL M F, MARSHALL R T and ANDERSON M E, 'Microbiological and sensory tests of beef treated with acetic and formic acids', *J Food Protect*, 1986 **49** 207–10.
61. BRACKETT R E, HAO Y-Y and DOYLE M P, 'Ineffectiveness of hot acid sprays to decontaminate *Escherichia coli* O157:H7 in beef', *J Food Protect*, 1994 **57** 198–203.
62. GILL C O and BRYANT J, 'Decontamination of carcasses by vacuum-hot water cleaning and steam pasteurizing during routine operations at a beef packing plant', *Meat Sci*, 1997 **47** 267–76.
63. GILL C O, BRYANT J and BEDARD D, 'The effects of hot water pasteurizing treatments on the appearances and microbiological conditions of beef carcass sides', *Food Microbiol*, 1999 **16** 281–9.
64. NUTSCH A L, PHEBUS R K, RIEMANN M J, SCHAFER D E, BOYER J E Jr, WILSON R C, LEISING J D and KASTNER C L, 'Evaluating of a steam pasteurizing process in a commercial beef processing facility', *J Food Protect*, 1997 **60** 485–92.
65. DAVEY K R and SMITH M G, 'A laboratory evaluation of a novel hot water cabinet for the decontaminating of sides of beef', *Int J Food Sci Technol*,

1989 **24** 305–16.
66. NUTSCH A L, PHEBUS R K, RIEMANN M J, KOTROLA J S, WILSON R C, BOYER J E Jr and BROWN T L, 'Steam pasteurization of commercially slaughtered beef carcasses: evaluation of bacterial populations at five anatomical locations', *J Food Protect*, 1998 **61** 571–7.
67. GILL C O and BAKER L P, 'Assessment of the hygienic performance of a sheep carcass dressing process', *J Food Protect*, 1998 **61** 329–33.
68. GILL C O, BEDARD D and JONES T, 'The decontaminating performance of a commercial apparatus for pasteurizing polished pig carcasses', *Food Microbiol*, 1997 **14** 71–9.
69. GILL C O, JONES T and BADONI M, 'The effects of hot water pasteurizing treatments on the microbiological conditions and appearances of pig and sheep carcasses', *Food Res Int*, 1999 **31** 273–8.
70. PEARSON A M and DUTSON T R (eds), *Edible Meat By-Products*, London, Elsevier Applied Science, 1988.
71. GILL C O, McGINNIS J C and JONES T, 'Assessment of the microbiological conditions of tails, tongues and head meats at two beef-packing plants', *J Food Protect*, 1999 **62** 674–7.
72. GILL C O and DeLACY K M, 'Microbial spoilage of whole sheep livers', *Appl Environ Microbiol*, 1982 **43** 1262–6.
73. STEWART A W, LANGFORD A F, HALL C and JOHNSON M G, 'Bacteriological survey of raw and soul foods available in South Carolina', *J Food Protect*, 1978 **41** 364–7.
74. GILL C O, 'Microbiology of edible meat by-products', in *Edible Meat By-Products*, eds PEARSON A M and DUTSON T R, London, Elsevier Applied Science, 1988.
75. GREER G G, GILL C O and DILTS B D, 'Predicting the aerobic growth of *Yersinia enterocolitica* on pork fat and muscle tissue', *Food Microbiol*, 1995 **12** 463–9.
76. GILL C O and HARRISON J C L, 'Evaluation of the hygienic efficiency of offal cooling procedures', *Food Microbiol*, 1985 **2** 63–9.
77. GILL C O and JONES S D M, 'Evaluation of a commercial process for collection and cooling of beef offals by a temperature fraction integration technique', *Int J Food Microbiol*, 1992 **15** 131–43.
78. GILL C O and PENNEY N, 'The shelf life of sheep livers packed in closed tubs', *Meat Sci*, 1982 **11** 73–7.
79. GILL C O, TAYLOR C M, TONG A K W and O'LANEY G B, 'Use of a temperature function integration technique to assess the maintenance of control over an offal cooling process', *Fleischwirtsch*, 1995 **75** 682–4.
80. REICHEL M P, PHILLIPS D M, JONES R and GILL C O, 'Assessment of the hygienic adequacy of a commercial hot boning process for beef by a temperature function integration technique', *Int. J Food Microbiol*, 1991 **14** 27–42.
81. JAMES S J and BAILEY C, 'Chilling of beef carcasses', in *Chilled Foods, the State of the Art*, ed GORMLEY T R, London, Elsevier Applied Science, 1990.

82. JAMES S J and BAILEY C, 'Process design data for beef chilling', *Int J Refrig*, 1989 **12** 42–9.
83. NOTTINGHAM P M, 'Microbiology of carcass meats', in *Meat Microbiology*, ed BROWN M H, London, Applied Science Publishers, 1982.
84. GILL C O and BRYANT J, 'Assessment of the hygienic performances of two beef carcass cooling processes from product temperature history data or enumeration of bacteria on carcass surfaces', *Food Microbiol*, 1997 **14** 593–602.
85. GILL C O and JONES T, 'Assessment of the hygienic performances of an air cooling process for lamb carcasses and a spray cooling process for pig carcasses', *Int J Food Microbiol*, 1997 **38** 85–93.
86. GILL C O, JONES S D M and TONG A K W, 'Application of a temperature function integration technique to assess the hygienic adequacy of a process for spray chilling beef carcasses', *J Food Protect*, 1991 **54** 731–6.
87. GILL C O, HARRISON J C L and PHILLIPS D M, 'Use of a temperature function integration technique to assess the hygienic adequacy of a beef carcass cooling process', *Food Microbiol*, 1991 **8** 83–94.
88. BROWN A D, *Microbial Water Stress Physiology: Principles and Perspectives*, Chichester, Wiley, 1990.
89. STRYDON P E and BUYS E M, 'The effects of spray chilling on carcass mass loss and surface associated bacteriology', *Meat Sci*, 1995 **39** 265–76.
90. DRANSFIELD E, 'Tenderness of meat, poultry and fish', in *Quality Attributes and their Measurement in Meat, Poultry and Fish Products*, eds PEARSON A M and DUTSON T R, London, Blackie Academic, 1994.
91. DAVIES R and OBAFEMI A, 'Responses of micro-organisms to freeze–thaw stress', in *Microbiology of Frozen Foods*, ed ROBINSON R K, London, Elsevier Applied Science, 1985.
92. KASPROWIAK R and HECKLMANN H, 'Weak points in the hygiene of slaughtering, cutting and processing firms', *Fleischwirtsch Int*, 1992 (2) 32–40.
93. GILL C O, BADONI M and McGINNIS J C, 'Assessment of the adequacy of cleaning of equipment used for breaking beef carcasses', *Int J Food Microbiol*, 1999 **46** 1-8.
94. GILL C O, BAKER L P and JONES T, 'Identifications of inadequately cleaned equipment used in a sheep carcass breaking process', *J Food Protect*, 1999 **62** 637–43.
95. LELIEVELD H L M, 'The EC machinery directive and food-processing equipment', *Trends Food Sci Technol*, 1993 **4** 153–4.
96. GILL C O, JONES T, BRYANT J and BRERETON D A, 'The microbiological conditions of the carcasses of six species after dressing at a small abattoir', *Food Microbiol*, 2000 **17** 233–9.

6

HACCP in primary processing: poultry

G. C. Mead, Royal Veterinary College, University of London

6.1 Introduction

Since 1990, poultrymeat consumption has increased worldwide at a mean annual rate of approximately 4.4% and it has been estimated that, in 1998, more than 40 000 million chickens were reared and slaughtered for meat purposes.[1] Such a large output is only possible with the highly intensive systems of production and processing that are characteristic of the modern poultry industry. In relation to processing, almost every operation is now capable of being mechanised, the main objectives being to reduce labour costs and increase productivity.[2] It is currently possible to kill and defeather birds at a rate of 12 000 per hour and to eviscerate, i.e. remove internal organs from, 9000 birds per hour on a single processing line. The type of process used to produce oven-ready carcasses and much of the machinery involved are similar in all the major poultry-producing countries. The main stages in a typical chicken processing line are shown in Fig. 6.1. Some modification of the process or the machines themselves is necessary for other poultry species, such as turkeys and ducks, but this aspect is not considered here.

6.1.1 Presence of foodborne pathogens and spoilage bacteria

Poultrymeat can become contaminated with a variety of foodborne human pathogens, including *Salmonella* and *Campylobacter* spp., *Listeria monocytogenes*, *Clostridium perfringens* and *Staphylococcus aureus*. Some of these organisms may be acquired by the birds during the hatching and growing stages, and are mostly carried asymptomatically. When the birds are sent for slaughter, there is a high risk that any foodborne pathogens present will be transmitted

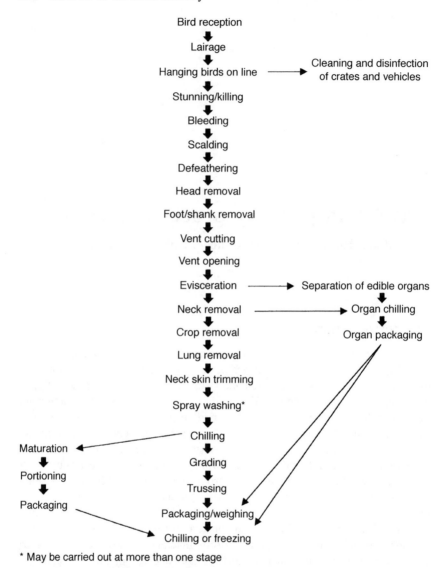

Fig. 6.1 Stages in a typical processing operation.

from carrier birds to the carcasses being processed. This underlines the need for HACCP principles to be applied in hatcheries, feed mills and on farms, as well as in processing plants, to minimise product contamination. In the case of salmonellas, levels of intestinal carriage in the live bird are generally low at slaughter, and contaminated carcasses typically carry fewer than 100 cells, although higher numbers can occur.[3] Experience has shown that the proportion of salmonella contaminated carcasses in a batch can vary from 0 to 100%. With regard to *C. jejuni*, on the other hand, both the proportion of contaminated carcasses and the numbers of viable cells present are consistently higher, so that finished carcasses may carry up to 10^6 colony-forming units of this organism,[3] reflecting the relatively high levels of intestinal carriage in the live bird. *Cl. perfringens* is another common intestinal organism in poultry, which is sometimes associated with human food poisoning. Surveys have indicated that 10–80% of carcasses may be contaminated,[3] but numbers are generally below 10 per gram of skin or per cm^2. In one study,[4] the organism was shown to be mainly present as spores. Like the other enteric pathogens referred to above, control of this organism in the processing plant depends on the extent to which faecal contamination in general can be prevented or reduced.

A different type of problem arises with *L. monocytogenes*, which is a pathogen that can multiply under chill conditions (psychrotrophic). The organism is relatively common on processed poultry, but is rarely found in the growing birds and these are not regarded as a significant reservoir.[5] Instead, carcass contamination is derived from the presence of listerias on processing equipment, especially that involved in the evisceration stages of the process. The strains present are likely to have originated from the incoming birds, but the problem can be exacerbated by inadequate cleaning and disinfection of the machinery. Some studies[6,7] have also shown an unusually high incidence of *L. monocytogenes* on the hands and gloves of operatives, suggesting that some multiplication had occurred. However, the numbers present on finished carcasses appear to be low, with a mean of 4.3 per cm^2 being reported.[8]

Most strains of *S. aureus* found on poultry carcasses do not produce the toxin that causes human food poisoning, and illness normally follows from contamination of the cooked meat by an infected food handler. Nevertheless, high levels of staphylococcal contamination can lead to the rejection of meat intended for product manufacture and therefore are to be avoided. The organism frequently occurs on the skin and in the nasopharynx of healthy birds and is usually present on carcasses at levels up to 10^3 per gram of skin.[8] Although *S. aureus* is generally regarded as a poor competitor in the presence of other microorganisms, there is much evidence to suggest that it is capable of colonising certain items of processing equipment, particularly the defeathering machines, from which contamination of carcasses occurs. The conditions leading to the colonisation of defeathering machines with *S. aureus* will be considered below.

In addition to controlling foodborne pathogens, there is a need for processors to minimise carcass contamination with psychrotrophic spoilage bacteria in order to ensure an adequate shelf-life for raw, chilled products. The main spoilage organisms of aerobically stored poultry meat are *Pseudomonas* spp. These and other spoilage bacteria originate in the live-bird environment and are brought into the processing plant on the skin and feathers of the birds. Also, they can be present in the plant water supply, if the water is not superchlorinated. Most pseudomonads on the carcasses are destroyed during the scalding process, but recontamination occurs at later stages of processing and the organisms can multiply on any wet surface in the plant. Their presence on selected items of processing equipment, etc., is shown in Table 6.1.

It is possible to draw some broad conclusions about the behaviour of microorganisms in poultry processing. Although birds entering the plant carry a large microbial load, especially in the alimentary tract, numbers on carcass surfaces are progressively reduced during processing, when appropriate hygiene controls are used. For example, in a study of five large processing plants in the UK,[9] total viable counts from skin samples were reduced by up to 100-fold. It is much more difficult, however, to control the spread of minority organisms such as salmonellas, and cross-contamination of carcasses can occur at virtually every stage of the process, even though the overall microbial load is being simultaneously reduced. A further complication for the processor is the fact that microbial contaminants soon become attached to the carcass surface or entrapped in the skin and abdominal cavity, so that washing the carcass does not readily remove them. Once attached or entrapped, the organisms are protected to some degree against the lethal effects of the scalding process or contact with superchlorinated wash water. It is important, therefore, to wash the carcasses as soon as possible after contamination has occurred.[10] For this reason, modern poultry processing utilises relatively large amounts of water and carcasses are often washed on more than one occasion during the evisceration stages. It is also a common practice to use eviscerating machines that allow continuous spray washing of the functional parts.

In many countries, the HACCP system is being applied to abattoirs, including those that deal with poultry. The main purpose is to improve the control of microorganisms, especially those of public health significance, by developing a strategy that targets resources in a concerted effort to maximise the available control options. This chapter will be concerned with the role of the HACCP system in meeting the desired objective, and attention will be given to the hazards associated with each of the main stages in the process. As a key part of any HACCP programme, the use of Good Manufacturing Practices (GMPs) will be described, together with the requirements for post-processing plant sanitation.

Table 6.1 Counts[a] of psychrotrophs and pseudomonads from processing equipment ($\log_{10}$cfu/cm^2 or swab)

Sampling site	Psychrotrophs	Pseudomonads
Plucking machine (internal) (A)[b]	2.7	2.7
Head puller (B)	1.7	1.7
Vent cutter (B)	3.7	2.5
Gloves of operative loading air chiller (B)	5.2	4.3
Conveyor from chiller (A)	3.1	2.5
Automatic grader (B)	4.4	3.1

[a] Counts obtained 4 h after start of processing
[b] A, cfu/cm^2; B, cfu/swab

The steps involved in establishing a HACCP programme will be considered in relation to both primary carcass processing and the production of cut portions and deboned meat. Future trends in processing technology and the use of carcass decontamination treatments will be discussed. The chapter concludes with some information on other publications that are intended to facilitate the application of HACCP principles to poultry processing.

6.2 Hazard analysis in the slaughter process

6.2.1 Scope of the HACCP plan

The plan deals with the application of HACCP principles at those stages in the production chain that begin with the arrival of live birds at the processing plant and end with the loading of finished, oven-ready carcasses or other raw products onto refrigerated vehicles prior to distribution. The loading stage is the point at which the processor's responsibility for handling the product often ceases, since the use of specialist haulage companies is becoming more common, particularly in the United Kingdom. In a fully integrated company that may own breeder and broiler farms, hatcheries and feed mills, as well as processing and possibly further-processing facilities, the HACCP system can be applied throughout the entire operation in a fully coordinated manner. Clearly, such a situation is more favourable for controlling microbial and other hazards than one in which ownership of the different sectors is fragmented, but the application of HACCP principles on the farm is still in its infancy and present control measures are largely based on Good Hygienic Practices rather than true CCPs. It is highly important, however, that carriage of foodborne pathogens in the live bird is suitably controlled, because, inevitably, conditions in the processing plant are conducive to cross-contamination, and elimination of the organisms at this stage is currently impossible, for reasons that will be discussed below.

6.2.2 Introducing the HACCP system

Developing and implementing the HACCP programme is a company responsibility and its initiation must come from senior management, who will need to give the project their full and continuing support if it is to succeed. The first step is to select the team and its coordinator, and these individuals will usually need to combine management experience with sound technical knowledge that is relevant to the processes and products under consideration. The team should be multidisciplinary and comprise no more than four or five individuals. The members could include, for example, the quality assurance manager, the production manager, a process engineer and the company microbiologist. An engineer is particularly relevant here, because hygiene control in poultry processing is dependent on the type of machinery used and its mode of operation. The coordinator will need to be well versed in the operation of each process and able to communicate well with staff at all levels. Some familiarity with the HACCP concept would be an advantage for all members of the team, but training is likely to be required, and each individual must also acquire a basic knowledge of the nature and properties of the hazardous microbes associated with poultrymeat production, as well as the means available to control them. In addition, the team must familiarise itself with the wide range of chemical and physical hazards that could arise. The former will include cleaning agents, disinfectants, lubricants, etc., while the latter will involve mainly foreign bodies such as metal, glass and plastic. As well as being responsible for developing the HACCP programme, the team members will be required to train other individuals as necessary. Some operations, especially at the smaller processing plants, may need help from an outside expert, who should be an experienced food microbiologist. The expert will provide advice and guidance to the HACCP team and give training in HACCP principles. He or she should be prepared to act as the team coordinator.

Once the team is chosen, one of the first tasks is to establish the types of product being produced, their distribution and the ways in which they will be handled and used by the consumer groups for which they are intended. Such information is needed in assessing the problems that could arise at any stage and in identifying any particularly susceptible consumers who may be more at risk in relation to foodborne illness. It is also important for the team to be aware of the possibility of product abuse and its microbiological consequences. This may even lead to better labelling of the product with regard to instructions for storage, handling and cooking.

It is clear that production practices at the farm level have an important influence on the contamination of raw poultry products with foodborne pathogens. Unless specific eradication measures for key pathogens are successfully implemented, involving high standards of husbandry hygiene and use of pathogen-free chicks, it is likely that most raw poultry will be contaminated with one or more agents of foodborne disease. Moreover, poultrymeat is a good substrate for microbial growth and, under appropriate conditions, will readily support the multiplication of both pathogens and spoilage bacteria. In primary processing and portioning, as well as in product distribution and retail sale, microbial growth is controlled by keeping the product chilled or frozen. There is also the option of using modified atmosphere packaging for chilled products as a means of extending their shelf-life. The packs contain an atmosphere that is enriched with carbon dioxide and inhibits the growth of aerobic spoilage organisms. This type of pack is becoming more common for retail presentation and raises the question of whether the extended shelf-life would provide greater opportunity for the multiplication of psychrotrophic pathogens such as *Listeria, Aeromonas* and *Yersinia* spp. In practice, however, no additional hazard is considered likely[11] and at CO_2 concentrations above 75% the pathogens, should they be present, tend to be inhibited.

6.2.3 Constructing the process flow-chart

Prior to carrying out any hazard analysis, it is necessary to make a careful examination of the entire process and to produce a comprehensive flow-chart on which to base the study. This may be overlaid onto a plant layout diagram, so that routes of products and personnel can be indicated to show, for example, sites where there is an increased risk of cross-contamination. The flow-chart should include every step in the operation from reception of the live birds, through to distribution of the final product, and must be agreed by the HACCP team before proceeding further. All relevant technical data should be made available to the team, including practices and procedures that may be different during night shifts or at weekends: for example, use of the air-chilling system to hold carcasses over from one shift to another. Some of the relevant information needed is shown in Table 6.2.

6.2.4 Carrying out the hazard analysis

The purpose of this exercise is to identify the points at which hazards can occur during processing and handling of the product up to the point of consumption. It is also necessary to identify the types of hazard that can arise, whether microbiological, chemical or physical, although in the present chapter only microbiological hazards are considered. Hazards may be associated with organisms that survive a particular processing step or are introduced from contaminated equipment, neighbouring carcasses, the processing environment or plant personnel. The team should also consider the impact of any atypical conditions, such as breakdowns in processing or temporary holding of the product while awaiting reworking. Once all the hazards have been identified, they need to be characterised in relation to their severity, i.e. human consequences, and likely frequency of occurrence (high, medium or low risk). The hazards that can arise at each of the main stages of the process will now be considered separately.

Birds arrive at the processing plant, either in fixed-crate vehicles or in lorries carrying some form of loose-crate system. During transit from farm to plant, the

Table 6.2 Examples of technical data relating to poultry processing and needed by the HACCP team

Floor plans and equipment layout
Equipment design details
Physical separation of high/low risk areas
Segregation of plant personnel
Conditions for product storage and distribution
Time/temperature history of products during processing
Arrangements for reworking faulty products
Types of packaging materials used
Consumer-use instructions
Efficacy of plant cleaning and disinfection

birds in any one crate remain in close proximity to each other and in contact with faecal material from the group. The period of fasting prior to slaughter inevitably reduces subsequent defecation, but there is evidence that the stress of transportation may result in more systemic invasion of the birds by salmonellas,[12,13] which could have consequences for contamination of hearts and livers as giblet components. Also, longer holding times in the crates can cause higher levels of faecal contamination on carcasses.[13] This may increase contamination with campylobacters, although transport stress does not appear to promote the shedding of these organisms.

After the birds have been unloaded, both crates and vehicles should be thoroughly cleaned and sanitised before returning to collect other batches of birds. Various studies have shown that, in practice, the crates are rarely cleaned properly and therefore serve as a source of flock-to-flock transmission of enteric pathogens.[14–16] The unloading process, which involves manual hanging of birds on the processing line, results in some struggling and wing-flapping. This scatters dust and microorganisms, and the atmosphere of the hanging-on bay can be heavily contaminated. The birds pass rapidly to the stunning area, where they are either stunned electrically to render them unconscious or killed by electrocution. The type of equipment most commonly used for the purpose is a water-bath stunner in which the head of the bird should be only partly immersed to avoid inhalation of contaminated water. Because the birds in any single load will vary in size, some go slightly deeper into the water bath and can inhale a small amount of water that may reach the lung cavity.[17] At the next stage, the neck is cut to allow the bird to bleed out and, in modern, high-rate processing plants, this is done by means of a rotating knife-blade, which cannot be heat-treated between birds and is therefore a possible source of cross-contamination. Organisms present on the slaughter knife may be carried into the body of the bird by residual blood circulation.[18]

The freshly slaughtered birds are passed through the scald tank, which contains hot water in a state of agitation, in order to loosen the feathers. This process removes large numbers of microbial contaminants from the carcass surface and the tank becomes heavily contaminated with faecal material. Each carcass is thought to contribute about 10^9 viable bacteria to the scald water as it passes through the tank.[19] Despite such a large input, conditions are such that, following an initial build-up, the number of viable bacteria present remains relatively constant over the working period, due to the influx of fresh water to replace that removed by outgoing carcasses. Survival of microorganisms is also influenced by the temperature of the scald water, which may vary from 50 °C to 63 °C, depending on the type of bird being processed and whether it will be water-chilled and frozen or air-chilled for the 'fresh' market. The latter product requires carcasses to be scalded at 50–52 °C to avoid damage to the cuticle and subsequent discoloration during air chilling. The persistence of many organisms in the scald-water at such low temperatures means that cross-contamination of carcasses occurs readily. The problem also arises during the defeathering stage and, initially, is due to aerial dispersion of bacteria from the scouring or flailing of carcasses to remove the feathers. Equally, carcasses may acquire microbial contamination from the plucking machines themselves, since the internal atmosphere is warm and moist and favourable for microbial growth. Also, the flexible rubber 'fingers' that are in contact with the carcasses soon become worn and cracked and can be penetrated by organisms that will then survive post-processing cleaning and disinfection. The most common colonisers are certain types of staphylococci, including *S. aureus*.[8] The colonising strains of this species usually have an unusual clumping morphology and may produce extracellular slime. They tend to be more resistant to chlorine than other strains and some have been shown to produce enterotoxins associated with human food poisoning, mainly types C and D toxins. Growth of *S. aureus* in the machines is such that the normally low levels of carcass contamination with this organism increase during the plucking process.

Before the birds are eviscerated, both the head and feet are usually removed and the head-puller, in particular, may be a site for cross-contamination of the neck skin. Further contamination and spread of faecal bacteria are likely to occur during all stages of evisceration (Fig. 6.1), especially if there is any rupture of the intestines and spillage of gut contents. In low-throughput processing plants, evisceration is carried out manually, but the same problems can arise, unless special care is taken by the operatives. With some evisceration machinery, gut breakage can occur because of natural variation in bird size, and the design of the machines is such that this cannot be entirely avoided. During the evisceration stages, continuous water-sprays are often used in the machines to avoid any accumulation of debris and the carcasses are spray-washed during and after evisceration to ensure that they are visibly clean, i.e. free from blood splashes and minor faecal contamination, although the effect is limited because of microbial attachment to carcass surfaces. In practice, there is a need to wash both carcasses and equipment with care in order to minimise direct splashing and formation of water droplets that can themselves be a vehicle for spreading contamination.[20]

After the final washing stage, the warm carcasses are chilled promptly to prevent growth of mesophilic pathogens and to limit multiplication of psychrotrophic bacteria such as spoilage pseudomonads. There are two main types of chilling system, one of which involves immersing carcasses in chilled water, the other requiring the use of cold air, with or without wetting the birds to maintain product yield and enhance the rate of cooling through evaporation (evaporative cooling). Both types of system chill the carcasses effectively, but carry some risk of cross-contamination between carcasses. In water chilling, this is likely to occur by carcass-to-carcass contact and via the cooling medium, even though there is an overall reduction in carcass contamination due to the washing effect.[21] Some air-chilling systems also allow contact between carcasses, but, equally, the transmission of microorganisms could occur via the air currents. When water-sprays are used, organisms may be spread by the droplets that are formed. However, any cross-contamination occurring at the chilling stage must be seen in the context of the greater hygiene problems at earlier stages of the process, especially during scalding, plucking and evisceration.

The stages that follow chilling are grading, trussing and packaging, and they involve further handling of the product, with the likelihood of additional cross-contamination from hands and equipment. Growth of psychrotrophs may also occur if there are delays in transferring the product to the secondary chilling or freezing stage.

6.2.5 Role of GMPs and plant sanitation programmes

The application of GMPs is a pre-requisite of any HACCP programme. GMPs can be defined as 'all basic preventive measures that are needed to produce food under acceptable, hygienic conditions'. The choice of appropriate measures is largely subjective and based on experience, and since there may be no quantifiable improvement in the hygiene status of the product as a result of their application, the effects of using GMPs are not usually apparent. Not all GMPs are even related to food safety issues but, when they are relevant, they may be regarded as an integral part of the HACCP programme. Examples of GMPs are the use of protective clothing, disinfectant footbaths and hand-washing by plant personnel, and the design and layout of the processing plant in relation to hygiene, including the separation of 'clean' and 'dirty' parts of the process. In this context, it is particularly important to separate the hanging-on bay and the scalding and plucking processes from the remaining stages. These are legal requirements in the EU and other GMPs can be in the same category. Essentially, GMPs create the environment in which HACCP principles can be successfully applied.

A comprehensive and effective plant sanitation programme is another necessary part of the HACCP approach and will cover not only the cleaning and disinfection of all the equipment and processing environment, but also maintenance of the fabric of the working areas and the hygiene of the plant as a whole, including the control of rodents and other pests. For all food handling premises, plant sanitation is a CCP and should be carried out only by designated, trained personnel. The requirements should be fully documented and include the exact procedures to be used, the frequency of cleaning, the equipment and chemical agents needed, the quantities of chemicals involved and the methods of preparing the required solutions, as well as any precautions necessary in handling the relevant substances. One person should be made responsible for the day-to-day management of the entire cleaning programme and different individuals designated to carry out particular tasks.

For poultry processing plants, a full programme of cleaning and disinfection should be carried out at least once per day, although some items, such as knives and gloves, require more frequent attention. In addition, consideration should be given to intermediate cleaning during the main break periods. The basic stages in the cleaning and disinfection process are shown in Table 6.3, bearing in mind

Table 6.3 Stages in the cleaning and disinfection process[22]

Stage	Aim	Items used	Application
Pre-clean	Removal of gross soil	Scrapers, brooms, shovels, brushes or squeegees	
Rinse	Removal of minor soil	Potable water	Medium pressure (30 bar)
Clean	Provide a visually clean surface	Detergent	With air as a foam or gel
Rinse	Remove soil and cleaning agent	Hot potable water (40–50 °C)	Medium or high pressure (30–60 bar)
Disinfect	Kill most microorganisms	Chemical disinfectant	Low pressure (large droplets)
Rinse	Remove dead micro-organisms and chemical residues	Potable water	Medium pressure
Dry	Prevent regrowth of viable microorganisms	(Air)	(Equipment should be designed to drain)

Reproduced with permission of the editors.[22]

that special treatment may be needed for some items of equipment, e.g. the defeathering machines, where worn rubber 'fingers' should be replaced regularly. Adequate technical knowledge is essential in selecting the appropriate chemicals for each type of application. Detergents are chosen according to the type of soil, nature of the surface to be cleaned, water hardness and method of cleaning. The choice of disinfectant depends upon the nature of the target surface and its accessibility. Surfaces must be cleaned thoroughly, if the disinfectant is to be fully effective, and a visual check is needed before disinfection is carried out. Afterwards, it is necessary to determine the efficacy of the disinfection process and this is usually done by microbiological testing (see below). Tests should not be confined to flat surfaces that are easy to sample, but should include crevices and other possible niches for microbial survival. Each establishment should fix limit values for colony counts, above which the standard of cleaning and disinfection is considered unacceptable. A more rapid method of evaluation is the ATP assay, but this is relatively expensive for routine use.

Both GMPs and plant sanitation practices should be subjected to a formal audit and a report prepared before the HACCP study is carried out.

6.3 Establishing CCPs

Because of the nature of the process, little can be done at present to prevent the spread of microorganisms in poultry processing and there are no CCPs in the process at which any foodborne pathogens on the carcasses can be eliminated. The development of carcass decontamination systems that would be effective and feasible to use under commercial conditions is still in its infancy and will be discussed later. To optimise hygiene control under present circumstances, the main objectives must be (i) to establish an adequate and effective plant sanitation programme, as described above; (ii) to ensure that GMPs are properly applied at all stages of the process; and (iii) to apply HACCP principles as a means of ensuring the control of *overall* carcass contamination. Failure to meet these objectives could result in finished carcasses of poor microbiological quality.[23]

Although the microbiological hazards in processing are well known and the process itself is similar across the world, no two processing plants are identical in all respects and, in some situations, the severity and probability of the hazards are likely to differ from one plant to another. Therefore, a separate HACCP programme should be developed for each processing line. As a part of GMPs, the process should be checked to ensure that the general standard of hygiene control is as high as possible. In countries where the use of chlorinated water is permitted, e.g. the United Kingdom and USA, there may be scope to introduce low-pressure chlorinated water sprays that can be targeted within specific items of equipment to control microbial contamination and hence transmission of organisms among the carcasses. This point is illustrated in Table 6.4, which shows the spread of a 'marker' strain of *Escherichia coli* from two inoculated carcasses to others that followed through the head puller.[24] The sprays should be used in such a way that generation of aerosols is minimal. At many processing plants, there are also points in the process where carcasses unnecessarily come into contact with soiled surfaces and small adjustments can be made to avoid such contact. When taken together, modifications of this kind can have a significant effect on levels of carcass contamination.[25]

Table 6.4 Controlling the transmission of a 'marker' strain of *Escherichia coli* by using chlorinated water sprays in the head-puller[24]

Carcass number[a]	No sprays	With sprays
1	5.3[b]	5.7
10	2.7	2.5
50	2.1	0.9
100	2.1	—[c]
250	1.4	—
500	0.9	—

[a] Mean of three trials in each case
[b] Log_{10}cfu/g of neck skin
[c] — not found

With regard to establishing CCPs, there are limited options in present processing systems and, in any case, the product is a raw one that will always be cooked before consumption. At the level of the processing plant, the rate of production in the larger premises and the close proximity of carcasses on the line allow little opportunity for the use of hygiene intervention measures after dealing with one carcass at any particular stage and before another arrives. This situation is conducive to the spread of any hazardous organisms present among the carcasses being processed. Also, the carcasses come into contact with equipment that rapidly becomes soiled and therefore contributes to the cross-contamination problem. A further factor is the aerial dispersion of microorganisms that is greatest in the unloading bay, but also occurs at other stages of the process,[26] and can contribute to the spread of pathogens. If no method of carcass decontamination is available, the net effect of processing is usually to increase the proportion of carcasses that carry minority organisms such as salmonellas.

In the absence of any specific CCP to prevent or reduce the proportion of contaminated carcasses, dependence must be placed on plant sanitation and the use of GMPs in which non-critical control points (CPs) can be recognised. A CP has been defined as 'any point in a specific food system where loss of control does not lead to an unacceptable health risk'.[27] Examples range from the maintenance of disinfectant footbaths to the use of chlorinated water sprays at key points in the process, as mentioned earlier. Thus, the CP concept is more realistic in the context of poultry processing; it allows the processor to focus on the principal control measures and it avoids the pitfall of setting too many CCPs, which could weaken their impact. The overall aim is to minimise microbial contamination of carcasses and the spread of foodborne pathogens. There are no universally accepted microbiological criteria for poultrymeat, and processing practices may vary according to national and regional requirements. In the USA, for example, current regulations do not permit any visible faecal contamination of carcasses and include microbiological performance standards for *Salmonella* that are related to the introduction of HACCP programmes. By contrast, meat found to be contaminated with *Salmonella* is regarded as unfit for human consumption in some Scandinavian countries.

6.3.1 Control measures at different stages of processing

This section considers the control measures currently available at various stages of processing, whether or not these should be regarded as GMPs, CPs or CCPs.

To prevent the spread of any pathogenic organisms from the live-bird reception area to the remainder of the processing plant, it is essential that the reception area is well separated and that doors are kept closed when not in use at all points of access. Scalding and plucking, too, must be physically isolated from other parts of the process. Facilities for cleaning and sanitising bird delivery crates and vehicles are usually located close to the arrival area. For loose crates, systems are available for soaking and washing the crates to remove all visible soiling. Unless this is done thoroughly, no sanitising agent is likely to be fully effective. The fixed-crate system is more difficult to clean because of the lack of a soaking stage to facilitate the removal of dried-on droppings. Cleaning is usually accomplished by hosing down the entire vehicle with superchlorinated water, but the efficacy of the process depends on the time spent on each vehicle and this can be inadequate.[24]

The rotating knife-blade of the automatic slaughtering equipment is one of a number of situations in processing where microbial transmission can be reduced by spraying key surfaces continuously with superchlorinated water, and the same is true of the head-puller and any conveyor belts used for carrying carcasses.[24] Chlorinated water can also be used in evisceration machinery to prevent a build-up of contamination. Chlorine concentrations will vary from one processing plant to another, but need to be at least 20 mg/l. The use of chlorine would not be appropriate for the scald tank because of the high level of organic loading in the water. In this situation, there are no additional control options but, as long as the system operates at the required temperature and with the necessary input of fresh water, combined with water agitation, microbial contamination of the carcasses will be reduced, while numbers of organisms in the scald water remain relatively stable after an initial build-up. The scald temperature has more effect on organisms circulating in the water than on those present on carcasses,[19] especially when bacterial attachment has occurred.[28] However, because of their relatively low heat-resistance, pseudomonads are rarely isolated from either scald-water or freshly scalded carcasses, even when scalding is carried out at 50°C. On the other hand, the higher heat-resistance of salmonellas can be reduced by adjusting the pH value of scald-water to 9.0 ± 0.2, through the constant addition of sodium bicarbonate or sodium hydroxide.[29] The treatment reduces microbial contamination of both carcasses and scald water, but tends to make the carcasses slippery and has not found favour with the industry.

The nature of the defeathering process does not lend itself to any specific control measures, although conditions that allow an excessive accumulation of feathers or a rise in the environmental temperature inside the machines are to be avoided. For example, when new machines were installed at a commercial plant, a plastic canopy was fitted and the chlorine concentration in the water was reduced, contamination of carcasses with staphylococci and other organisms increased.[30] Because the rubber 'fingers' become worn and cracked during use, thereby harbouring bacteria, regular replacement of the 'fingers' is essential.

Evisceration of carcasses is carried out in several stages and includes opening of the abdominal cavity and exposure of the viscera prior to carcass inspection. The damage that occurs to some viscera and spillage of gut contents can only be controlled by careful setting of the machines and resetting when birds of a different weight are being processed. This problem is less evident with the newer kind of evisceration machinery described below. The introduction of automatic line transfer, taking carcasses from the killing process to the evisceration line and then on to the chilling line, has also been an improvement, because product handling and therefore carcass contamination are reduced. The spray-washing that is carried out both during and after the evisceration stages can be expected to reduce carcass contamination by about ten-fold, provided that washer design and water usage are adequate. In the EU, there is a requirement to use 1.5 litres per carcass for carcasses up to a weight of 2.5 kg and correspondingly higher amounts for larger birds. Superchlorination of the water may also be beneficial, using a free available chlorine concentration of 20 mg/l or more. However, the use of an 'inside–outside' washer, as required in the EU, does not guarantee any better removal of organisms than that achieved with other types of washer.[31] For washing at intermediate stages of evisceration, the siting of the washers is important if contamination is to be removed effectively. Likely sites are after opening of the abdominal cavity and following exposure of the intestines.

Within the EU, control of water immersion chilling systems is prescribed by a code of practice that is part of the poultry hygiene regulations in each of the Member States. The aim is to prevent a build-up of microorganisms in the chill water during the working day and to cool the carcasses rapidly to prevent microbial growth. The regulations specify the amount of water to be used for different sizes of carcass, how the water should be divided between chill tanks, if more than one is used, the temperature of the water at the carcass entry and exit points, and the maximum dwell-time for carcasses in the first part of the system. It has been shown that properly controlled immersion chilling systems lead to a reduction in both aerobic plate counts and counts of coliform bacteria.[32,33] There are no specific requirements for air chilling in the legislation and, without any washing effect, levels of microbial contamination are likely to remain virtually the same, as long as the chilling process is effective in cooling the carcasses.

The stages that follow chilling are grading, trussing and packaging. These involve further handling of the product and possible cross-contamination from hands and equipment. Any delays in refrigerating or freezing the carcasses should be avoided to prevent multiplication of psychrotrophic spoilage bacteria.

As indicated above, some items of processing equipment are particularly difficult to clean and disinfect properly and may require special attention. The live-bird delivery systems are one example and defeathering machinery is another, although the machines are now more accessible for cleaning than was once the case. It is also important that the eviscerating machines receive adequate attention because there is evidence that they can contribute to the carry-over of microorganisms from one processing period to another.[5]

6.3.2 Establishment and operation of CCPs

CCPs will be determined by the HACCP team following detailed consideration of both the product and the process, and this will involve the setting of limit values. In doing so, it is important that all the critical components of each CCP are identified. For example, the cleaning and disinfection of live-bird delivery crates requires a process that will remove all visible signs of residual droppings and subsequent application of a sanitising agent that will destroy most viable organisms. The consequences of not meeting these requirements (flock-to-flock transmission of pathogens) are clearly important in relation to food safety. If scalding is considered to be a CCP in controlling overall carcass contamination, then control of water input and temperature is essential. It is also necessary to ensure that the water in the tank is agitated continuously to facilitate circulation and the removal of microorganisms from the skin and feathers of carcasses. All three parameters are included in Table 6.5. The table shows the minimum acceptable temperature for 'soft' scalding of poultry. Below the minimum, growth of some microorganisms could occur in the scald tank. However, the normal upper limit of 52 °C does not appear in the table because it does not relate to food safety and is, in fact, a product-quality criterion for air-chilled poultry. Similarly, with spray washing at any stage, the use of chlorine must not exceed the concentration that is acceptable to operatives working in the area and is largely determined by the effectiveness of plant ventilation. There is little risk of tainting the product, even with chlorine levels that are several times higher than those normally used. However, the water pressure should be such that any generation of aerosols capable of spreading bacteria is minimised.

It is debatable whether carcass evisceration is really a CCP because the efficacy of this process depends upon the type of machinery being used and cost considerations usually dictate the timing of any replacements that are deemed necessary. Even with the older, less efficient machines, visible contamination of carcasses with gut contents can be minimised by proper attention to setting the machines, and this needs to be checked regularly. In the case of chilling, which is the most important CCP from the microbiological viewpoint, the upper limit of 10°C for chicken carcasses is significant in relation to growth rates of foodborne pathogens. Salmonellas would double in number only about once at this temperature in an eight-hour working shift. Further chilling is, of course, necessary at the end of the process to ensure an adequate shelf-life for the product and to meet the EU legislative requirement of 4°C. Any air cooling process must avoid crust-freezing of the product, because this is unacceptable for quality reasons.

In all instances, monitoring of the necessary parameters is carried out by relatively simple tests, from which any necessary corrective action can be readily taken. Measurements can be made continuously or at appropriate intervals, as required. The temperature of the scald water is usually measured automatically, while simple visual checks are sufficient to determine whether transport crates are being adequately cleaned or eviscerating machines are functioning properly. Chlorine concentrations can be determined rapidly and at regular intervals using colorimetric tests. Whatever corrective action is needed, the person responsible must be clearly identified. For some purposes, such as any necessary adjustment of chilling conditions, the relevant individual may need to be an engineer. Details of any action taken must always be fully recorded.

Table 6.5 Examples of microbial hazards and their control at different stages of processing

Process step	Aspect	Hazard	Monitoring			Control requirement	
			Procedure	Frequency	Critical limits		Action
Cleaning and disinfection of delivery crates	(i) Cleanliness of crates	Flock-to-flock contamination	Visual check	Continuous	Visibly clean		Reclean or use extra cleaning
	(ii) Use of sanitising agent		Check dosage	Hourly	Not below recommended concentration		Adjust dose
Scalding	(i) Water temperature	Inadequate reduction in carcass contamination	Check temperature reading	Hourly	Not below e.g. 50°C		Adjust temperature
	(ii) Water level		Visual check	Hourly	Not below specified level		Adjust water input
	(iii) Agitation of water		Visual check	Hourly	Constant		Stop line and repair
Evisceration (mechanical)	Mode of operation	(i) Faecal contamin. of carcasses	Visual check	For each batch	Not below agreed standard		Adjust machines
		(ii) Microbial build-up on machines	Check water usuage/chlorine concentration	Hourly	Not below set limits		Adjust flow/chlorine concentration
Post-evisceration spray wash	Efficacy	Inadequate carcass cleaning	Check (i) water usage	Hourly	Not below e.g. 1.5 litres per carcass		Adjust flow
			(ii) chlorine level		Not below e.g. 20 mg/l		Adjust dosage
Air chilling	Efficacy	Microbial growth on carcasses	Measure deep carcass temperature	Hourly	Not above e.g. 10°C		Modify chilling conditions

6.3.3 Validation and verification

Validation of CCPs and their associated controls may involve microbiological testing to ensure that the hazards at particular stages are reduced to an acceptable degree. Microbiological tests may also be relevant in verifying the effectiveness of the HACCP programme, when all the controls are in place and operational. Verification should be carried out regularly by at least two individuals who were not directly involved in the development or implementation of the HACCP plan. Apart from microbiological testing, verification will involve visual checks on the entire process and an examination of production records to ensure compliance with stated CCPs, procedures, corrective actions, etc. The plan itself will also need to be evaluated. Verification is particularly important when a change in the product or the process is introduced, or when new information relating to product safety becomes available.

Although processing of poultry should aim to minimise the spread of hazardous microbes, direct testing for pathogens is laborious, time-consuming and costly when searching for organisms that are present only sporadically and in low numbers. A more practical alternative is to carry out trials using a non-pathogenic and readily identifiable bacterium that can be artificially introduced into the processing environment so that its dissemination, and any proposed control measures, can be evaluated. For this purpose, a suitable organism is a nalidixic acid-resistant strain of *Escherichia coli* K12, which can be isolated directly and specifically on a medium containing nalidixic acid as one of the selective agents. The resistance is chromosomal and not plasmid-borne. The organism was referred to previously (Table 6.4) in relation to hazard identification and control in chicken processing.[24] Its use is also relevant to verification, and experience has shown that the results obtained are reproducible. Because the techniques involved are simple and yield results relatively rapidly, the 'marker' organism could also be used in the training of plant personnel and in demonstrating microbiological hazards.[34]

Another type of test that is relevant to HACCP verification is to determine the shelf-life of the end product, i.e. time taken for 'off' odour formation during chill storage. Pseudomonads are the principal spoilage bacteria of poultry, and contamination of carcasses with these organisms occurs mainly after the scalding stage. Since shelf-life is related to the level of *Pseudomonas* contamination on the product, a long shelf-life would indicate that the organisms have been adequately controlled. In a sense, therefore, pseudomonads can also serve as hygiene indicators.

6.3.4 Microbiological testing of products and equipment

The need to use microbiological testing in validation and verification raises the question of how best to sample carcasses and other raw products. Most of the microbial load on the carcass is present on the skin, but is not evenly spread, and some sites, such as the thigh and neck skin, tend to be more heavily contaminated than others, such as the breast area.[35] Also, the fact that some organisms will be attached to the skin or entrapped means that they are difficult to remove without the use of a destructive sampling method. On the other hand, certain foodborne pathogens, including salmonellas, may be present infrequently or in low numbers and could be missed if only a part of the carcass is sampled.

The two most popular methods of sampling have advantages and disadvantages. To obtain total viable counts or counts of faecal indicator or spoilage bacteria, it is common to sample the neck skin. Since a sample of only about 5 g is needed for the purpose, this can be taken without damaging the carcass or removing it from the processing line, when samples are obtained post evisceration. Maceration of the skin or use of a 'stomacher' provides a suspension for examination and tends to give a higher recovery of organisms than other sampling methods. An alternative is to use a whole-carcass rinse method[36] and, although it does not recover all of the organisms present, this method is the best means of detecting minority pathogens, such as salmonellas, because all surfaces of the carcass are rinsed. The technique involves placing the carcass in a waterproof plastic bag and adding an appropriate amount of diluent, equivalent to at least half the weight of the carcass. The bag is closed and held in such a way that it can be shaken vigorously for, say, 30 seconds. A portion of the rinse water is then examined for the required organisms. The method is unsuitable for larger carcasses, especially those of turkeys, and, obviously, it requires carcasses to be removed from the processing line. Rinse sampling can be tiring to carry out on a large scale and a means of mechanising the shaking process has been developed.[37] Giblets, cut portions and mechanically recovered meat are usually sampled by taking an appropriate amount of material for maceration or stomaching. For portions, both skin and cut muscle can be included in the sample.

Another method is swab sampling, but this is of limited value for carcasses or other products, because the recovery of organisms from the skin is relatively low. However, this method is highly suitable for sampling equipment and working surfaces, since it can be used to reach those parts of the process that are most difficult to clean effectively. Swabbing and other methods of testing processing equipment, including use of agar contact plates and ATP and impedance measurement, have been reviewed recently.[38]

For monitoring of equipment to determine the effectiveness of cleaning and disinfection, it is usually sufficient to obtain total viable counts (otherwise known as 'aerobic plate counts'). An incubation temperature of 30°C is advisable for the agar plates, since this permits the recovery of both faecal bacteria and spoilage organisms, the latter being inhibited at 35°C or above. It is both advisable and convenient to use pre-prepared spread plates, because spoilage bacteria are relatively heat-sensitive and can be inactivated by molten agar, as used in pour-plates.

For some purposes, it may be relevant to use indicator organisms as a measure of faecal contamination of carcasses, and possible candidates are *Escherichia coli*, coliforms and Enterobacteriaceae. However, faecal contamination of carcasses post scalding can be difficult to distinguish from that occurring initially on the skin, especially when 'soft' scalding at about 50°C is used. Coliform counts of 10^3 per gram of neck skin are not uncommon after scalding.[9] Although *E. coli* is a more specific indicator of faecal contamination and most coliforms isolated from carcasses during processing belong to this species,[39] there is a preference in Europe for the Enterobacteriaceae test, involving a pour-plate method, because of its wider scope. The coliform test, which is aimed only at lactose-positive strains, has been used for surface inoculation of pre-prepared plates. In this form, the test is less specific but more convenient and appears to be adequate in the context of the processing plant. Whatever test is used, there is no predetermined correlation between levels of carcass contamination and the presence of salmonellas, but if salmonellas are present in the incoming birds, control of faecal contamination will clearly be a factor in limiting the spread of these organisms.

The prevalence of salmonellas in raw end-products can be an important issue with respect to customer requirements and export–import opportunities, and the HACCP team will need to establish close liaison with their counterparts in other sectors of the production chain to maximise control measures.

6.4 Other processing operations

6.4.1 Giblets

Giblets comprise the gizzard, heart, liver and neck and, since these organs are pooled after harvesting, it is almost certain that each of the organs within a giblet pack will have originated from a different bird. The inclusion of giblets in chilled carcasses is rare now in the UK and, although liver may be sold separately or used for pâté manufacture, most of the edible offal produced in poultry processing is utilised in pet food. This situation followed the recognition that giblets were more frequently contaminated with salmonellas than the carcasses themselves and a recommendation of the UK National Advisory Committee on the Microbiological Safety of Food that incorporation of giblet packs in retail poultry be discontinued.[20] Even packaging of the organs cannot be relied on to prevent the transmission of contaminants to the carcasses. In some cases, the giblet packs are not impermeable to water, while in others the packaging material is impermeable but there can be a high rate of failure in sealing. The method of production is such that there is abundant opportunity for microbial cross-contamination between individual organs, from giblets to carcasses via leaking packs and from the outsides of the packs to carcasses due to contact with contaminated surfaces during and after the packaging process. The main microbiological hazards in giblet handling and their control are shown in Table 6.6. While it is evident that cross-contamination cannot be prevented in this type of operation, the table shows that much can be done to reduce the problem, mostly by giving attention to staff training and the observance of GMPs. Control of faecal contamination is largely related to the general standard of carcass evisceration. As with carcasses, chilling in cold air or water is a CCP that effectively controls microbial growth, while chlorination of all water that

HACCP in primary processing: poultry

Table 6.6 Microbial hazards associated with giblet handling and their control[a]

Processing stage	Source of contamination	Control of hazard
Harvesting	Handling	Regular hand washing
	Contact surface	Cleaning and disinfection
	Intestinal contents	Related to control of evisceration
Fluming	Process water	Chlorination of water
	Contact surface	Cleaning and disinfection
Chilling	Contact surface	Cleaning and disinfection
	Process water	Chlorination of water
	Microbial growth	Control of water temperature
Transfer	Handling	Regular hand washing
	Contact surface	Cleaning and disinfection
Packaging	Leaking packs	Staff training
	Surface contact after packing	Cleaning and disinfection
Transfer	Handling	Regular hand washing
	Contact surface	Cleaning and disinfection
Insertion of packs into carcasses	Handling	Regular hand washing
	Pack damage	Staff training

[a] Based on information given by Chappell[40]

comes into contact with the giblets is likely to be beneficial in reducing microbial contamination. The table also highlights the importance of cleaning and disinfecting the relevant surfaces and equipment. This should be done by hosing down with chlorinated water at break times and more comprehensively at the end of the working period, as part of the regular plant cleaning programme.

6.4.2 Portioning and deboning operations

There is a high demand for cut portions of poultry from both retail and catering sectors of the industry and for deboned meat used in product manufacture. Increasingly, these processes are being carried out by mechanical or semi-mechanical means, which allow relatively high line-speeds and throughputs, although manual portioning operations still exist, especially for turkeys, and are highly labour intensive. A plan of a typical mechanical line is shown in Fig. 6.2. It might be expected that microbial contamination would increase during the manipulations involved, because of the exposure of newly cut surfaces of the meat and contact with conveyor belts, cutting equipment and other soiled items. The extent to which microbial counts increase depends upon the degree of hygiene control being exercised and the length of time that particular meat surfaces are exposed.[41] When hygiene standards are high, portioning does not significantly increase microbial contamination. However, contamination can increase during the carcass hanging-on stage and some items of equipment may

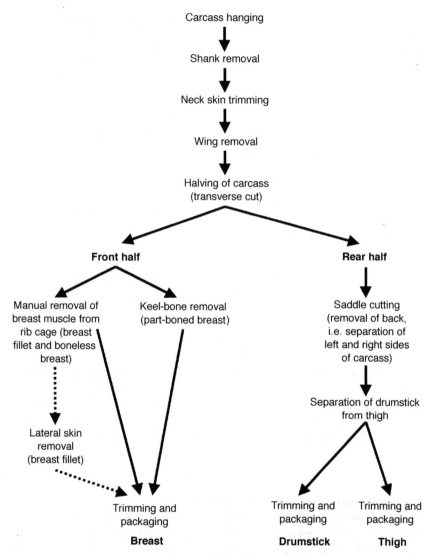

Fig. 6.2 A mechanised process for carcass portioning.[41]

become excessively contaminated with *S. aureus*, thus underlining the importance of thorough cleaning and disinfection. Also, psychrotrophic bacteria, such as *Pseudomonas* spp. and *Brochothrix thermosphacta*, are able to multiply and it is important, therefore, that equipment and working surfaces are kept as dry as possible and the processing environment cool to retard microbial growth. As with the carcass processing plant, aerial contamination occurs, but is likely to be at a low level.[42] In these circumstances, only visibly clean carcasses should be portioned and they should be adequately chilled beforehand. Subsequent control measures depend on a variety of GMPs, including the use by operatives of full protective clothing, hand-washing facilities and hot-water baths for knives, etc., as well as general control of the cutting environment. As before, proper staff training and cleaning and disinfection of equipment, surfaces, gloves, etc., are vital and will feature in CCPs. Maintaining microbial quality depends upon rapid transfer of freshly cut portions to a chiller or blast freezer and holding the final product at the required temperature, again a CCP.

The machines used to harvest residual meat from bones are conducive to microbial cross-contamination. They are relatively difficult to clean and some of the older machines can cause a rise in product temperature during operation. The bones themselves must be refrigerated and stored hygienically before use, and the recovered meat needs to be chilled promptly or frozen. The meat is in a finely divided state and highly favourable for microbial growth.

Any kind of raw product that is used in the manufacture of value-added products is likely to be tested microbiologically to ensure that it meets company specifications for that purpose.

6.5 Future trends

6.5.1 Development of more hygienic processing equipment

The spread of microorganisms in poultry processing has been exacerbated over the years by increasing intensification of the process. Understandably, hygiene control has not been the highest priority for poultry processors and improving product yield and uniformity have been the principal goals in developing new process machinery. Within the EU, legislation on poultrymeat hygiene makes no specific demands on the design of processing equipment, other than requirements for cleanability and avoidance of any interference with the product. This does not mean that equipment could not be developed to function more hygienically but, unless there is a financial incentive to do so, processors are unlikely to invest in costly new machinery.

Despite the lack of incentives, some attention has been given to developing systems that would reduce microbial contamination of poultry and diminish opportunities for cross-contamination of carcasses. An EU-funded research project known as EUROVOL, involving some of the leading equipment manufacturers, has allowed a study of conditions in processing that affect carcass contamination and has been associated with the development of new equipment.[43] In one case, the design of live-bird transport crates has been improved, so that birds arriving for slaughter have less faecal contamination. A more effective crate washer has also been developed and this incorporates a soaking stage to loosen dried-on faeces and inversion of the crates so that loose, solid material falls away. The scalding process has been redesigned to include a series of immersion tanks that provide a progressive dilution of microbial contaminants in the scald water. By changing to a counter-flow system, carcasses meet the cleanest water as they leave the final tank. During processing, any of the individual tanks can be emptied and refilled, as required, without disrupting the process. It is claimed that the newer system reduces microbial contamination of carcasses by 60%. Studies involving counter-flow scalding followed by a hot-water wash have shown small but statistically significant reductions in microbial counts,[44,45] when compared with the conventional scalding process. Other changes have also been beneficial, including the use of superchlorinated water at selected stages of the process as a whole. However, in relation to pathogens such as *Salmonella*, *Campylobacter* and *Listeria* spp., the effects of process changes have varied from one processing plant to another, sometimes with no apparent benefit, and the reasons for this are unknown.

A notable improvement in carcass hygiene has been evident with the newer carcass opening and eviscerating systems that are now widely used. The greater degree of control provided by these machines reduces gut breakage and hence faecal soiling of carcasses, and, instead of leaving the exposed viscera in contact with the carcasses, the organs are transferred to a separate but synchronised line that facilitates on-line inspection. The new approach also reduces the amount of handling needed in harvesting giblets. Further developments that are applicable to both rotating eviscerating machines and transfer belts are cleaning-in-place systems that allow the equipment to be cleaned and sanitised during break periods. Unfortunately, these can spread contamination via the formation of droplets of contaminated water and are rarely used.

Despite the improvements described above, some parts of the process still require attention from the hygiene viewpoint. The defeathering stage is an example and, although the modern machines are more accessible for cleaning, they are still a major cause of cross-contamination and often favour microbial colonisation. Preliminary studies have shown that the cross-contamination problem could be significantly reduced by some means of shielding individual carcasses, especially in the first plucker unit (D.B. Tinker, Silsoe Research Institute, pers. comm.). Again, however, there is little incentive for processors to change the existing machines, because there is no financial advantage in doing so. An alternative is to remove contaminants during or after the plucking process, as in the so-called 'closed-loop' system, which involves spraying the carcasses with hot water that is then partly recycled.[46]

Individually, the newer developments in processing equipment have only a relatively small effect on carcass contamination and, in practice, must be part of a properly integrated approach to hygiene control, involving the full range of GMPs. The EUROVOL project recognised the importance of controlling the processing environment with respect to temperature and humidity, in order to limit microbial growth and surface attachment. Clearly, future reductions in the growth and spread of microorganisms in poultry processing will need to take more account of environmental control.

6.5.2 Predictive modelling of microbial contamination in processing

Because poultry processing is now highly mechanised and can be relatively closely controlled, there appears to be scope for developing mathematical models that would take account of microbial behaviour at particular stages of the process and could be used to provide information for risk analysis and to predict conditions for optimum hygiene control. So far, deterministic models have been developed for only two stages of the process: scalding[47] and water immersion chilling.[48] These situations are highly appropriate for modelling, because both involve immersion of carcasses in water, which provides a homogeneous environment, with stable temperatures, pH values, etc. In relation to through-flow water chilling, the model showed that, with appropriate controls, microbial numbers in non-chlorinated chill water would be unlikely to exceed 10^5 per ml. The prediction was subsequently confirmed in a study of commercial chillers.[49] The modelling approach could be applied to other stages of the process, but would require detailed knowledge of microbial survival, growth and dissemination under the relevant conditions, and this information is not yet available.

6.6 Decontamination of carcasses

Treatments to reduce microbial contamination of carcasses are available and new systems are being developed, but the EU is reluctant to accept this approach because of fears that too much reliance will be placed on decontamination and not enough on general plant hygiene. Methods available for both red meat and poultry have been reviewed[50] and the requirements discussed. The ideal method would have no adverse effect on the appearance, smell, taste or nutritional properties of the meat. Also, treatment of carcasses would leave no residues, pose no threat to the environment and raise no objections from consumers or legislators. The treatment would be low in cost and easy to apply. Processors would expect not only effective destruction of foodborne pathogens, but an extension of product shelf-life through the inactivation of spoilage bacteria.

In reality, it is unlikely that any one type of treatment would completely satisfy all the above criteria. Many different chemical treatments have been investigated and some would appear to be unsuitable for use in a food processing environment or would carry a risk of tainting the product. For example, dipping chicken carcasses in a solution of hot succinic acid was found to be highly effective in destroying salmonellas, but had an adverse effect on the colour of the meat.[51] However, physical and microbiological treatments have also been developed and evaluated, as shown in Table 6.7. Physical methods have the advantage of avoiding any chemical residues or problems of waste disposal, but may still affect the appearance of the product unless the treatment is carefully controlled. Combinations of treatments are also feasible, but have yet to be fully investigated.

Washing carcasses with hot water has been used successfully for red meat and could be applied to poultry. Since temperatures of 80 °C or above are needed to obtain reductions in microbial contamination of up to 99.9%, there is a danger of partly cooking the meat. The same is true of steam treatment, although steam

Table 6.7 Possible treatments for reducing microbial contamination of poultry carcasses

Physical	Chemical	Microbial
Cold or hot water	Organic acids	Lactic acid bacteria
Steam	Inorganic acids	Bacteriocins
Steam-vacuum	Alkalis	Microbial parasites
Ultrasound	Sorbates	
Electromagnetic radiation	Chlorine and chlorine compounds	
Ionising radiation	Chlorine dioxide	
Electrical methods	Phosphates	
High pressure	Ozone	
Air ions	Hydrogen peroxide	
Freeze–thaw cycling	Miscellaneous salts	
	Lysozyme	
	Disinfectants	
	Antibiotics	
	Lactoperoxidase system	
	EDTA	
	Ethanol	

has a higher heat capacity than the equivalent amount of water and can inactivate bacteria in a much shorter time than that required to cook the flesh. In order to prevent cooking, the steam must be allowed to condense and then re-evaporate quickly from the surface to provide a cooling effect. A method involving re-evaporation under vacuum has been used for repeated cycles of very short treatment times and, with artificially inoculated carcasses, can achieve a 99.99% reduction in carcass contamination.[52] Among other possible physical treatments, ultraviolet (UV) light has been considered but is known to have poor powers of penetration, and bacterial cells in shadows or crevices are likely to be shielded. Although salmonellas and campylobacters are relatively sensitive to UV light, its effectiveness is a function of intensity and time of exposure, which may limit its applicability to rapid, on-line processing.

While physical treatments are still undergoing development, chemical methods are more readily available and, in some cases, there is both laboratory and field experience to draw upon. Of the possible options (Table 6.7), certain organic acids are among the most promising substances to use, including acetic and lactic acids. These are applied at a concentration of 1.2–3.0% to avoid any lasting discoloration of the carcasses, and are best used when the carcasses are still warm. Reductions in carcass contamination are of the order of 90–99%. Lactic acid may be preferred to acetic because it is a natural metabolite of muscle. However, both acids are used in food manufacture and are regarded as safe in relation to human health.

Another possible decontaminant is trisodium phosphate (TSP)[53] and a commercially applicable method of treating poultry carcasses with this compound is available. The mode of action involves removal of attached bacteria and appears to depend on the high pH (>12.0 for an 8% solution), which results in lysis of Gram-negative organisms. TSP can be applied as an immersion treatment immediately after water chilling or prior to air chilling. The degree of bacterial kill is again 90–99%.

Although chlorine is included in Table 6.7 as a carcass decontaminant, it is not directly effective for this purpose, because it is rapidly inactivated in contact with carcasses. Chlorine is most effective in controlling microbial contamination of the processing environment, including equipment, working surfaces and process water. In this way, it has an indirect effect on carcass contamination and is best regarded as an aid to hygienic processing.

6.7 Sources of further information and advice

Among the numerous publications that cover both theoretical aspects and practical applications of HACCP principles, there are some that deal specifically with poultrymeat production and products. Certain of these include generic plans that are intended to provide guidance to processors prior to developing their own HACCP programmes and not as ready-made systems for blanket application. One such generic plan for processing is provided by the US National Advisory Committee on Microbiological Criteria for Foods[54] and includes six CCPs that cover venting/opening/evisceration, final carcass wash, chilling, cut-up/boning/packing/product chilling, labelling and refrigerated storage. The inclusion of labelling seems to be logical, because the label should provide a clear means of identifying the source of the product, should it become necessary to trace and withdraw a particular lot from sale. The label also gives instructions to the consumer on product handling and cooking, which are relevant to food safety. This generic plan differs in some respects from one published earlier by the International Commission on Microbiological Specifications for Foods,[55] where scalding is among the proposed CCPs. The US committee rejected scalding as a CCP, partly because of the apparent difficulty in establishing a scientifically valid critical limit. The difference of opinion in this respect confirms the fact that the HACCP concept is still somewhat subjective in its application and there can be other disagreements over what constitutes a hazard. Nevertheless, both publications give useful information and enable the reader to become HACCP-orientated.

Further material on HACCP applications in the poultry industry is given in *HACCP in Meat, Poultry and Fish Processing*,[56] which covers farms and ranches as well as processing and further-processing establishments. However, the book deals more with theoretical considerations and, for a more practical approach of a general kind, use can be made of *HACCP: A Practical Guide*.[57]

6.8 References

1. ANON, Watt poultry statistical yearbook 1998, *Poultry Int*, 1998 **37** (9) 3.
2. HUPKES H, Automation and hygiene in relation to poultry processing, in *Factors Affecting the Microbial Quality of Meat 2. Slaughter and Dressing*, pp 95–8, Langford, UK, University of Bristol Press, 1996.
3. WALDROUP A L, Contamination of raw poultry with pathogens, *World Poultry Sci J*, 1996 **52** 7–25.
4. GIBBS P A, The incidence of clostridia in poultry carcasses and poultry processing plants, *Br Poultry Sci*, 1971 **12** 101–10.
5. OJENIYI B, WEGENER H C, JENSEN N E and BISGAARD M, *Listeria monocytogenes* in poultry and poultry products: epidemiological investigations in seven Danish abattoirs, *J Appl Bacteriol*, 1996 **80** 395–401.
6. GENIGEORGIS C A, DUTULESCU D and GARAYZABAL J F, Prevalence of *Listeria* spp in poultry meat at the supermarket and slaughterhouse level, *J Food Protect*, 1989 **52** 618–24.
7. GENIGEORGIS C A, OANCA P and DUTULESCU D, Prevalence of *Listeria* spp in turkey meat at the supermarket and slaughterhouse level, *J Food Protect*, 1990 **53** 282–8.
8. MEAD G C and DODD C E R, Incidence, origin and significance of staphylococci on processed poultry, *J Appl Bacteriol Symp Suppl*, 1990 81S–91S.
9. MEAD G C, HUDSON W R and HINTON M H, Microbiological survey of five poultry processing plants in the UK, *Br Poultry Sci*, 1993 **34** 497–503.
10. NOTERMANS S, TERBIJHE R J and VAN SCHOTHORST M, Removing faecal contamination of broilers by spray-cleaning during evisceration, *Br Poultry Sci*, 1980 **21** 115–21.
11. GARCIA DE FERNANDO G D, NYCHAS G J E, PECK M W and ORDONEZ J A, Growth/survival of psychrotrophic pathogens on meat packaged under modified atmospheres, *Int J Food Microbiol*, 1995 **28** 221–31.
12. SELIGMAN R and LAPINSKY Z, Salmonella findings in poultry as related to conditions prevailing during transportation from the farm to the processing plant, *Ref Vet*, 1970 **27** 7–14.
13. PEZZOTTI G, MIGLIORATI G, TOSCANI T, SEMPRINI P and CONTE A M, Influence of transportation stress on *Salmonella* spp contamination in broiler carcasses, in *Prevention of Contamination of Poultry Meat, Eggs and Egg Products*, eds FRANCHINI A and MULDER R W A W, COST Action 97, Luxembourg, European Communities, pp 7–9, 1998.
14. BOLDER N M and MULDER R W A W, Faecal materials in transport crates as source of *Salmonella* contamination of broiler carcasses, *Proceedings of the Sixth European Symposium on Quality of Poultry Meat, Ploufragan*, eds LAHELLEC C, RICARD F H and COLIN P, 1983, pp 170–6.
15. GOREN E, DE JONG W A, DOORNENBAL P, BOLDER N M, MULDER R W A W and JANSEN A, Reduction of salmonella infection of broilers by spray application of intestinal microflora: a longitudinal study, *Vet Quart*,

1988 **10** 249–54.
16. RIGBY C E, PETTIT J R, BAKER M F, BENTLEY A H, SPENCER J L, SALOMONS M O and LIOR H, The relationships of salmonellae from infected broiler flocks, transport crates or processing plants to contamination of eviscerated carcasses, *Can J Comp Med*, 1982 **46** 272–8.
17. GREGORY N G and WHITTINGTON P E, Inhalation of water during electrical stunning in chickens, *Res Vet Sci*, 1992 **53** 360–2.
18. MACKIE B M and DERRICK C M, Contamination of the deep tissues of carcasses by bacteria present on the slaughter instruments or in the gut, *J Appl Bacteriol*, 1979 **46** 355–66.
19. MULDER R W A W and DORRESTEIJN L W J, Hygiene beim Brühen von Schlachtgeflügel, *Fleischwirtsch*, 1977 **57** 2220–2.
20. *Report on Poultry Meat*, Advisory Committee on the Microbiological Safety of Food, London, HMSO, 1996.
21. BARNES E M, Microbiological considerations in the chilling of poultry meat, *Proceedings of the Fourth European Poultry Conference, London*, 1973, pp 339–45.
22. HINTON M H, MEAD G C and ROWLINGS C (eds), *Microbial Control in the Meat Industry 5. Cleaning and Disinfection of Equipment and Premises*, EU Concerted Action CT94–1456, Bristol, University of Bristol Press, 1996.
23. ABU-RUWAIDA A S, SAWAYA W N, DASHTI B H, MURAD M and AL-OTHMAN H A, Microbiological quality of broilers during processing in a modern commercial slaughterhouse in Kuwait, *J Food Protect*, 1994 **57** 887–92.
24. MEAD G C, HUDSON W R and HINTON M H, Use of a marker organism in poultry processing to identify sites of cross-contamination and evaluate possible control measures, *Br Poultry Sci*, 1994 **34** 345–54.
25. MEAD G C, HUDSON W R and HINTON M H, Effect of changes in processing to improve hygiene control on contamination of poultry carcasses with campylobacter, *Epidemiol Infect*, 1995 **115** 495–500.
26. ELLERBROEK L, Airborne microflora in poultry slaughtering establishments, *Food Microbiol*, 1997 **14** 527–31.
27. NATIONAL ADVISORY COMMITTEE ON MICROBIOLOGICAL CRITERIA FOR FOODS, *HACCP Principles for Food Production*, USDA-FSIS Information Office, Washington DC, 1990.
28. NOTERMANS S and KAMPELMACHER E H, Heat destruction of some bacterial strains attached to broiler skin, *Br Poultry Sci*, 1975 **16** 351–61.
29. HUMPHREY T J, LANNING D and BERESFORD D, The effect of pH adjustment on the microbiology of chicken scald-tank water, with particular reference to the death-rate of salmonellas, *J Appl Bacteriol*, 1981 **51** 517–27.
30. PURDY J, DODD C E R, FOWLER D R and WAITES W M, Increase in microbial contamination of defeathering machinery in a poultry processing plant after changes in the methods of processing, *Lett Appl Microbiol*, 1988 **6** 35–8.
31. MULDER R W A W and BOLDER N M, The effect of different bird washers on

the microbiological quality of broiler carcasses, *Vet Quart*, 1981 **3** 124–30.
32. MEAD G C and THOMAS N L, The bacteriological condition of eviscerated chickens processed under controlled conditions in a spin-chilling system and sampled by two different methods, *Br Poultry Sci*, 1973 **14** 413–19.
33. SURKIEWICZ B F, JOHNSTON R W, MORAN A B and KRUMM G W, A bacteriological survey of chicken eviscerating plants, *Food Technol*, 1969 **23** 1066–9.
34. MEAD G C, HUDSON W R and HINTON M H, Use of 'marker' organisms for hygiene assessment and control in poultry processing, in *Factors Affecting the Microbial Quality of Meat 2 Slaughter and Dressing*, pp 13–18, Langford, UK, University of Bristol Press, 1997.
35. BARNES E M, IMPEY C S and PARRY R T, The sampling of chickens, turkeys, ducks and game birds, in *Sampling – Microbiological Monitoring of Environments*, eds BOARD R G and LOVELOCK D W, Society for Applied Bacteriology Technical Series No. 7, London, Academic Press, pp 63–75, 1973.
36. SIMONSEN B, Methods for determining the microbial counts of ready-to-cook poultry, *World Poultry Sci J*, 1971 **27** 368.
37. DICKENS J A, COX N A, BAILEY J S and THOMSON J E, Automated microbiological sampling of broiler carcasses, *Poultry Sci*, 1985 **64** 1116–20.
38. RUSSELL S M, COX N A and BAILEY J S, Microbiological methods for sampling poultry processing plant equipment, *J Appl Poultry Res*, 1997 **6** 229–33.
39. MEAD G C, ADAMS B W and HAQUE Z, Studies on the incidence, origin and spoilage potential of psychrotrophic Enterobacteraceae occurring on processed poultry, *Fleischwirtsch*, 1982 **62** 1140–4.
40. CHAPPELL A, Microbial hazards associated with handling and packaging of chicken giblets, *Technical Manual No. 41*, Campden Food & Drink Research Association, 1993.
41. HOLDER J S, CORRY J E L and HINTON M H, Microbial status of chicken portions and portioning equipment, *Br Poultry Sci*, 1997 **38** 505–11.
42. ELLERBROEK L, JANSSEN T, KRAUSE P and WEISE E, Process control of poultry cutting plants, in *Factors Affecting the Microbial Quality of Meat 3 Cutting and Further Processing*, pp 23–32, Langford, UK, University of Bristol Press, 1996.
43. ZWANIKKEN R, Improved hygiene in the slaughtering process, in *Prevention and Control of Potentially Pathogenic Microorganisms in Poultry and Poultry Meat Processing 2. Contamination with Pathogens in Relation to Processing and Marketing of Products*, eds LÖPFE J, KAN C A and MULDER R W A W, Flair No. 6, Beekbergen, Spelderholt Centre for Poultry Research and Information Services, pp 45–9, 1993.
44. JAMES W O, PRUCHA J C, BREWER R L, WILLIAMS W O, CHRISTENSEN W A, THALER A M and HOGUE A T, Effects of countercurrent scalding and postscald spray on the bacteriologic profile of raw chicken carcasses, *J Am Vet Med Assoc*, 1992 **201** 705–8.

45. WALDROUP A L, RATHGEBER B M, FORSYTHE R H and SMOOT L, Effect of six modifications on the incidence and levels of spoilage and pathogenic organisms on commercially processed postchill broilers, *J Appl Poultry Res*, 1992 **1** 226–34.
46. VEERKAMP C H and PIETERSE C, Cleaning of broiler carcasses using a closed loop washing system during plucking, *Proc Eleventh European Symposium on the Quality of Poultry Meat, Tours, France*, 1993, pp 576–9.
47. VEERKAMP C H, PIETERSE C, BOLDER N M and VAN LITH L A J T, Model experiments for cleaning broiler carcasses during scalding, *Proc Tenth European Symposium on the Quality of Poultry Meat, Doorwerth, The Netherlands, Vol 1 Poultry Meat*, 1991, pp 79–86.
48. JARVIS B and BLOOD R M, The relationship between water usage and the level of bacterial contamination of water in spin-chillers, Paper no. A7, *Proc Poultry Meat Symposium, Roskilde, Denmark*, 1973.
49. BLOOD R M and JARVIS B, Chilling of poultry: the effects of process parameters on the level of bacteria in spin-chiller waters, *J Food Technol*, 1974 **9** 157–69.
50. CORRY J E L, JAMES C, JAMES S J and HINTON M, *Salmonella, Campylobacter* and *Escherichia coli* O157:H7 decontamination techniques for the future, *Int J Food Microbiol*, 1995 **28** 187–96.
51. COX N A, MERCURI A J, JUVEN B J, THOMSON J E and CHEW V, Evaluation of succinic acid and heat to improve the microbiological quality of poultry meat, *J Food Sci*, 1974 **39** 985–7.
52. MORGAN A I, GOLDBERG N, RADEWONUK E R and SCULLEN O J, Surface pasteurization of raw poultry meat by steam, *Proc Second Annual Meeting of EC COST Working Group No 2, Zaragoza, Spain*, 1995, pp 215–22.
53. KIM J W, SLAVICK M F, PHARR M D, RABEN D P, LOBSINGER C M and TSAI S, Reduction of *Salmonella* on post-chill chicken carcasses by trisodium phosphate (Na_3PO_4) treatment, *J Food Safety*, 1994 **14** 9–17.
54. NATIONAL ADVISORY COMMITTEE ON MICROBIOLOGICAL CRITERIA FOR FOODS, Generic HACCP application in broiler slaughter and processing, *J Food Protect*, 1997 **60** 579–604.
55. INTERNATIONAL COMMISSION ON MICROBIOLOGICAL SPECIFICATIONS FOR FOODS, *HACCP in Microbiological Safety and Quality*, Oxford, Blackwell, 1988.
56. PEARSON A M and DUTSON T R (eds), *HACCP in Meat, Poultry and Fish Processing*, Advances in Meat Research Series, Vol 10, London, Blackie, 1995.
57. LEAPER S (ed), *HACCP: A Practical Guide* (Second Edition), Technical Manual No 38, Campden & Chorleywood Food Research Association, 1997.

Part 3

HACCP tools

7

Microbiological hazard identification in the meat industry

P. J. McClure, Unilever Research, Sharnbrook

7.1 Introduction

This chapter describes the main microbiological hazards associated with meat and meat products. Meat is associated with a variety of pathogenic microorganisms some of which we are relatively familiar with. Worryingly, the past two decades has seen the emergence of 'new' microbial agents capable of causing disease. These recent developments and the continued increase in the number of cases of foodborne illness have resulted in widespread concern for consumer safety. Many of the recently emerging foodborne pathogens are associated with meat from poultry, cattle and other animals, and they do not necessarily cause overt signs of illness in these animals. The appearance of these pathogens is, generally speaking, a global trend and is not restricted to particular geographic locations. The reasons for their emergence and spread are poorly understood and it is suspected that the shift to a global economy, international trade, and changes in the livestock industry may have contributed to these recent developments. No doubt, some of this is also due to improved surveillance, reporting and methods of detection.

The first principle of HACCP (conducting a hazard analysis) includes determination of the food safety hazards likely to occur and these may come from a variety of different sources. The list of hazards associated with meat and meat products includes protozoal parasites, helminths, arthropods, viruses, prions and bacteria, arguably the most important of these categories. Many bacteria are common inhabitants of animal intestines and their presence may be transient or long term. In addition to livestock being a source of infection, through internal carriage or hide contamination, pathogens may be introduced at any point in slaughter, processing, packaging, distribution and preparation of food. The bacteria and main protozoal parasites considered to be hazards in meat and meat products are discussed.

Understanding the origin of these different pathogens and their fate during processing is essential for control of the hazards and managing the risk posed by their presence. The analytical methods used to detect the presence of many of these pathogens have advanced significantly in recent years. These improvements in detection and characterisation methodologies now allow for the tracking of different pathogens through processing, enabling identification of the origin of these agents. These developments also allow links to be made between apparently unrelated (e.g. sporadic) cases. The specific methodologies used for enumeration and detection of particular pathogens are not within the scope of this chapter, but the general approaches and recent advances will be discussed. A number of future trends likely to impact on the hazards associated with meat and meat products are also discussed in this chapter. Genetic evolution will continue to contribute to the appearance of new pathotypes or pathovars of microorganisms and this will result in pathogens that possess new combinations of known and unknown virulence factors.

7.2 The main hazards

7.2.1 *Salmonellae*

The genus *Salmonella* is subdivided into over 2000 serotypes or serovars, based on unique antigenic structure. Further subdivision is possible through phage- and biotyping. Salmonellae are primarily intestinal parasites of humans and many animals, including rodents, wild birds and domestic animals. Recently, the nomenclature of salmonellae has been revised since modern taxonomic methods suggested that all serotypes of *Salmonella* probably belonged to one DNA-hybridisation group. *S. enterica* was originally subdivided into seven subgroups, *S. enterica* subspp. *enterica, salamae, arizonae, diarizonae, houtenae, bongori* and *indica*. *S. enterica* subspp *bongori* has since been elevated to species level. Only serotypes of subsp. *enterica* are still named (e.g. *S. enterica* subsp. *enterica* serotype Typhimurium or *S.* Typhimurium or simply Typhimurium) indicating that the named serotype is a member of subsp. *enterica*.

Although many salmonellae are potentially pathogenic in animals, the response to infection by the same serotype in different animals may be different. Although a large number of serotypes have been identified, less than 10% have been isolated from man and other animals. Salmonellae are most often isolated from cattle and poultry. Serotypes are classified as either host-adapted or non-host-adapted, depending on their host range and the majority show no host specificity. Host adapted serotypes rarely cause disease in other hosts. *S.* Dublin is traditionally host adapted to cattle but in some case has shown a tendency to spread to swine and was originally isolated from a child. This serotype can cause severe disease (septicaemia, osteomyelitis, and meningitis) in some individuals.

Salmonellosis in animals
Generally speaking, young animals are more susceptible to salmonellosis than older ones. There are a number of factors that predispose animals to clinical salmonellosis and these include poor sanitation, overcrowding, parturition, transportation and concurrent infections with other pathogens (e.g. parasites, viruses, etc). Many animals, particularly swine and poultry are fed contaminated feed without developing any apparent clinical symptoms. Feed is usually contaminated through meat and bone meal, fish meal or soybean meal with organisms entering these materials during or after processing. Wild birds and rodents also provide a source of contamination from faeces contaminating feed or buildings, and other possible sources include contaminated poultry litter and water courses.

Salmonellosis in cattle usually begins as an enteric infection, commencing with colonisation of the intestine and invasion of the intestinal epithelium. This can be followed by septicaemia, abortion, meningitis, pneumonia or arthritis, after entry into the bloodstream. The two most important serotypes in cattle are *S.* Typhimurium and *S.* Dublin. Typhimurium is found worldwide and Dublin is found mainly in Europe, western US and South Africa. Antibiotic resistant strains of Dublin are now spreading to the north-eastern US.[1] Persistently shedding carrier animals are thought to be the primary reservoir of Dublin, with most infections occurring when animals are on pasture. Unlike Dublin, disease caused by Typhimurium is usually self-limiting, since persistent shedders are not the norm. Typhimurium is known for primarily enteric disease states whereas Dublin causes primarily septicaemia. The most important mode of transmission for Typhimurium is the faecal-oral route. Both serotypes cause serious disease with mortality rates sometimes as high as 50–75%. Other serotypes that have caused infection in cattle include Anatum, Montevideo, Newport and Saint-paul.

In sheep, serotypes associated with disease include Abortus ovis, Dublin, Montevideo and Typhimurium. Infection in flocks results from introduction of infected sheep and ingestion of the organism. Dublin and Typhimurium cause enteritis, septicaemia and abortion, similar to the conditions observed in cattle. For sheep and goats, Typhimurium infection can come from a variety of environmental and animal sources whereas cattle are the usual source of Dublin.

The serotypes most frequently associated with disease in pigs are Choleraesuis and Typhimurium and other serotypes that can cause disease in susceptible animals include Anatum, Derby, Heidelberg, Newport and Panama. Choleraesuis causes paratyphoid, Dublin causes enteritis and meningoencephalitis and Typhimurium and other serotypes cause enteritis and septicaemia. Heidelberg can also produce severe catarrhal enterocolitis.

In poultry, serotypes causing disease include Agona, Bareilly, Hadar, Oranienburg, Typhimurium, Gallinarum and Pullorum. The last two of these cause fowl typhoid and bacillary white diarrhoea. Pullorum is now rarely isolated in the United States and northern Europe, due to successful eradication programmes, but is of increasing importance in Latin America, the Middle East, the Pacific Rim, Africa and some parts of southern Europe. The incidence of Gallinarum has also been reduced due to changes in husbandry and through eradication of Pullorum, with which it shares common antigens. In the areas where Pullorum has been eradicated, Typhimurium is often found, causing paratyphoid. Typical conditions of paratyphoid in poultry include enteritis, diarrhoea and septicaemia.

During 1985/1986, *S.* Enteritidis PT4 emerged as a 'new' problem in poultry in Europe. In 1993, the first outbreak of PT4 occurred in the US, and the number of isolations from eggs and the farm environment of laying flocks suggests that eggs have had a major contribution to the dramatic increase in associated human illness. Enteritidis is an invasive serotype and has achieved prominence because of its association with poultry eggs. Although eggs have been recognised as a source of infection for Typhimurium, the incidence of food poisoning cases from this source has always thought to have been low.

S. Typhimurium DT104 has recently emerged in cattle populations in particular parts of the world and causes severe diarrhoea, with an associated mortality rate of 50–60%. Long-term carriage (up to 18 months following an outbreak) has been observed in many species including cattle.[2]

Salmonellosis in man
Non-typhoidal *Salmonella spp.* are one of the most commonly reported foodborne pathogens in industrialised countries. Symptoms of human salmonellosis include nausea, vomiting, diarrhoea, abdominal cramps and fever, with illness lasting for 3–12 days.[3] Associated clinical conditions also include reactive arthritis, Reiters syndrome, septic arthritis and septicaemia.

Certain serotypes are being increasingly reported as the cause of salmonellosis. In 1989, Typhimurium, Enteritidis, Heidelberg, Hadar and Agona accounted for 57.9% of all serotypes isolated from human infections and accounted for 46.5% of isolations obtained from poultry, in the US. One serotype that has increased significantly in recent years is *S.* Enteritidis. Before 1990, Typhimurium was the most common cause of reported salmonellosis in a number of geographic regions. In 1990, this serotype was overtaken by Enteritidis and is now a major cause of human food poisoning in many countries.[4] In recent years in the UK and western Europe, the predominant phage type responsible for egg-borne salmonellosis is PT4 whereas in the US, although there is no predominant phage type associated with egg-borne infection, PT8 and PT13a are the most commonly isolated phage types.[5] The emergence of other phage types, such as PT6 in the UK, continues to occur, as does the emergence of other types such as *S.* Typhimurium DT104, which is now appearing in Europe, north America and elsewhere.[6] The main reservoir of this pathogen, which often exhibits resistance to multiple antibiotics, is thought to be cattle, but there are reports of increasing incidence in poultry, sheep, pigs and goats. This is in contrast to *S.* Enteritidis, which is mainly associated with eggs and poultry. The invasiveness of DT104 in humans does not appear to be any different to other salmonellae, but an increase in occurrence of severe illness

7.2.2 Escherichia coli

Like salmonellae, the primary habitat of *E. coli* is the intestinal tract of man and other warm-blooded animals. Many *E. coli* are commensal organisms and cause no harm but there are some types that are pathogenic to man and other animals and these are not regarded as part of the normal flora of the human intestine. *E. coli* is also commonly found in external environments (e.g. soil and water) that have been affected by human and animal activity. *E. coli* is divided into more than 170 serogroups based on the somatic (O) antigens, and over 50 flagellar (H) and 100 capsular (K) antigens allow further subdivision into serotypes. Serogrouping and serotyping are used with biotyping, phage typing and enterotoxin production to distinguish strains able to cause infectious disease in man and animals.

There are many types of disease caused by *E. coli* and these depend on the virulence factors present. The known virulence factors include adhesins and colonisation mechanisms, haemolysin, ability to invade epithelial cells and production of a number of toxins including heat labile enterotoxins, heat stable enterotoxins, cytotoxic necrotising factors and vero cytotoxins (or Shiga toxins, Stx1 and 2). The adhesion and colonisation factors include fimbrae, haemaglutinnins and specific adhesins such as the F4 (K88) antigen. The encoding genes of these and other virulence factors may be carried on transmissible plasmids or on the chromosome. There are currently six recognised virulence groups comprising enteropathogenic *E. coli* (EPEC), enterotoxigenic *E. coli* (ETEC), enteroinvasive *E. coli* (EIEC), verotoxigenic *E. coli* (VTEC or Shiga-like toxin producing *E. coli* or SLTEC, which include enterohaemorrhagic *E. coli* or EHEC), enteroaggregative *E. coli* (EAggEC) and diffusely adherent *E. coli* (DAEC).

Disease in animals

E. coli infections occur frequently in many farm animals, including poultry.[7,8] In younger animals, there are principally two types of disease which are systematic colibacillosis, caused by a range of O-serogroups and enteric colibacillosis, caused by a few host-specific enteropathogenic strains. In older animals, a third group of diseases, caused by a number of O-serogroups, causes mastitis in cows and sows. Other, sporadic infections, such as urinary tract infections, can also occur.

Enteric colibacillosis involves oral infection, followed by site specific adhesion to intestinal mucosa, allowing colonisation and release of toxins, which causes damage to intestinal cells or other organs. Colibacillary diarrhoea is an acute disease and occurs most frequently in calves, lambs and piglets, soon after birth and is mainly caused by ETEC. The OK groups in calves and lambs tend to be the same, whilst the OK groups associated with pigs are rarely isolated from other species. In cattle, common serotypes include O8, O9 and O101. In pigs, the most common serotype is O149 and other commonly isolated serotypes include O8, O138, O147 and O157. There appears to have been little change in the serotypes that cause colibacillary diarrhoea or their virulence factors, in recent years. Other forms of enteric colibacillosis are colibacillary toxaemia in pigs, associated with a few serotypes that cause shock in weaner sydrome, haemorrhagic colitis and oedema disease. These forms of disease are thought to relate to production of enterotoxin, endotoxin or a neurotoxin.

Systemic colibacillosis is caused by invasive strains and involves their survival and multiplication in extra-intestinal sites. This occurs frequently in calves, lamb and poultry, but not in pigs, and develops by passage of *E. coli* from the alimentary or respiratory mucosa to the bloodstream. From there, a localised infection, such as meningitis or arthritis in calves and lambs or air sacculitis and pericarditis in poultry, or a generalised infection (colisepticaemia) can develop. O78 and O2 are commonly isolated from poultry, and these serotypes are rarely observed in human isolates. Strains causing colibacillosis in calves belong to relatively few serotypes such as O15:K, O35:K, O137:K79, and O78:K80. The last of these is the most frequently isolated and is also associated with similar conditions in lambs and poultry.

Bovine mastitis is still an important disease and can range from mild forms, which cause clots, milk discoloration and udder swelling, to severe illness that can result in death of the affected animal. Mastitis is caused by a large number of serotypes that are not easily distinguished from strains in normal faeces. Endotoxin and necrotising cytotoxin are thought to play significant roles in this disease.

It is generally thought that VTEC do not cause overt disease in animals, but there is increasing evidence of illness caused by some VTEC in neonatal calves and older animals.[9,10] Bovine VTEC strains share many of the virulence markers with VTEC strains causing infection in man, but in Germany, the intimin-positive strains (those strains causing attaching and effacing lesions, encoded by the *eae* gene) are thought to be restricted to the stx_1 genotype, only capable of producing Stx1. In Brazil, however, a recent study has shown that 60% of the VTEC strains isolated from cattle possess both stx_1 and stx_2. Serogroups O5 and O118 are mainly associated with disease in calves, serogroups O26, O103 and O111 cause disease in calves and humans, but O157 is generally considered to be carried by healthy animals and only associated with disease in humans.

Disease in man
Worldwide, the importance of diarrhoeal and other diseases caused by *E. coli* is immense, particularly in children in developing countries. In developed countries, the incidence of foodborne illness associated with *E. coli* is also significant and appears to be increasing. More worryingly, recent years have seen the emergence of particularly virulent *E. coli*, such as *E. coli* O157:H7 (the predominant VTEC serotype), that are able to cause serious illness in man, with low infectious doses, e.g. fewer than 100 cells. The severity of

disease caused by VTEC can vary from asymptomatic carriage to haemorrhagic colitis (HC), to life-threatening conditions such as haemolytic uraemic syndrome (HUS) in children and thrombotic thrombocytopaenic purpura (TTP) in adults. HUS is the most common cause of acute renal failure in children. For the other pathotypes of *E. coli*, such as EPEC, ETEC and EIEC, clinical studies suggest that more than 10^5 EPEC are necessary to produce diarrhoea, 10^8 ETEC are necessary for infection and diarrhoea and 10^8 EIEC are required to produce diarrhoeal symptoms in healthy adults. EPEC cause a bloody diarrhoea in infants (commonly referred to as infantile diarrhoea), which in some cases may be prolonged; ETEC cause self-limiting diarrhoea, vomiting and fever, and travellers' diarrhoea; EIEC cause shigella-like dysentery; EAggEC cause persistent diarrhoea in children, particularly in developing countries; and DAEC cause childhood diarrhoea.

Even though there are many *E. coli* responsible for disease in animals, most of the *E. coli* pathogenic in man are not the same as those causing illness in animals. Indeed, the principle reservoir for many human pathogenic *E. coli* is believed to be man. However, this is not true of VTEC, including *E. coli* O157:H7, where the main reservoirs are thought to be cattle and other ruminants.[11] Dairy cattle, particularly young animals within herds, have been identified as a reservoir of *E. coli* O157:H7 and other VTEC, and this serotype has also been isolated from other ruminants such as sheep and goats. Hence raw foods of bovine or ovine origin are likely to be vehicles of *E. coli* O157:H7 and other VTEC through faecal contamination during slaughter or milking procedures. In one survey, four per cent of cattle were contaminated prior to slaughter, and after processing, 30% of the carcasses were contaminated.[12] The most frequently implicated vehicle of infection for *E. coli* O157:H7 is undercooked ground beef. Surveys of raw meats for sale have revealed *E. coli* O157:H7 in 2–4% of ground beef, 1.5% of pork and poultry, and 2% of lamb.[13,14] Other studies suggest contamination rates for VTEC in some raw foods of between 16 and 40%. The incidence of *E. coli* O157:H7-related illness is worldwide.

Human infections with VTEC O157:H7 are under nationwide surveillance in a number of countries, but detection of other non-O157 VTEC types is more difficult and performed only by specialist laboratories. Humans are likely to be more exposed to non-O157 VTEC because these strains are more prevalent in animals and as contaminants in foods. The growing number of non-O157 serogroups associated with human disease now include O26, O103, O111, O118 and O145.[15] It is thought that both horizontal gene transfer and intragenic combination are important for evolution of VTEC. Particular regions of the globe show patterns of emergence that appear to be unique, e.g. 20–25% of *E. coli* O157 isolates in Germany are sorbitol +ve. The most common non-O157 serotypes in Germany are O26:H$^-$, O103:H$^-$, O111:H$^-$ and O145. In Italy, the HUS cases caused by O26 strains now outnumber those caused by O157:H7. Studies in animals demonstrate that some of these serogroups, such as O118, are also prevalent in farm animals, and are a likely reservoir.

Human pathogenic strains of VTEC vary in their ability to cause illness, and this depends on virulence attributes and other unknown factors.[16] Pathogenic VTEC O26, O103 and O111 belong to their own lineages and possess unique profiles of virulence determinants that are different from the virulence profile of *E. coli* O157:H7, which is said to contain a more complete repertoire of virulence traits. This may explain why *E. coli* O157:H7 is the predominant VTEC serotype.

One of the problems that is becoming increasingly recognised is that the terminology used to describe diarrhoeagenic *E. coli* is complex and by no means definitive. Since it was first recognised that *E. coli* could cause diarrhoea, an array of virulence factors have been discovered and a number of categories of diarrhoeagenic *E. coli* have been proposed, generally based on the presence of non-overlapping virulence factors. However, EPEC strains and EHEC strains are often regrouped under the name of attaching/effacing *E. coli* (AEEC) on the basis of the ability to produce common attaching and effacing lesions in their hosts.

There are already a number of documented studies describing isolates that do not fit neatly into any of the recognised categories of diarrhoeagenic *E. coli*. This should not be surprising considering that the virulence factors are encoded on 'pathogenicity islands', bacteriophage, transposons and transmissible plasmids. Some of these elements have also been found in other members of the *Enterobacteriaceae*. Therefore, we should anticipate that there will be other combinations of known and currently unknown virulence factors appearing in the group of organisms we currently call *E. coli*, and other members of the *Enterobacteriaceae*.

7.2.3 *Campylobacter jejuni*

C. jejuni was not recognised as a cause of human illness until the late 1970s but is now regarded as the leading cause of bacterial foodborne infection in developed countries.[17] It is one of 20 species and sub-species within the genus *Campylobacter* and family *Campylobacteriaceae*, which also includes four species in the genus *Arcobacter*. Despite the huge number of *C. jejuni* cases currently being reported, the organism does not generally trigger the same degree of concern as *E. coli* or salmonellae, since it rarely causes death and is rarely associated with newsworthy outbreaks of food poisoning. It is among the most common causes of sporadic bacterial foodborne illness. *C. jejuni* is associated with warm-blooded animals, but unlike salmonellae and *E. coli* does not survive well outside the host. *C. jejuni* is susceptible to environmental conditions and does not survive well in food and is, therefore, fortunately relatively easy to control. Food associated illness usually results from eating foods that are re-contaminated after cooking or eating foods of animal origin that are raw or inadequately cooked. The organism is part of the normal intestinal flora of a wide variety of wild and domestic animals, and has a high level of association with poultry.[18] The virulence of the organism, as suggested by the relatively low infectious dose of a few hundred cells and its widespread prevalence in animals, are important features which explain why this relatively sensitive organism is a leading cause of gastroenteritis in man.

Campylobacteriosis in animals
C. jejuni is a commensal organism of the intestinal tract of a wide variety of animals.[19] In cattle, young animals are more often colonised than older animals, and feedlot cattle are more likely to be carriers than grazing animals. Colonisation of dairy herds has been associated with drinking unchlorinated water. Day-old chicks can be colonised with as few as 35 organisms, and most chickens in commercial operations are colonised by four weeks. Reservoirs in the poultry environment include insects, unchlorinated drinking water and farm workers, but probably not feeds, since these are thought to be too dry for survival of campylobacters. It has been proposed that *C. jejuni* is a cause of winter dysentery in calves and older cattle, and experimentally infected calves have shown some clinical signs of disease such as diarrhoea and sporadic dysentery. Nevertheless, the aetiology of naturally occurring disease in animals remains unconfirmed. *C. jejuni* is, however, a known cause of bovine mastitis, and the organisms associated with this condition have been shown to cause gastroenteritis in persons consuming unpasteurised milk from affected animals. Other campylobacters, such as *C. fetus*, are known to cause abortions in sheep and cattle and some strains of *C. sputorum* are known to cause porcine intestinal adenomatosis and regional ileitis in pigs, but these appear to be host-specific diseases.

Campylobacteriosis in man
C. jejuni and *C. coli* are the most common campylobacters associated with diarrhoeal disease in man and are clinically indistinguishable. Also, most laboratories do not attempt to distinguish between the two organisms. It is thought that *C. coli* constitute 5–10% of cases reported as caused by *C. jejuni* in the US. Campylobacteriosis in man is usually characterised by an acute, self-limiting enterocolitis, lasting up to a week. A small proportion (5–10%) of affected individuals suffer relapses. Symptoms of disease often include fever, abdominal pain and diarrhoea, which may be inflammatory, with slimy/bloody stools, or non-inflammatory, with watery stools and absence of blood. Reactive arthritis and bacteraemia are rare complications and infection is also associated with Guillain-Barré syndrome, an autoimmune peripheral neuropathy causing limb weakness. This condition is thought to be associated with particular serotypes (e.g. O:19, O:4 and O:1) capable of producing structures that mimic ganglioside motor neurons. There are a number of pathogenicity determinants that have been suggested for *C. jejuni*, including motility, adherence, invasion and toxin production, but little is known about the mechanism causing disease in man.

There is considerable evidence that poultry is the main vehicle for transmitting *Campylobacter* enteritis in man. Poultry typically has populations of 10^4–10^8 *C. jejuni* per gram of intestinal content and more than 75% of chickens and turkey often carry the organism in their intestinal tract. It is estimated that 30% of retail poultry is contaminated with *C. jejuni* at levels of 10^2–10^4 per gram. Also, serotypes associated with poultry are also frequently associated with illness in humans.

Prior to 1991, *Arcobacter butzleri* and *A. cryaerophilus* were known as aerotolerant *Campylobacter*. These organisms have been associated with abortions and enteritis in animals and enteritis in man. Although both species are known to cause disease in man, most human isolates come from the species *A. butzleri*. There is very little known about the epidemiology, pathogenesis and real clinical significance of Arcobacters, but it is thought that consumption of contaminated food may play a role in transmission of this group of organisms to man. Although Arcobacters have never been associated with outbreaks of foodborne illness, they have been isolated from domestic animals, poultry, ground pork and water.

7.2.4 *Yersinia enterocolitica*

Surveillance data suggest that *Yersinia enterocolitica* is an increasing cause of gastroenteritis in man in Europe and the US.[20] The main cause of yersiniosis during the 1970s and 1980s was thought to be milk and the main serotype associated with disease was O:8. Since then, O:3 has become the predominant serotype in developed countries. The main reservoir of this serotype and other important serotypes, such as O:9, is pigs and consumption of pork is an important risk factor for infection. *Y. enterocolitica* is a component of the intestinal flora of red meat animals, particularly pigs. Poultry is known to carry significant levels of yersinias. Although meat and meat products from goats and sheep have never been implicated in outbreaks of foodborne yersiniosis, small ruminants can harbour the pathogen.

Y. enterocolitica causes gastroenteritis in man and can also cause persistent arthritis. Infection does not, however, always result in diarrhoea. Yersiniosis is usually characterised by abdominal pain, accompanied by fever, with or without diarrhoea. Because of its ability to multiply at refrigeration temperatures, *Y. enterocolitica* is of special interest to particular areas of the food industry. There is relatively little known about the mechanisms of pathogenicity but the genes for invasion of mammalian cells lie on the chromosome and all the other known pathogenicity determinants are found on a plasmid. The other member of this genus that can cause gastroenteritis is *Y. pseudotuberculosis* and large outbreaks of gastroenteritis caused by this organism have been reported in Japan. Disease associated with *Y. pseudotuberculosis* resembles typhoid and is often fatal. *Y. enterocolitica* is not known to cause disease in animals. *Y. pseudotuberculosis* is rarely associated with infections in cattle and sheep, with those in cattle manifesting as pneumonia or abortion.

7.2.5 Staphylococcus aureus

Meat or meat products are not thought to be a major source of *S. aureus* infection in man even though *S. aureus* is an important pathogen in animals. The principle source of transmission between animals and man is unpasteurised milk and cheese made from unpasteurised milk. Outbreaks of staphylococcal food poisoning in man are frequently associated with improper food handling and temperature abuse of foods of animal origin, but it is generally believed that the main source of contamination is food handlers. Nevertheless, strains of *S. aureus* can become endemic in food processing plants and meat can be contaminated from animal or human sources. *S. aureus* has been isolated from cattle carcasses and is also found in raw beef. There is a high correlation between coagulase production and production of enterotoxins, of which there are at least seven heat-stable types associated with food poisoning. In animals, *S. aureus* causes a number of different diseases. The most relevant disease for transmission of the organism to man is bovine and ovine mastitis.

7.2.6 Listeria monocytogenes

Listeriosis is an atypical foodborne disease that has attracted a great deal of attention since the early 1980s mainly because of the severity, high mortality rate and non-enteric nature of the disease. Listeriosis is caused by *L. monocytogenes*, which is found in many environments and is frequently carried in the intestinal tract of many animals, including man. *L. monocytogenes* is often found in healthy animals and humans, with a carrier rate of 10–50% in cattle, poultry and swine. The organism has been isolated from a variety of foods, at levels of 13% in raw meat, 3–4% raw milk and 3–4% of dairy products.[21] Some of the major outbreaks in man have been attributed to meat products such as pork tongue and meat paté. Foods associated with outbreaks have largely been refrigerated, processed and are ready-to-eat. The disease in man is commonly associated with meningitis, septicaemia and abortion. Recent outbreaks, however, have been associated with a milder form of disease characterised by gastroenteritis and flu-like symptoms. In these recent outbreaks, serogroup 1/2 has been implicated whereas many of the human strains isolated previously belong to serovar 4b and to one major ribovar. Serogroup 1/2 accounts for most of the food and environmental isolates and together, serotypes 1/2a, 1/2b and 4b account for up to 96% of the isolates in man. Host factors are likely to play an important role in the susceptibility to listeriosis, together with presence of virulence factors in the organism. Many individuals frequently ingest *L. monocytogenes* without any apparent ill effects. Although listeriosis is a severe disease, the number of cases, compared to some of the other foodborne diseases, is relatively low.

Since *L. monocytogenes* is widely distributed in soil, vegetation and faeces, most animals are exposed to it during their lifetime. *L. monocytogenes* is also commonly found in large numbers in poor-quality silage, and ruminants fed this material are more likely to develop listeriosis. As in humans, predisposing factors are important for disease in animals. The clinical conditions associated with animal listeriosis are similar to the human disease, and include septicaemia, abortion, enteritis and meningoencepahalitis. Interestingly, the isolates associated with processed meats more often originate from the processing environment than from the animal itself.

7.2.7 Clostridium perfringens

Strains of *C. perfringens* are classified into 5 types, A–E, according to the extracellular toxins that are formed. Type A is responsible for almost all cases of foodborne disease in humans. Type C very rarely causes foodborne disease and results in necrotic enteritis, but is only a concern in individuals who are nutritionally impaired or whose intestinal proteolytic enzyme activity is reduced. Type A *C. perfringens* is usually present in the soil at concentrations of 10^3–10^4/g. The other types are obligate parasites of domestic animals and do not persist in the soil. Type A strains occur widely in raw and processed foods, but at numbers too low to cause infection. The organism is found in the alimentary tract of nearly all species of warm-blooded animals.

C. perfringens is primarily associated with outbreaks of food poisoning involving handling problems and meat, meat products and poultry are frequently implicated in outbreaks. Illness usually results from ingestion of heavily contaminated food and typical symptoms are diarrhoea and severe abdominal pain. Occasional reports of illness within 2 h of ingestion indicate ingestion of preformed toxin. Sporulation of ingested bacteria is also associated with production of enterotoxin, which is destroyed by heating (e.g. 60°C for 10 min).

In animals, type A causes yellow lamb disease in sheep and a similar illness (toxin produced in the small intestine) in goats. Type B is known to cause lamb dysentery, and haemorrhagic dysentery in sheep, goats and calves. Type C causes enterotoxaemia in a variety of animal species and type D causes the same disease, but apparently only in sheep. Type E is believed to cause haemorrhagic necrotic enteritis in calves.

7.2.8 Clostridium botulinum

Strains of *C. botulinum* are classified into several types (A–G) depending on the antigenic properties of the toxin produced. Types A, B, E and F are responsible for most cases of human botulism, whereas types C and D cause illness in animals. The outbreaks of foodborne botulism associated with meats, such as home-cured hams, tend to occur mainly in Germany, France, Poland and Italy. The incidence of foodborne botulism is extremely low, but the severity of disease and its heat resistance mean that it is the target microorganism of many preservation processes used for foods. Spores of *C. botulinum* are present in the soil and environment, but to a lesser extent than *C. perfringens*. Spores may be present in meat, but this is usually at levels between 0.1–10 spores/kg. The disease in man is an intoxication and causes general weakness of limbs and

Microbiological hazard identification in the meat industry 169

respiratory muscles, and often nausea and vomiting. Like humans, botulism in animals almost always arises from ingestion of food contaminated with preformed toxin. There is evidence that animals carry spores and this may lead to internal contamination and contamination of meat processing environments. In Europe, occurrence of spores is generally infrequent but when it occurs, levels can reach 7/kg of sample whereas incidence in meats in the US is much lower, probably reflecting the incidence of meat-associated botulism is the two areas.

7.2.9 Other bacteria

Other bacteria associated with meat animals include brucellae (e.g. *Brucella melitensis*) and *Bacillus anthracis*, which can cause disease in man but are regarded as a relatively low risk from meat and meat products. *B. cereus* is a ubiquitous organism and has been found in raw beef and milk, and the organism is directly linked to dairy cows, being incriminated in abortions and mastitis. Therefore, contamination of carcasses of dairy cows is possible but is not thought to constitute a significant risk in foods of animal origin. Foodborne illness caused by *B. cereus* generally results from improper handling of foods. Other organisms, such as *Corynebacterium pseudotuberculosis*, *Mycobacterium paratuberculosis* and *Pasteurella spp.* are responsible for diseases in animals and have been linked to disease in humans but transmissibility from animals to man has yet to be proven.

7.2.10 Parasites

Giardia duodenalis and G. lamblia
G. duodenalis is one of the most common protozoal infections in man, causing diarrhoeal disease in infants and young children, in both industrialised and developing countries. The parasite is also found in many domestic animals including cattle, sheep and goats, particularly young animals. There is evidence of zoonotic transmission, but the major sources of contamination are thought to be water or food contaminated with water that has been in contact with faecal material.[22]

Cryptosporidium parvum
C. parvum has a wide spectrum of animal hosts, including cattle, goats, other farm animals and man. It is an intracellular parasitic protozoan responsible for self-limiting diarrhoeal illness in its hosts.[23] Symptoms include watery diarrhoea, nausea, anorexia, abdominal cramps, fever and weight loss. The life-cycle is completed within one host and large numbers of oocysts are then transmitted, in faeces, to the environment, where they may survive for long periods of time. In man, if individuals are young or immunocompromised, more serious gastroenteritis can occur and this can be fatal. In diarrhoetic young goats and sheep, there is a high prevalence of *C. parvum*, suggesting a strong association between infection and disease. In surveys looking for presence of *C.*

parvum in animals, oocysts were found in calves at levels of up to 22% of animals tested. The reservoirs and routes of transmission suggest that meat and meat products may be a source of infection in humans. Sausage and tripe have been shown to contain *C. parvum* oocysts.[23] Contaminated water is known to be the cause of large outbreaks of disease. Poor diagnosis of disease in man and the small numbers of oocysts (100s) necessary to cause infection mean that many cases of cryptospordiosis may go undetected.

Other parasites
Toxoplasma gondii is a protozoal parasite well known for causing abortions in sheep. The organism is also known to cause acute primary infection in man and is a particular risk to pregnant women. Consumption of raw or undercooked mutton is thought to be responsible for transmission to man.

Trichinella spiralis is responsible for trichinellosis, which, in man, begins as an acute gastrointestinal condition and is followed by fever and myalgias. Chronic illness may result since 10–20% of cases develop neurologic or cardiac symptoms. Illness in man results from consumption of raw or undercooked pork, wild boar or horse meat, with most cases occurring in Europe. *Taenia saginata* also causes outbreaks of disease in Europe through consumption of infected beef.

Cyclospora spp. cause very similar disease to *C. parvum* and are also similar in other respects such as biology and pathogenesis, but there is only one species, *Cyclospora cayetanensis*, known to cause illness in man. This species is not known to have any other animal host. Other members of the genus cause disease in other animals.

Echinococcus granulosus is another parasite that can cause infection in man. The larval stage is found in sheep, goats, cattle, pigs and man. The final host for the parasite is the dog. Contamination of meat is not thought to occur directly; the main route of infection is through contamination of eggs from dogs.

7.2.11 Other agents

Transmissible spongiform encephalopathies (TSEs)
Scrapie has been prevalent in sheep and goats in particular parts of the world for many years. This disease is regarded as the prototype of TSEs, found in humans and other animals. These TSEs cause progressive degenerative disorders of the nervous system and result in death. There is no doubt that these are infectious diseases but the nature of the infectious agents remains elusive. Theories about the causative agent vary and there is continuous debate about the presence of nucleic acids and the importance of a protease resistant protein (prion theory), derived from a normal host protein. In the early 1980s, an epidemic of bovine spongiform encephalopathy (BSE) began in the UK, and the recycling of infected cattle material is thought to have continued driving this epidemic.

The recent emergence of a new variant of Creutzfeldt-Jakob disease (vCJD) in humans in the UK has led to the belief that this new disorder is related to the transmissible agent causing BSE. The working hypothesis is that transmission has occurred through contaminated material entering the food chain. This, in turn, has focussed attention back on scrapie as a potential source of infection, despite the fact that large quantities of contaminated material must have been consumed without any apparent ill effects. It is not known how many vCJD cases are likely to emerge as a consequence of the BSE epidemic and there is still much to learn about all aspects of this group of diseases.

Viruses
Viruses are not generally considered to be transmitted to man via meat and meat products, although caliciviruses infect humans and other animals. Within the family *Caliciviridae*, there are four distinct genera comprising vesiviruses and lagoviruses, which contain a broad range of animal viruses and Norwalk-like viruses (NLV or small round structured viruses) and Sapporo-like viruses, which until recently have only been associated with man. NLVs are the main cause of gastrointestinal illness in restaurants and institutions. Recent data suggest that NLV infections often occur in calves and sometimes in pigs.[24] The significance of this recent finding is unknown at the present time.

7.3 Analytical methods

This section of the chapter provides a brief overview of the types of methods available for detection of foodborne pathogens. Detailed description of methods for each of the pathogens discussed above are not included. Conventional methods for the detection and characterisation of bacteria associated with foods rely on specific media. These methods tend to be relatively cheap, sensitive and can provide both quantitative and qualitative information. However, they can be lengthy procedures, are labour intensive, rely on multiplication of the target organism and do not use genetic information, which can be used to discriminate between closely related organisms. Nevertheless, there have been advances in recent years that facilitate some of these conventional procedures such as the introduction of chromogenic or fluorogenic media, removing the need to do further sub-culturing and biochemical steps. Modifications to particular media have also been made to improve performance and cut down some of the other steps involved in conventional culture methods. Other improvements include availability of automated colony counting, using image analysis, and availability of automated biochemical identification systems. These advances provide results directly comparable to conventional tests but make testing much more convenient.

Reliance on particular methods for the detection of pathogens can lead to problems where atypical types or responses are evident. For example, *E. coli* O157:H7 isolates are routinely distinguished from other *E. coli* because of their inability to ferment sorbitol. This means that sorbitol +ve *E. coli* O157:H7 strains, such as those found in Germany, would go undetected during routine testing. Selective media, because of their inclusion of inhibitory agents, may also underestimate target organisms if they are injured. In such cases, inclusion of a recovery stage is critical to the detection procedure.

Alternative approaches for the detection of specific microorganisms have also been developed in recent years and these include flow cytometry, impedimetry, immunological techniques and nucleic acid based assays. Flow cytometry is an optically-based approach that can detect low numbers of cells (e.g. 10^2–10^3 bacteria) rapidly (within minutes), but food matrices can interfere with the technique and distinction between live and 'dead' cells can be problematic. It has been used for the enumeration of viruses in water and is also used to enumerate *Cryptosporidium* oocysts. Impedimetry is based on changes in the electrical conductivity of liquid media caused by growth of the target organism. Although this method is not 'rapid', it is convenient for high throughput since it is fully automated and can deal with multiple samples simultaneously. Specificity is dependent on the media used to grow the target organisms.

Immunological methods are based on the specific binding of an antibody to an antigen. The advent of monoclonal antibodies now provides a consistent and reliable source of characterised antibodies. Immunoassays are divided into homogeneous and heterogeneous assays. There is no need for markers with homogeneous assays, since the antibody-antigen complex is directly measurable and the test time is short. Examples of this type of assay are agglutination reactions, immunodiffusion and turbidimetry, and tests are available for most pathogens. Heterogeneous assays are more complex procedures and use immobilised antibodies on a variety of supports and reporting systems. These precedures can be carried out without the need for special equipment. Detection limits are between 10^3–10^5 cell/ml for most pathogens. Direct detection in foods is not possible and enrichment is required. Immunoassays can also detect bacterial toxins. Automated immunoassays are also now commercially available.

Developments in genetically-based techniques in recent years provide a step-change in analytical capability for detection and characterisation of pathogens. These techniques are based on the hybridisation of target DNA or RNA with a specific DNA probe. The specificity of this probe is dependent on its nucleotide sequence. When hybridisation has occurred, detection can be via a number of methods, similar to those used in immunoassays. Commercial assays are now available for a number of pathogens. The detection limit for bacteria is 10^3 cells, so enrichment is sometimes required. Alternatively, an amplification step may be used. Examples of this are polymerase chain reaction, involving denaturation of the target DNA and annealing of primers to the single strand, followed by extension of the primers using a thermostable polymerase, or RNA amplification through the concerted action of enzymes (NASBA®). Use of amplification methods requires clean samples, and availability of commercial kits now enables routine laboratories to carry out procedures which until recently were regarded as complex and only carried out in specialist laboratories. Because these methods are based on genetic elements, results only indicate the potential to produce toxin or express virulence. There are also problems with false positives (e.g. 'dead' cells) and negatives (polymerase inhibitors or accessibility to the target organism).

Molecular typing is also possible now, allowing identification to sub-species level, aiding epidemiological and taxonomic studies. These techniques are often referred to as fingerprinting methods. They include restriction fragment length polymorphism (RFLP), random amplified polymorphic DNA (RAPD), pulsed field gel electrophoresis (PFGE) and AFLP® which combines PCR and RFLP. Whilst conventional methods still have an important role to play, molecular methods are likely to become more commonly used. The next breakthrough in diagnostic methodology is likely to come from 'DNA chip' technology, which combines semiconductor manufacturing with molecular techniques. This technology will allow rapid and cheap analysis of multiple sequences, using large arrays of nucleotides, making it possible to detect and type different organisms in the same food sample. There are, however, significant hurdles to be overcome, with viruses and parasites posing their own particular problems.

7.4 Future trends

Foodborne pathogens that have emerged in recent years share a number of characteristics. Nearly all of these have an animal reservoir from which they spread to man, i.e. they are foodborne zoonoses, but unlike established zoonoses, they do not often cause illness in the animal host. Another worrying trend is that these pathogens are able to spread globally in a short period of time. Many of the emerging pathogens are becoming increasingly resistant to antibiotics and this has been attributed, partly, to the use of antibiotics in animals. The practice of using antibiotics in animal production is coming under increasing pressure and there have been recent legislative changes that address this issue in particular parts of the world. Unfortunately, it is likely that some of these practices will continue in those areas that are not properly regulated or policed.

New food vehicles have been identified in recent years. These new vehicles include foods that were once thought to be 'safe' such as eggs, apple juice, fresh fruit, fresh vegetables and fermented meats. With consumer preferences for fresher, less heavily processed foods likely to continue, it is possible that new food vehicles for foodborne disease will continue to emerge. Alternative processes, if incorrectly assessed, may also provide an additional source of infection. Continued consolidation within the food industry is likely to lead to increasingly large markets and wider distribution from centralised manufacturing operations. With increasing demand from increasing populations, we are likely to see more re-use and recycling of water and waste, and this may have an impact on the microbiological hazards we have to face.

Fortunately, improved epidemiological capability, provided through better detection methods and better cooperation/coordination between different surveillance networks, is likely to allow quicker detection of geographically widespread outbreaks of foodborne disease. Molecular methods are transforming taxonomy and our understanding of the genomes of particular pathogens and groups of pathogens, such as the *Enterobacteriaceae*. This has already let us gain some insight into evolutionary processes and should allow us to better anticipate the potential of microorganisms to incorporate new genetic material and develop new virulence characteristics. Better understanding of pathogenesis of foodborne disease and colonisation of animals may also allow development of new intervention strategies.

With anticipated increases in the average life expectancy, through improved medical treatment of chronic disease and other advances, there is likely to be an increase in the proportion of persons with age-related susceptibility to foodborne disease. Also, there is likely to be a continuing increase in the number of immuno-suppressed individuals, due to infection with HIV and other chronic illnesses.

At the present time we are seeing a decrease in the number of cases of some common foodborne pathogens, such as salmonellae, in developed countries like the US, UK and other parts of Europe. This is encouraging and suggests that some disease prevention strategies may be beginning to take effect. Despite this, the incidence of foodborne illnesses and deaths caused by unsafe food are increasing. The genetic plasticity of the microorganisms poses a serious threat for the future, and will undoubtedly lead to the emergence of novel infectious diseases. At the genetic and molecular level, the virulence traits of pathogens clearly show us that pathogenicity does not arise by slow adaptive evolution but rather by step changes.

7.5 Sources of further information and advice

General articles describing members of the *Enterobacteriaceae*, such as salmonellae, *E. coli* and *Y. enterocolitica* are available.[3,20] *Enterobacteriaceae* and *E. coli* infections in animals have been reviewed in a number of articles.[7,8] Specific articles describing foodborne listeriosis, campylobacteriosis and the emergence of *E. coli* O157:H7 are also available.[11,18,21] Foodborne parasites are reviewed in several articles.[22,23,25] A review of analytical methods used in microbiology has been published recently.[26]

7.6 References

1. McDONOUGH P L, FOGELMAN D, HIN S J, BRUNNER M A and LEIN D H, '*Salmonella enterica* serotype Dublin infection: an emerging infectious disease for the Northeastern United States', *J Clin Micro*, 1999 **37** 2418–27.
2. BARLEY J P, '*S. typhimurium* DT104 in cattle in the UK', *Vet Rec*, **140** 75.
3. D'AOUST J, '*Salmonella* species' in *Food Microbiology – fundamentals and frontiers*, ASM Press, Washington, 129–58, 1997.
4. RODRIGUE D C, TAUXE R V and ROWE B, 'International increase in

Salmonella enteritidis: a new pandemic?' *Epid Inf*, 1990 **105** 21–7.
5. MISHU B, KOEHLER J, LEE L A, RODRIGUE D, BRENNER F H and BLAKE P *et al.*, 'Outbreaks of *Salmonella enteritidis* infections in the United States, 1985–1991', *J Inf Dis,* 1994 **169** 547–52.
6. CENTRES FOR DISEASE CONTROL AND PREVENTION, 'Multi-drug resistant *Salmonella* serotype Typhimurium – United States', *Mor Mort Wkly Rep*, 1997 **47** 308–10.
7. LINTON A H and HINTON M H, 'Enterobacteriaceae associated with animals in health and disease', *J App Bact Sym Supp*, 1988 71S-85S.
8. WRAY C and WOODWARD M J, '*Escherichia coli* infections in farm animals' in Escherichia coli*: mechanisms of virulence*, ed SUSSMAN M 1997, 49–84.
9. DEAN-NYSTROM E A, BOSWORTH B T and MOON H W, 'Pathogenesis of O157:H7 *Escherichia coli* in neonatal calves' in *Mechanisms in the Pathogenesis of Enteric Diseases*, Plenum Press, New York, 47–51, 1997.
10. PEARSON G R, BAZELEY K J, JONES J R, GUMMING R F, GREEN M J, COOKSON A and WOODWARD M J, 'Attaching and effacing lesions in the large intestine of an eight-month-old heifer associated with *Escherichia coli* O26 infection in a group of animals with dysentery', *Vet Rec*, 1999 **25** 370–2.
11. ARMSTRONG G L, HOLLINGSWORTH J and MORRIS J G, 'Emerging foodborne pathogens: *E. coli* O157:H7 as a model of entry of a new pathogen into the food supply of the developed world', *Epid Rev*, 1996 **18** 29–51.
12. CHAPMAN P A, WRIGHT D J, NORMAN P, FOX J and CRICK E, 'Cattle as a possible source of verocytotoxin-producing *Escherichia coli* O157:H7 infections in man', *Epid Inf*, 1993 **111** 439–47.
13. DOYLE M P and SCHOENI J L, 'Isolation of *Escerichia coli* O157:H7 from retail fresh meats and poultry', *Appl Env Micro*, 1987 **53** 2394–6.
14. SEKLA L, MILLEY D, STACKIW W, SISLER J, DREW J and SARGENT D, 'Verotoxin-producing *Escherichia coli* in ground beef – Manitoba', *Can Dis Weekly Rep*, 1990 **16** 103–5.
15. WORLD HEALTH ORGANISATION, 'Report on a WHO working group meeting on shiga-like toxin producing *Escherichia coli* (SLTEC) with emphasis on zoonotic aspects', Bergammo, Italy, Rep no WHO/CDS/VPH/94.136, 1994.
16. NATARO J P and KAPER J B, 'Diarrheagenic *Escherichia coli*', *Clin Rev Micro*, 1996 **34** 2812–14.
17. TAUXE R V, 'Emerging foodborne diseases: an evolving public health challenge', *Emer Infect Dis*, 1997 **3** 425–34.
18. KETLEY J M, 'Pathogenesis of enteric infection by *Campylobacter*', *Micro*, 1997 **143** 5–21.
19. STERN N J and KAZMI S U, '*Campylobacter jejuni*' in *Foodborne Bacterial Pathogens*, Marcel Dekker, New York, 71–110, 1989.
20. OSTROFF S '*Yersinia* as an emerging infection: Epidemiologic aspects of yersiniosis', *Contrib Micro Immunol*, 1995 **13** 5–10.
21. FARBER J M and PETERKIN P I, '*Listeria monocytogenes*, a foodborne pathogen', *Micro Rev*, 1991 **55** 476–511.

22. TREES A J ,'Zoonotic protozoa', *J Med Micro*, 1997 **46** 20–4.
23. HOSKIN J C and WRIGHT R E, '*Cryptosporidium*: an emerging concern for the food industry', *J Food Prot*, 1991 **54** 53–7.
24. VAN DER POEL W H M, VINJÉ J, VAN DER HEIDE R, HERRERA M-I VIVO A and KOOPMANS M P G, 'Norwalk-like calicivirus genes in farm animals', *Emerg Inf Dis*, 2000 **6** (1).
25. GOODGAME R W, 'Understanding intestinal spore-forming protozoa: cryptosporidia, microsporidia, isospora and cyclospora', *Ann Intern Med*, 1996 **124** 429–41.
26. DE BOER E and BEUMER R R, 'Methodology for detection and typing of foodborne microorganisms', *Int J Food Micro*, 1999 **50** 119–30.

8

Implementing HACCP in a meat plant

M. H. Brown, Unilever Research, Sharnbrook

8.1 Introduction

The primary purpose of HACCP is to help processors and suppliers to identify and to control potential hazards in meat and meat products and to ensure that the finished product will be safe for consumers (Tompkin, 1990). HACCP is not a zero-risk system, but is designed to reduce the risk of manufacturing hazardous food to a minimum. How effective it will be is determined by its scope and the effectiveness of its implementation. It is a practical system that requires proper management and maintenance. HACCP cannot stand alone within a production plant; any system that is out of line with the other management systems in a plant will fail. Therefore integration with other management activities, especially Quality Assurance, is essential. Production staff and supervisors have responsibility for the everyday operation of the plan, and production management should have overall responsibility, including review and updating of the plan.

To allow successful implementation, the HACCP study must identify realistic hazards and critical control points, critical limits, control measures and corrective actions (Jouve, 1994). Implementation will focus around these activities and provide for their continuing review, maintenance and updating.

Codex Alimentarius (1997) provides the principles for validation and verification. Principle 6 of its guide to the management of HACCP asks that 'procedures are established for verification, to confirm that the HACCP system is working correctly'. And this in turn relies on Principle 7, which asks food producers to 'establish documentation concerning all procedures and records appropriate to those principles and their application'. Apart from these principles there is very little formal regulatory guidance on the presentation

178   HACCP in the meat industry

and content of HACCP plans, their degree of detail or the means and extent of implementation (Gombas, 1998). This chapter explains how the principles of validation and verification can be effectively applied to plans using the Codex approach.

Many businesses and employees in the meat industry are still not familiar with the concept of HACCP and have difficulty understanding it. Because of this they cannot understand what is required for implementation. A major factor hindering effective implementation can be the requirement for management to change from a simple QC or QA approach, often based on final product testing, to the preventive, hazard and process control-based approach of HACCP. The task is made more difficult as there is no global consensus on the right coverage for a HACCP plan that management can follow. Uncertainty exists in the following areas:

- The optimum scope for a HACCP study, so that it remains practical and manageable and provides good protection
- Which hazards should be covered
- The expected output of the HACCP study
- Which controls should be implemented
- How a plan for implementation can be developed
- The means available to ensure successful implementation
- Whether the study is line specific, or can be more generic, so that trained workers can move around a plant, or outline plans can be used to cover different products made on the same line
- How performance indicators can be derived from the HACCP plan
- How the performance of the plan can be monitored and success reported.

Although the meat industry is well covered by general industry guides published by research institutes, regulatory and industry bodies (e.g. CCFRA, 1997; ILSI Europe, 1998; NACMCF, 1993), there is a scarcity of published guidance on strategies for implementation in meat plants that have a diversity of hygiene requirements and types of process (see Sandrou and Arvanitoyannis, 1999). This may be because implementation is the transfer of ownership of the plan from the HACCP team to the management structure and workforce of a plant and is likely to be different from company to company. Although implementation will require tuning to suit different types and sizes of businesses, the principles and stages involved are likely to be similar.

8.2   The elements requiring implementation

HACCP is regarded as being based on seven principles (Bauman, 1995); the study, its implementation and management have been broken down into 14 steps (CCFRA, 1997). For the purpose of implementation the seven principles may be used as a guide. These elements and the sections of this chapter in which they are discussed are as follows:

- Hazard identification (8.2.1)
- Critical Control Points (8.2.2)
- Targets and critical limits (8.2.3)
- Monitoring (8.2.4)
- Corrective actions (8.2.5)
- Recordkeeping (8.2.6)
- Verification (8.2.7)

8.2.1 Hazard identification

This involves determining the likely hazards to consumers that could be presented by a meat product. These hazards and their seriousness must be communicated to the workforce and its management, or supervision, so that they understand the risks they run as a business and the consequences of selling unsafe products. Realistic hazards include microbiological contamination, natural toxins, any toxins formed in meat by decomposition, chemical contamination, drug residues and the presence of foreign bodies or other specified materials (such as offals) that may cause injury.

Hazards can occur or be introduced before, or after, a processor receives a raw material, on the premises where the study was done or elsewhere in the supply chain; therefore different groups of employees will be involved in implementation and control. The HACCP plan should be used to identify the employees at all levels likely to be involved in control, so that they can take part in implementation. All the process stages, personnel and controls within the scope of the study should be covered. The major stages in the production of meat and meat products are shown in Fig. 8.1.

To aid implementation and identify the personnel that should be involved in implementation, hazards can be divided into:

- **'Material-related'** that occur before raw or part-processed materials are received by the processor. Typically these hazards have their origins in the environment or in the animal, or are introduced during harvest, pre-processing or transportation to the processor. For effective control, implementation should extend upstream of the factory gate, through the manufacturing stages to the retailer's shelf.
- **'Process-related'** that occur during processing, packaging or distribution. Implementation can start at the factory gate and should cover the downstream supply chain to the retailer's shelf or beyond.

Clearly different groups of personnel will contribute to control and monitoring of the identified hazards, and therefore need to be involved in implementation and may require training. For the 'material-related' hazards, farmers, buyers, suppliers and the lairage and transport personnel may control safety. For the 'process-related' hazards, production, packaging, distribution, hygiene and maintenance personnel have to support the objectives of the plan and be involved in implementation, whether they are on or off site.

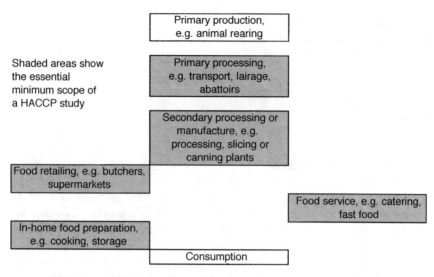

Fig. 8.1 Major stages in the production of meat and meat products.

The degree of risk and severity of the hazard arising from contamination or survival of microorganisms will also depend on the type of product being made. This should determine the efforts made to involve personnel at the extremes of the supply chain, such as animal rearers, transporters, etc., who may not be directly under the control of the plant and therefore may be more difficult to train and monitor (Grotelueschen, 1995). The more severe the hazard the greater the efforts that should be made. The following examples illustrate the point:

- Contamination of meat (or animals) with salmonella or *E. coli* O157 may occur before the processor sees the raw material. Effective control prior to entry to the factory may, or may not, be possible. Depending on the scope of the study, implementation could involve only personnel from the processing and butchery part of the supply chain or additionally extend back to the farm. Because of the severity of the hazard from infectious pathogens, every effort should be made to extend control back onto the farm. The extent of control (e.g. the starting and finishing points) must be defined at the start by the scope of the HACCP plan. Challenging whether these are sufficient, or not, to deliver a safe product is not part of implementation.
- Meat products, such as steaks or burgers, may be eaten partially cooked. Normal butchery and manufacturing techniques cannot eliminate any pathogens present on meat. The minimum degree of control should ensure that numbers and the incidence of infectious pathogens are not increased by slaughter, butchery and manufacture and that any cooking instructions provide a safe, good quality product. Who needs to be involved in providing this protection should be clear from the scope of the HACCP plan, and although the customer is an essential part of the safety system, it is not realistic to involve them directly in implementation (other than through pack labelling).

- Raw meat may also be brought into the plant for further processing (e.g. cooking for making pie fillings). In these cases, depending on the process involved, contamination with infectious pathogens before processing poses a less serious hazard, because factory cooking can reliably eliminate it. The control measures, procedures and personnel required will be different from the example above, as the objective of manufacturing personnel is to ensure absence in the cooked material, rather than preventing contamination of the raw material. In this example only the contribution of manufacturing personnel will be essential to product safety.
- In contrast, the presence of harmful microorganisms in a cooked, ready-to-eat bulk meat product bought from an outside factory for slicing and vacuum packaging is a serious hazard. Contamination may be due to either a failure of the cooking process (to achieve a temperature high enough to kill the microorganisms) or contamination by poor handling or hygiene after cooking. It is essential that the scope of the HACCP plan identifies critical process stages and personnel in the original producer, to give reliable control of safety in the whole supply chain. The implemented HACCP plan should extend from the slicing plant upstream to the producer of the bulk material to ensure the correct controls and monitoring are present in both the supplier and the slicing factory. Small producers of such materials should not be treated any differently from large ones, as the risks are no less.

8.2.2 Critical Control Points

By completion, the HACCP study should have

- identified the CCPs and
- verified that Good Manufacturing and Hygiene Practices are followed in the plant.

If verification of this has not been done, the effectiveness of the plan should form the subject of an inspection or audit prior to implementation of the HACCP plan. Establishment of GMP should not be included in the implementation plan, but is an essential pre-requirement (Sperber *et al.*, 1998).

It is common to identify too many CCPs in a HACCP plan. The result can be that implementation becomes unmanageable and important control issues cannot be distinguished from matters that are either Good Manufacturing Practice or have no real impact on control. Before starting implementation, the team should be confident that CCPs have been separated from other important aspects of processing. However failure to control the real CCPs within the production process will cause, or contribute to, the presence of a hazard in the product.

Each CCP will depend on employees and equipment that can eliminate or reduce an existing hazard, or prevent or minimise the chances of one being introduced. Therefore, to allow implementation, the following need to be identified for each CCP, based on the HACCP plan:

182　HACCP in the meat industry

	Control			Monitoring			Corrective actions		
	Location	Equipment covered	Responsibility	Location and technique	Frequency of measurement	Responsibility	Location	Equipment required	Responsibility
Process stage – CCP									

Fig. 8.2 Summary chart of management information on requirements for control, monitoring and corrective actions. Typically targets and limits will be included elsewhere in process specifications.

- personnel
- equipment
- locations.

The key information can be summarised as shown in Fig. 8.2.

The process of implementation should help the individuals or production teams to identify their responsibilities, consider the tools at their disposal and focus on the critical contributors to product safety – CCPs. For example, if raw meat is the product, control of the supply chain back to the farmer, including veterinary and agricultural practice, may be the only way to prevent animals carrying harmful bacteria. On the other hand, for a processor slicing and packing cooked meats, the receiving personnel, or QA, can exert effective control of product safety by ensuring that the cooking step, post-cook handling and hygiene and handling temperatures are well controlled and monitored in their suppliers. This should not involve direct testing of each batch, but should rely on auditing and examination of process or test records from the supplier.

8.2.3　Targets and critical limits

The HACCP study should have established critical limits for each CCP. Critical limits, such as a temperatures, hygiene level, cooking time or residue concentration, draw the line between acceptable and unacceptable operation of a plant or supplier, e.g. cooling to below 5 °C within 16 hours, or salmonella absent in five samples of 25 g per batch.

Targets are the working limits, related to the capability of the equipment and personnel. They are usually set within the critical limits to ensure that process or operator variability does not cause the limits to be exceeded. Setting or validation of targets is an essential part of implementation; for example, if a temperature limit for a slicing area is 12 °C, the room may be operated at 8–10 °C to allow for temperature rises during door opening, etc. Such targets and limits must be consistently achievable by the personnel, equipment and procedures involved at the CCP. If not, action will be required before the plan can be satisfactorily implemented. This may involve changes to equipment, layouts, suppliers or training of personnel, or even to the payment or productivity structure for a particular processing operation.

Some limits may only be used for verification purposes, rather than on a day-to-day basis for control (Bautista *et al.*, 1997). These may be limits where real-time control methods do not exist and analytical results may not be available until the opportunity for altering the processing conditions has passed. For example, microbiological testing may be used as a means of checking how well a CCP has been implemented. Such tests will be valuable for verifying how well controls on temperature, hygiene and handling practices are operated. For example, carcasses may be examined for *E. coli* to determine the extent of contamination from hides or gut contents, or cooked meats may be examined for the presence of Enterobacteriaceae or *Listeria* after cooking and cooling, to monitor hygiene or the extent of cross-contamination in clean areas (see Table 8.1).

8.2.4 Monitoring

Production personnel will not usually be responsible for assessing their own control of CCPs. This is done by monitoring, which is a planned sequence of observations or measurements, preferably made independently (by QA, not production), to assess whether a CCP is under control or not. Monitoring may classify whether a process stage is operating well within, near to or outside its limit or may produce numerical data for trend analysis. For example, the monitoring procedure for a cooking step leading to a cooked ready-to-eat product should show that

- product type and size and heating conditions are matched,
- the operating temperature of the oven is achieved and maintained,
- the proper length of process time is achieved by every batch,
- products enter the oven at or above a specified minimum temperature,
- the correct internal temperature is achieved within the product units when cooking is complete,
- cooling is satisfactorily achieved and re-contamination prevented,
- control systems are calibrated,

and will be predominantly numerical data.

Monitoring measurements may be similar to or less extensive than the control measurements used by production staff. For example, the temperature of ovens may be automatically controlled on a continuous basis and performance of the control system monitored from time to time by inspection of records or an independent thermometer. Monitoring may also include additional microbiological measurements, or be limited to review of the records, but should always be done with a specified frequency, as circumstances require.

Independent monitoring by QA will sometimes initiate corrective actions by production. Under these circumstances the frequency of monitoring will determine the extent of out-of-specification product made. During implementa-

Table 8.1 Examples of the training/communication material used for implementation of a HACCP study

Kitchen team – CCP 1				
Personnel – Ingredient handlers & Hygiene operators share responsibility for the CCP				
Equipment – chill rooms and docking bay				
Related process stages 1, 5, 8 & 9				
Refrigerated storage		Receipt, Expiry and removal from store dates must be entered by the FLT driver on the raw material HACCP doc 5.71. Date Code is 4-digit no. ÷ 3 = day of year		
Preventive measure	Target	Critical limits	Monitoring procedure	Corrective actions
Time control	Storage time no longer than shelf-life	Storage time exceeds shelf-life.	J-I-T & FIFO inventory policy used by receiving personnel and fork-lift drivers precludes extended storage	If recommended shelf-life exceeded Receiving personnel notify plant QA & adjust order quantity
Temperature control	5°C	7°C	Check room thermometer daily, check between-pack temps on entry – every batch & exit 2/d	If recommended temperature exceeded Receiving or handling personnel notify plant QA & & maintenance & block product

Kitchen team – CCP 3				
Personnel – Cooks, Ingredient handlers & Hygiene operators share responsibility for the CCP				
Equipment – heated + cooled tanks, colloid mill, dicer and connecting pipework				
Related process stages 1, 3, 4, 5.				
Emulsion preparation & storage		Linked CCP's 3 & 4		
Preventive measure	Target	Critical limits	Monitoring procedure	Corrective actions
% salt in protein slurry	10%	$\geq 9.5\%$	QA technician takes sample of pre-mix slurry. Salt level measured by titration. After salting if level <10%, each tote bin is sampled accept, remix or reject.	Hold, remix or reject product by forklift driver
Colloid mill exit temperature	25°C	30°C	Check temp. on discharge	Redistribute in totes and place in 2°C chill for 8h
Final salt content	6.5%	$\geq 6.0\%$	QA technician takes sample of final mix	If post-emulsion salt <6%, exit temp >30°C or operating time >6h, block & check batch for *S.aureus* (QA procedure 8.2)
Tank scrape down after every mix	Removal of all visual material	Presence of residues after scrape down	Supervisor inspects empty, scraped tank	Wash down tank

Table 8.1 continued

HACCP Procedure 12.3.
THERMOMETER CALIBRATION
Issue date 12. Nov 1998. Supercedes HACCP Proc. 12.2.
2 oil baths are set-up to maintain a constant temperature with suspended certified thermometers
One bath (blue) is set-up to a constant temperature of 10°C & one (red) at 95°C. DO NOT CHANGE ADJUSTMENT
Insert the thermometer and allow time for the reading to stabilize. When a constant reading is obtained (approx. 1 minute or until the reading is stable) record both temperatures, date and your initials on the thermometer Daily Check List in the specified column.
Check the reading of the certified thermometer and indicate if it is OK on the thermometer Daily Check List in the specified column
If adjustment is required to co-ordinate your thermometer with the certified thermometer, turn the screw adjustment to obtain the desired reading. (The holder is built with a recess to protect the adjusting screw and to make adjusting easier)
Ensure that after adjustment BOTH temperatures are accurate, if not record the offset in the Daily Check List and notify Maintenance
Record results along with job site & date on the Thermometer Daily Check List
Calibrate your thermometer everyday. Remember that your data is only as good as the accuracy of your thermometer. A uniform stable temperature must be used to ensure accuracy when calibrating.

tion the economic impact of sampling frequency on possible losses should be considered. If the plant runs so that it relies on intervention by QA to control quality or safety, then it is unlikely that implementation has been effective, as the personnel controlling the process evidently do not understand or support the limits set out in the HACCP plan. Therefore reliable control cannot be achieved. Alternatively the limits set out in the HACCP plan may be unachievable by the business or unnecessarily restrictive to its operation. In either case further review is needed.

8.2.5 Corrective actions

Corrective actions should be prompted by failure to meet a critical limit. Responsibilities for them should be allocated to production and supervisory staff during implementation. These staff must understand the importance of, and be trained to carry out, corrective actions whenever process control or monitoring data indicates that a process step is outside its critical limit. Production has to be responsible for these actions, as they have the best information on process performance. For example:

- If a cooker does not reach its proper operating temperature, corrective action will be required to repair the equipment, but also to ensure that product is held at least until it can be determined whether it is safe for sale. This is necessary because undercooking may allow the survival of infectious pathogens in a ready-to-eat product.
- If a chiller is found to be above its critical temperature, then the department involved should initiate maintenance or repair work, and also block product until it has been determined whether safety has been impaired. If, as a result

of evaluating the length of time product was exposed to the higher temperature, QA determines there is no safety hazard, then the product can be released. Corrective action should then focus on finding out the cause(s) of the failure and ensuring that it does not happen again. On the other hand, if the product was found to be potentially unsafe, then safe disposal has to be undertaken.

For both these examples, personnel in addition to those directly concerned with control of the CCP will be involved in safe management of deviations and this should be covered during implementation.

8.2.6 Recordkeeping

The HACCP plan should recommend which information has to be recorded during control and by monitoring, along with procedure to be used, the frequency and who is responsible. Good recordkeeping is essential to the success of HACCP and will make implementation and verification much easier. It is important that good performance monitors are developed to allow tracking of improvements or problems with the plan. The following two types of record are required.

Recordkeeping of the HACCP system and its implementation
The output of the HACCP study (i.e. its CCPs, limits, targets, control and monitoring procedures and corrective actions plus the personnel involved, their authorities and responsibilities) should be documented. The format should follow the flow diagram of the HACCP plan, covering all the hazards it controls and any procedures and pro-formas for recording the results. Good recordkeeping will provide:

- An understandable presentation of the HACCP plan, covering the key aspects of the identified hazards, their control and monitoring
- An outline of the implementation plan covering activities and responsibilities
- A summary of key limits and targets and the procedures used for measurement or analysis
- Records of training and information sources (for all levels of employee)
- A record of responsibilities and authorities or reporting lines
- Templates for recording the results of control, monitoring and the management of process deviations and indicating any trends.

Recording the performance of the HACCP system
To show the performance of production at each CCP, records must include the results of control and monitoring. They may be based on fill-in-the-blank sheets in a standard format or may use more sophisticated methods, such as the output of software systems. Recordkeeping can be done with all levels of complexity, depending on the needs of the plant, its management and customers. At the most

basic level it will be a paper record of conditions at a CCP (e.g. a fridge temperature) at a particular time, shift records of hygiene checks or continuous records from automatic equipment. Sometimes only exceptions or out-of-control events (such as poorly cleaned equipment) may be recorded. Where records are required for CCPs that are not part of the 'core' manufacturing process, e.g. describing suppliers' performance, additional records should cover what must be done and achieved by them and how it is monitored, and the records sent to the in-factory QA function.

Systematic recording of results is necessary so that all interested parties can look when necessary, or over periods of time, at how critical aspects of the operation are functioning. The production of summaries can show trends and allow potential problems to be fixed before they become critical. HACCP records should also provide a means for management or supervision to find out quickly if something has gone wrong, identify the cause and keep track of what is being done to remedy it. Document control is an essential part of record-keeping to ensure that everybody works to the current specification and that results are not lost. This may be done using the ISO 9000 format (ISO, 1994).

8.2.7 Verification

During implementation, procedures to review that the HACCP system is working properly must be put in place. At a minimum this involves a record review, but ideally will include an audit of performance of the plant (Sperber, 1998; van Schothorst, 1998). On a routine basis verification must check the following:

- The critical limits and scope of the plan are adequate to control or eliminate the identified hazards. This can be done by inspection of QA or customer complaint data.
- Accurate and suitable records are being produced as agreed. There should also be evidence that these records are being compared with critical limits and used to manage the process.
- Where process deviations occur more than infrequently, management are reviewing any impact on product safety and also questioning the effectiveness of the plan, with the aim of seeing whether it needs to be revised.

8.3 The implementation process

Communication is the key to successful implementation. The importance of food safety to the business and the aims of HACCP must be understood and supported by all employees. In the meat industry this is critical, as the success or failure of HACCP in controlling safety lies mainly in the hands of those on the farm, handling, processing and distributing meat, rather than in the implementation of automatic controls. HACCP is a tool for reliability improvement, but it is only a tool – the important parts are the controls and their implementation, monitoring and review. The system has to be operated by the production management in the context of a supportive longer-term management strategy; if it is outside this strategy implementation will not work. Senior management commitment is essential, and implementation has to be done down to factory floor level through specific activities (e.g. training) and appropriate tools.

8.4 The differences between large and small businesses

Safety must be achieved through HACCP in a way that is compatible with the company's capability and meets the needs and expectations of customers. Training and changes in company or plant culture during implementation must address these needs (Kukay *et al.*, 1996). Matters of Good Manufacturing Practice (or pre-requisites) should be in place before implementation of a HACCP plan is attempted. The plant should already have basic hygiene training in place to meet the underlying requirements of legislation (e.g. EU directive 93/43; EC, 1993). HACCP may reinforce or change existing practices, systems and requirements, and is likely to produce improvements and raise organisational, cost or technical issues. In addition to human food safety hazards and controls, the principles may also be used to address quality issues.

Successful implementation of HACCP needs to be directed toward an endpoint that is clear to everyone involved. Implementation of HACCP will represent a major investment of employee time and the company's resources as it requires the active participation and continuing support of all members of the workforce. At the onset some personnel may be unsure where to start with HACCP and what is required. Management may be unwilling to spend time and money, especially if there is a risk that they may not succeed and may have wasted their time. In companies of all sizes, some members of the workforce may have problems expressing themselves both verbally and in writing; some may not even speak English. Successful implementation will require practical help and reassurance from the beginning to overcome these barriers. This may be obtained by using professional trainers or external consultants, or by training specified managers or supervisors in HACCP and by careful and realistic management of the HACCP team. Alternatively, regulatory authorities or trade associations in some countries may offer help. Some companies may have existing, strong QA or QC systems and HACCP implementation should build on these as much as possible, rather than challenge them.

When assessing implementation of HACCP system it is important to recognise and understand that there will be differences between large and small businesses in their approach to HACCP, because of the expertise and ability of the people involved and the controls they can implement and maintain. Slaughter facilities are likely to differ in their layout and practices and each is likely to require development of a different HACCP plan. Even in processors of similar size there may be differences between the control measures and corrective actions used or advocated for similar processes. In some cases, controls relevant to one processor may not apply to another, because of conditions specific to the type of meat, its origin, the processor or the supply chain. Whichever controls are used, they must assure an equivalent degree of product safety.

In some small businesses the output of a HACCP study may not be a written document, even though its output will require implementation. Lack of documentation can prove a problem, as clear documents provide a good basis for guiding change and give something to check progress against, making the task of implementation easier. Simple paperwork covering GMP, CCPs and controls should be produced by any HACCP study, as a minimum requirement for implementation; in some cases existing QA systems can be used for guidance.

In small companies there may be too few personnel to allow teams to be formed and taken away from day-to-day production for training. Under such circumstances training needs to be integrated with production in a planned way, to take account of the pressures of production. It should be directly supervised within the production framework by the existing management or by external consultants. Where consultants are used, plant management should actively approve and support their plans and ideas.

In large companies where there is familiarity with systems and procedures, often within the ISO 9000 framework (ISO, 1994), there is a temptation to produce HACCP systems of unmanageable complexity. All HACCP studies should be clear and concise and provide facts that the workforce can relate to, understand and implement. They must focus on the critical points and not become bogged down in trivia.

8.5 Where to start with implementation

To facilitate the implementation of the seven HACCP principles, they may be organised as 11 discrete stages or management activities, to be carried out in the plant by the HACCP team or other members of the workforce most involved in the activity or having the most appropriate knowledge and skills. These stages, with the sections in which they are discussed, are:

- Explanation of the reasons for HACCP (8.6)
- Review of food safety issues (8.7)
- Planning for implementation (8.8)
- Allocation of resources (8.9)
- Selecting teams and activities (8.10)
- Training (8.11)
- Deriving specific requirements and transferring ownership to production or manufacturing personnel (8.12)

190   HACCP in the meat industry

- Implementation on the line (8.12.1)
- Tackling barriers (8.13)
- Measuring performance of the plan (8.14)
- Auditing and review (8.15).

8.6   Explanation of the reasons for HACCP

As a starting point for implementation, the reasons for developing HACCP in the plant and its objectives should be explained to the workforce and management. To help this, the factory jobs and roles of those in the HACCP team should be explained to underline the practical basis and relevance of the plan. The explanation should be focused on the line or area selected for implementation and should include a description of the product, its principal raw materials and ingredients, the stages of manufacture, and product usage, for example whether it is intended to be cooked or ready-to-eat. The roles of Quality Assurance, production and engineering in implementing HACCP in manufacturing operations should be explained; and where necessary on-farm, transport and abattoir activities should also be included.

8.7   Review of food safety issues

Next there should be an explanation of food safety issues to the employees, related to the products being made. The information needed for this may have been assembled by the HACCP study team and it should be presented by them; if not, it can be collected from publications, external experts such as regulatory officials, or quality assurance data or customer complaints. It may need simplifying to make it accessible to management and the workforce. Based on this explanation, the reasons for prioritising the elements of the HACCP plan (especially the CCPs) can be explained to obtain the support of the workforce.

The scope of the plan for raw meat and meat products should cover on-farm control, although this is likely to be difficult to achieve. Livestock and poultry are a reservoir of infectious pathogens and farmers have limited ways of knowing which animals carry pathogens as part of their normal gut flora. In addition, these microorganisms can survive in soil and will be present in faeces. The effects of husbandry and transport practices on the carriage of infectious pathogens are largely unknown and therefore effective controls cannot be proposed. However, an essential part of control is the delivery of animals to the abattoir or poultry plant in a clean condition, carrying the minimum amount of soil or faecal matter. Farmers, commercial hauliers and livestock transporters should be aware of their responsibility to send animals to slaughter in a clean condition, and control and monitoring measures should be produced by the HACCP study and included in its implementation. Although abattoirs may differ in their practices, the following procedures should always be covered by implementation, with the aim of reducing the potential for contamination of the final product:

- Acceptance of animals, including minimising faecal or soil contamination
- Hide, fleece or feather removal
- Evisceration without contamination of meat
- Chilling
- Prevention of cross-contamination
- Enforcement of hygiene requirements.

8.8 Planning for implementation

It is unlikely that the plant will have sufficient resources concurrently to train all personnel, make changes and roll out the plan in all departments, therefore choices will have to be made.

8.8.1 Appointment of an implementation team
Management should appoint a HACCP implementation team drawn from production and QA plus additional departments (depending on the CCPs being covered). Its job is to decide on the timetable and means of implementation, taking account of critical issues and constraints on the plant and workforce. It may or may not benefit from inclusion of members of the HACCP study team, who should be included only if they bring specific technical or organisational skills to the team. Their documented output from the HACCP study should be good enough to stand without explanation.

8.8.2 Development of an implementation plan
The aim of the implementation team is to develop and cost a logical and structured outline plan for changes and improvements. The plan should assign responsibility for specific parts of the HACCP plan to specific departments and personnel, so that the workforce is able to take over its day-to-day running. Medium- or long-term changes in premises, personnel, equipment, supplies or controls should be identified as soon as possible, and the plan should cover the whole implementation exercise for a line or department and provide a timetable for changes with clear requirements and criteria for success covering the following aspects:

- Organisational (e.g. assignment of authority and responsibility)
- Personnel (e.g. management systems or procedures and training)
- Supply chain – raw materials and products (e.g. completion or amendment of specifications and deletion of risky or hazardous products)
- Process (e.g. improvement of process capability and modification of control systems or targets and limits)

- Layout (e.g. improved forward flow or segregation of raw and cooked product)
- Procedural changes addressing specific practices, resources or the sequence of activities relevant to the HACCP plan
- Procedures and lines of communication and reporting/authority during plant operation.

The implementation team should also identify any specific actions needed to implement the HACCP plan at least cost. Useful background information for such an exercise will include:

- A comparison of costings, time requirements and responsibilities between the existing QA plan and the HACCP plan proposals
- Any limitations imposed by numbers of personnel and current skills
- A review of past problems or shortcomings, and actions plus outcomes and their relationship with the requirements of the HACCP plan – are they covered?
- Regulatory requirements
- Legal or contractual requirements, liability and due diligence issues
- Coverage of normal and abnormal and emergency production conditions
- Specific priorities or actions to fit in with other management plans.

Examination of these topics should produce a review of the current situation and may indicate benefits from implementation, such as a reduction in the levels of defective or potentially unsafe product. This will lead to an estimate of the probable impact of HACCP on the current operation and may indicate the likely one-off and recurring costs for training and other requirements, such as equipment or layout changes.

8.8.3 Functions of the implementation plan

Senior management must approve the implementation plan, as they play a major role in facilitating its implementation. Preparation and publication of the implementation plan and its criteria for success should provide a means of gaining or keeping support at all levels. Clearly written, it will make management confident that they are not signing a blank cheque and the company and customers will benefit.

The receivers of the training need to be identified as work groups by the plan, so that their work responsibilities and training needs can be considered and included in production planning over the period of implementation. They must be provided with an effective means of feedback to the team.

The plan should show where implementation will be started, on one line with one product or, for larger companies, in one department with several lines, depending on the resources available. For preference, lines with the highest level of risk or most severe hazards should be tackled first. Experience from the early attempts at implementation should be fed back into the plan and used to help the development of procedures for other lines or implementation in other departments.

8.9 Allocation of resources

Management must ensure that there are sufficient resources for HACCP implementation available at the right time. They must approve

- the membership of the implementation team and any sub-teams dealing with implementation at specific process stages,
- release of the workforce from their normal duties or production targets for training and putting the plan into practice.

The resources for this must be in addition to essential on-going support for production, product development and quality assurance. This will normally cause conflicts, require compromises and determine the rate of progress. The largest part of implementation is likely to be devising and agreeing new procedures, practices and methods of recording performance. Layout and equipment changes are likely to figure to a lesser extent. Therefore the implementation team, in conjunction with line or department supervision, must ensure that the plan contains realistic timings, costings and priorities related to the specific line, its personnel and products. It may have to be revised by the implementation team after discussion with management, but a final plan must be agreed before proceeding any further.

8.10 Selecting teams and activities

Implementation sub-teams may be set up outside the main team and can include production and QA personnel plus additional departments, depending on the CCP being covered, e.g. buyers for raw materials, the head of the hygiene crew or the distribution manager for final product. This will reduce the time demands on the main team, but to be effective these sub-teams must be able to manage and lead their activities, so that everyone connected with the line or department is provided with the opportunity to participate. Activities may be specific to one part of the process (i.e. involving the personnel linked to a specific CCP – cooking temperatures or packaging) or cross-functional (i.e. covering CCPs controlled by more than one activity – receiving raw materials or cleaning).

8.11 Training

Staff training is an essential part of implementing the HACCP plan. It should provide staff with the capabilities and resources needed to carry out the necessary tasks in an efficient and competent fashion, time after time. Wherever possible it should be based on the strengths of the existing supply chain, production and QA systems and personnel. Training is likely to be required at all levels in the factory, ranging from ensuring that senior management are familiar with the concepts and objectives of HACCP, to the development of procedures and techniques with production staff and suppliers (Moy *et al.*, 1994).

8.11.1 Management responsibilities

Management should coordinate, set aims and ensure consistency of training to establish operational control within the HACCP framework under normal, abnormal and emergency situations. This will require the correct allocation of human, technical and financial resources to ensure motivation of the workforce by communication, recognition of work well done, and barriers to work well done and encouragement to make suggestions. Therefore the training of management needs to develop their awareness of requirements, importance and priorities within the HACCP plan (Mayes, 1993, 1994).

8.11.2 Training at the factory floor level

Training for production staff should be based on aspects of their job that affect product safety and be related to specified work practices or procedures. The type and extent of training will be determined by the extent to which suitable preventive or control measures already exist in the plant or whether they need to be developed. It can include on-the-job instruction, practical demonstration, and discussion in quality circles or formal discussion and evaluation. Where procedures or targets/limits are new, human resources or industrial relations personnel may need to be involved to ensure that the implications of changes are recognised.

Production staff may require new or additional technical, analytical or data interpretation skills, if new procedures or techniques are introduced at the factory floor level, e.g. temperature or time measurements or sampling procedures. Recording this data may involve either documentation or data entry by computer; such skills can be learnt by a mixture of training and experimentation.

Training for staff located at CCPs should provide them with:

- The location and a description of the CCP they are responsible for, including its role in ensuring product safety
- Knowledge of the hazards and the means of control at the CCP
- Work practices and limits of responsibility
- Targets, tolerances and critical limits
- Control and monitoring procedures
- Recordkeeping requirements

- Corrective actions
- Working contacts for review and discussion of problems.

Within the HACCP plan, production may share responsibility with QA for the interpretation of control or monitoring data showing whether or not targets are achieved.

8.11.3 After the initial training

The outputs of training should be put into practice as soon as possible after it is completed, which may give a progressive roll-out of the HACCP system to production areas. To ensure this is manageable, training should be timetabled against the needs of production. During rollout, problems may be found with some procedures or targets and changes may need to be made; the training team should check that any practical changes do not compromise the requirements of the HACCP plan. Teams should be asked to produce instructions that help them manage their CCPs; examples are given below.

Continuing training is an essential part of maintaining HACCP and at a local level this may be done by providing posters showing the procedures and requirements at CCPs. Regular review sessions with the teams or individuals at CCPs should be used to assess their ability to deliver against the requirements and the overall performance of the plan. The effectiveness of training may also be judged by the quality of proposals for improvement or change from the factory floor.

8.12 Transferring ownership to production personnel

To help transfer ownership of the HACCP plan, management, and especially Quality Assurance and Development Departments, should have developed and communicated positive attitudes and actions towards food safety and its control. The role, responsibilities and impact of employees and their activities, or practices, on food safety need to be emphasised by the overall plan. For implementation, some key areas need to be covered by line management or supervision with the production personnel at each CCP. This should:

- link specific personnel or teams to CCPs;
- provide them with knowledge of
  — why a hazard or procedure is important
  — what to do when a process stage moves out of control
  — whether the process stages up- and downstream of the CCP are important;
- lead to acceptance by personnel of responsibility for compliance with the limits at the CCP (i.e. for the process stages contributing to the CCP);
- define their authority (e.g. empowerment to stop the line or reject defective material).

It is essential that the employees agree that the explanation of these topics is appropriate to their jobs and that the requirements are realistic (see Table 8.1). If this is not the case and they do not have ownership of the HACCP plan, then implementation will fail. If the plant has a multi-shift operation or has a rapid turnover of staff, training and recordkeeping methods should ensure consistent understanding and standards of enforcement. This issue should be specifically addressed as part of the implementation plan. Information for suppliers, regulators and customers must also be defined during implementation.

8.12.1 Implementation on the line

When individual or team responsibilities and targets have been completed and agreed with the implementation team and management, then the plan should be put into action as soon as possible. The type of management on the production line is changed and implementation on the line starts. Planning should have covered the generalised stages shown for the production of raw meat and meat products, depending on the animal being processed (Fig. 8.1). Activities in shaded cells in the figure are likely to include CCPs or important aspects of GMP (see Fig. 8.3). If substantial training or equipment/layout modifications were needed, they should be substantially completed before implementation starts. This will be important if the control requirements necessary to meet HACCP targets without compromising quality or productivity cannot be met by existing process capabilities.

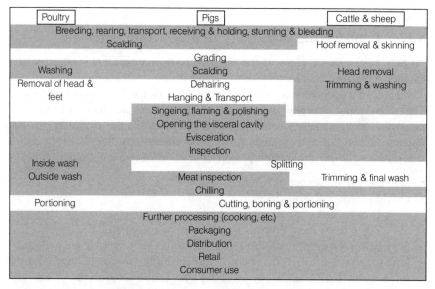

Fig. 8.3 The production of raw meat and meat products is likely to include the following generalised stages. The shaded areas show process stages common to production from all three types of animals/birds and meats.

New procedures and working practices for control and monitoring should be used, with support or help as necessary, and allowance should be made for any temporary increase in production costs or loss of line efficiency while learning progresses. The scale of the tasks involved and the amount of disruption will depend on the current knowledge, attitudes, facilities and responsibilities of the workforce concerning food safety and quality assurance, relative to the proposals of the plan.

The team should already have made allowance for this and its associated costs in the implementation plan. Based on their findings, implementation may be assisted by them, or a team with additional skills may be needed to help the workforce.

In many plants, implementation of the HACCP plan will only amount to a formalisation of existing controls and QA procedures and may only involve a change in the organisation or collection of control and monitoring data and the formal allocation of responsibilities to production staff. Where this is not the case, then the implementation sub-teams should plan to be heavily involved with the production line to ensure that operatives and production are not overwhelmed by changes.

8.13 Tackling barriers

At the end of the planning stage a number of barriers to implementation will have been identified. To judge their importance, the HACCP implementation team needs to separate what is acceptable and what is unacceptable with respect to food safety performance and practice in the plant and its supply chain. They should direct the organisation towards the best management practices and best practicable technology via the HACCP plan. Specific areas for improvement may be needed to deliver the HACCP plan as a workable system; they may only become evident when HACCP is working on the line and may include:

- Organisational barriers
- Inadequate resources or other cost constraints
- Necessary process, layout, procedural or product changes
- Improvements to the safety or quality of raw materials
- Cultural and language factors
- Technical or equipment factors – compatibility of HACCP requirements with plant, equipment, layout and any needs for improved controls, monitoring or control technology
- Distribution or marketing constraints
- Motivation and training needs
- Impacts of the requirements that may not have been estimable, e.g. pay/cost impact of the HACCP plan.

It is essential that a summary of these barriers and the actions necessary, plus the likely costs and benefits of overcoming them, is delivered to management by the team regularly during implementation. Further progress cannot be made without management's active support.

8.14 Measuring performance of the plan

Implementation should have translated the HACCP plan into a set of realisable actions focused on the CCPs covering the scope of the HACCP plan, including core and non-core processes. Performance indicators should be chosen to show whether implementation of the plan is complete, whether it is working or not, and the rate of progress (changes and improvements). These indicators can be quantitative or qualitative (e.g. rate of production, percentage of non-conforming product, material wastage or frequency of completing result sheets) and be included in recordkeeping. They will usually be derived from the specified control or monitoring measurements already noted in the HACCP plan. Performance indicators should be built up into quality systems that meet the needs and expectations of all parties and protect the company from the consequences of unsatisfactory performance during implementation and up to the first routine review. Different indicators may be needed to cover the period of implementation of the HACCP plan and its routine operation. Information produced by the implementation team must keep management up-to-date with the financial and human resources impact of the plan.

Effective recordkeeping procedures and measurement or analytical techniques must be put in place covering all areas critical to product safety. Records should always be referenced to the CCPs noted in the plan. The usefulness of any proposed measurements not related to CCPs, GMP or quality should be examined and they should be eliminated if they do contribute to control. Decisions on sampling or measurement frequency for control and monitoring at each CCP must be made and recorded and the period of retention of each record should be specified. Additional measurements or information needed for process deviations or emergencies should be noted. Records may include

- The outline and scope of the HACCP plan and the procedures being implemented and run, plus the data used in the review of food safety issues (see Section 8.7)
- The HACCP study and implementation team and sub-team members and their responsibilities
- Decisions and compromises made during implementation with the reasons
- CCP control records, trends (if available) and management actions
- CCP monitoring records, trends (if available) and management actions
- Records of process deviations and the actions taken to restore control and ensure product safety
- Training records (and plans)
- Audit and inspection records, actions and completions
- Regulatory input and necessary records, reviews, etc.

8.15 Auditing and review

Progress of the plan needs to be reviewed against the agreed timetable and the criteria for success. This is best done by a formal audit of the plan on the factory floor; a record review is unlikely to yield enough information and will not examine whether production personnel are fully in control and committed to the way of working demanded by the HACCP plan. If all is going well, this audit and review should lead to only minor adjustments and improvements. If there are major problems, then the barriers to success should be identified, so that remedies can be sought. These may not always lie in the implementation phase, but may result from unrealistic targets within the HACCP plan itself. Where this is the case, help from outside experts or regulatory agencies should be sought to confirm that the original requirements were realistic. Audit and recordkeeping together should provide sufficient evidence to allow an external inspector to decide whether problems have occurred, and if so how they were addressed.

8.16 Conclusions

HACCP plan implementation is more difficult and requires more time than the HACCP study itself. Without implementation the HACCP study is an empty document. The definition of successful implementation is still not available; many questions still remain about its relationship with GMP, ISO and other methods of quality management, not to mention legislation and the provision of evidence to show safe manufacture (Folstar, 1999). Wherever possible, implementation should build on existing strengths and systems, with training used to fill in the gaps and change attitudes. Problems with implementation are found not only on the factory floor; educating and training senior management is as important and may be more difficult. Without doubt their support is essential to the success of HACCP, because it will not survive in a business where it is at variance with the management culture. Effective implementation and maintenance of HACCP-based quality assurance systems, with control exercised by production staff rather than roving quality inspectors, leads to improved protection of consumers and businesses.

8.17 References

BAUMAN H E (1995) The origin and concept of HACCP, in *HACCP in Meat, Poultry and Fish Processing*, eds PEARSON A M and DUTSON T R, pp 1–7, London, Blackie Academic and Professional.

BAUTISTA D A, SPRUNG D W, BARBUT S and GRIFFITHS M W (1997) A sampling regime based on an ATP bioluminescence assay to assess the quality of poultry carcasses at critical control points during processing, *Food Res Int*, **30** 803–9.

CAMPDEN AND CHORLEYWOOD FOOD RESEARCH ASSOCIATION (CCFRA) (1997) *HACCP: A Practical Guide*, 2nd edn, Technical Manual No. 38.

CODEX ALIMENTARIUS COMMISSION (1997) Hazard Analysis and Critical Control Point (HACCP) System and Guidelines for its Application, Annex to CAC/RCP1 – 1969, Rev 3. FAO/WHO Food and Agriculture Organisation of the United Nations, Rome.

EUROPEAN COMMUNITY (EC) (1993) Council Directive 93/43/EEC (14 June) on the Hygiene of Foodstuffs, *Official Journal of the European Communities*, 19 July, No. L 75/I.

FOLSTAR P (1999) Editorial, *Food Control*, **10** 233.

GOMBAS D E (1998) HACCP implementation by the meat and poultry industry: a survey, *Dairy, Food and Environmental Sanitation*, **18** 288–93.

GROTELUESCHEN D M (1995) Impact of farm management practices on the incidence of human enteric pathogens in cattle, in *HACCP: An Integrated Approach to Assuring the Microbiological Safety of Meat and Poultry*, eds SHERIDAN J J, BUCHANAN R L and MONTVILLE T J, pp 9–14, Trumbull, CT, Food and Nutrition Press.

ILSI EUROPE (1998) *Food Safety Management Tools*, eds JOUVE J L, STRINGER M F and BAIRD-PARKER A C, ILSI Europe Task Force, 83 Avenue E. Mounier, B-1200 Brussels, Belgium.

INTERNATIONAL STANDARDS ORGANIZATION (ISO) (1994) ISO 9000 series, *Quality Management and Quality Assurance Standards: Guidelines for Selection and Use:*

ISO 9001, *Quality Systems – Model for Quality Assurance in Design/ Development, Production, Installation and Servicing*.

ISO 9002, *Quality Systems – Model for Quality Assurance in Production and Installation*.

ISO 9003, *Quality Systems – Model for Quality Assurance in Final Inspection and Test*.

JOUVE J L (1994) HACCP as applied in the EEC, *Food Control*, **5** 181–6.

KUKAY C C, HOLCOMB L H, SOFOS J N, MORGAN J B, TATUM J D, CLAYTON R P and SMITH G C (1996) Application of HACCP by small-scale and medium-scale meat processors, *Dairy, Food and Environmental Sanitation*, **16** 74–80.

MAYES T (1993) The application of management systems to food safety and quality, *Trends in Food Science and Technology*, **4** 216–19.

MAYES T (1994) HACCP training, *Food Control*, **5** 190–5.

MOY G, KAFERSTEIN F and MOTARJEMI Y (1994) Application of HACCP to food manufacturing: some considerations on harmonization through training, *Food Control*, **5** 131–9.

NATIONAL ADVISORY COMMITTEE ON MICROBIOLOGICAL CRITERIA FOR FOODS (NACMCF) (1993) Generic HACCP for raw beef, *Food Microbiology*, **10** 449–88. United States of America, Department of Agriculture.

SANDROU D K and ARVANITOYANNIS I S (1999) Implementation of Hazard Analysis Critical Control Point in the meat and poultry industry, *Food Rev Int*, **15** 265–308.

SCHOTHORST M VAN (1998) Introduction to auditing, certification and inspection, *Food Control*, **9** 127–8.
SPERBER W H (1998) Auditing and verification of food safety and HACCP, *Food Control*, **9** 157–62.
SPERBER W H, STEVENSON K E, BERNARD D T, DEIBEL K E, MOBERG L J, HONTZ L R and SCOTT V N (1998) The role of prerequisite programs in managing a HACCP system, *Dairy, Food and Environmental Sanitation*, **18** 418–23.
TOMPKIN R B (1990) HACCP in the meat and poultry industry, *Food Control*, **5** 153–61.

9

Monitoring CCPs in HACCP systems

J. J. Sheridan, TEAGASC (The National Food Centre), Dublin

9.1 Introduction

9.1.1 HACCP prerequisite programmes

According to Sperber[1] HACCP cannot be successfully applied in a vacuum but must be supported by a strong foundation of prerequisite programmes. Prerequisite programmes provide the basic environmental and operating conditions that are necessary for the production of safe food. The most common prerequisite programmes include Good Manufacturing Practices (GMPs), Standard Operating Procedures (SOPs) and Sanitation Standard Operating Procedures (SSOPs). These programmes should be established, managed and verified separately from the HACCP plan.

GMPs
GMPs are the correct processes and procedures to be followed in the preparation of food to prevent microbial, chemical or physical contamination of the finished product. The GMP programme includes activities such as maintaining the facilities, grounds, equipment and utensils, pest control, receiving and storage, process control (functions directly related to the manufacturing process such as employee hygiene, formulation control, labelling/code dating and reworking/ reconditioning), product recall and personnel training. An example of a GMP in beef slaughter is the immersion of knives and steels in water at 82 °C, for at least 30 seconds, between carcasses. This prevents the spread of bacteria from one carcass to the next and results in an overall reduction in bacterial contamination.[2] Hazards not addressed in the HACCP plan are usually covered as part of the GMP programme. This can lead to confusion, as the differences between a GMP and a CCP are not always obvious. GMPs are concerned with

hygiene and require action directed at the plant and equipment, while CCPs deal with food safety aspects and involve direct action on the carcass or meat products. These and other important differences between GMPs and CCPs are outlined in Table 9.1, while the relationship between GMPs, CCPs, SOPs and SSOPs is illustrated in Fig. 9.1.

In fresh meat slaughter the abattoir should be designed so that contact between carcasses and equipment, such as the hide puller or the evisceration table, is prevented. This should be documented in the GMP manual.

SOPs
SOPs are established or prescribed methods to be followed routinely for the performance of designated operations or in designated situations. The exact procedures for performing specific tasks are detailed as SOPs. These may be divided into three categories (Table 9.2). Routine procedures on the slaughter line, such as stunning and carcass splitting, are detailed in the SOP manual. The exact procedures to be followed at a CCP are also detailed as SOPs in the HACCP manual. Finally, SOPs concerned with cleaning the plant and equipment as well as personnel hygiene are referred to as SSOPs.

SSOPs
SSOPs are defined as those operations involved in providing a clean sanitary environment for the preparation, handling and storage of meat and meat products.

Table 9.1 Differences between GMPs and CCPs

GMP	CCP
Primarily concerned with hygiene A small number of GMPs are concerned with food safety: see Fig. 9.1	Concerned solely with safety
Covers area such as facilities, grounds, equipment, utensils, pest control, receiving and storage, process control, product recall and personnel training	Covers the carcass and related meat products
Does not address plant/product specific hazards	Addresses specifically identified hazards
No consideration of whether non-compliance poses an unacceptable health risk	Non-compliance always poses an unacceptable health risk
May not always be possible to monitor, establish critical limits and corrective actions	Must be monitored, have established critical limits and be capable of corrective actions.

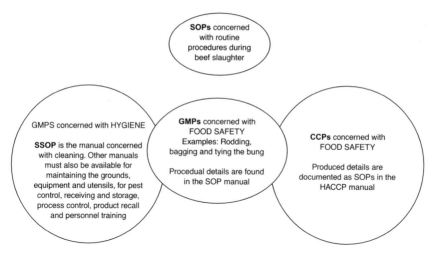

Fig. 9.1 Relationship between GMPs, CCPs, SOPs and SSOPs.

Table 9.2 Procedures covered in the SOP manual

	SOPs	
Routine procedures	CCP procedures	SSOPs
Examples: stunning carcass splitting	Examples: steam vacuuming steam pasteurisation	Examples: cleaning the plant and equipment personnel hygiene

9.1.2 Establishing critical control points

In a HACCP system one of the most important steps is the establishment of critical control points (CCPs). These are the steps at which control can be applied to eliminate or reduce food safety hazards. Loss of control at these points may result in the development of a food safety hazard. In relation to fresh meat it is important to distinguish between critical control points and control points. According to the International Commission on Microbiological Specifications for Foods,[3] control points in a HACCP system either assure control of a safety hazard as a result of elimination and are therefore critical, or they can reduce or prevent the risk. In terms of fresh meat processing, safety hazards cannot be eliminated but they can be prevented or reduced, while for some processed meats, especially those that involve a heating step, elimination is possible. In the present discussion both fresh and processed meat will be considered in relation to the use of HACCP.

9.1.3 The need for monitoring, validation and verification in HACCP systems

In a HACCP system the identified hazards are controlled using monitoring, validation and verification.

Monitoring

This involves carrying out a sequence of observations or measurements to determine if a CCP is under control. In any HACCP plan this is an essential process, since it provides proof that the system is working as designed. The data from monitoring must be carefully recorded, and used to ensure that process control is being maintained.

Validation

When a HACCP plan has been introduced into a meat plant it is necessary to validate the CCPs. Validation is a demonstration that the controls being introduced are capable of controlling the identified hazards. This is intended to be an active demonstration that the HACCP system is reducing or preventing a safety hazard, e.g. that faecal contamination on carcasses, as indicated by *E. coli* levels, is under control.

Verification

When a HACCP system is installed in a meat processing plant it is necessary to determine if it is operating as intended or needs modification. This is the process of verification and uses data from both monitoring and validation, as well as microbiological testing, to achieve this goal.

9.2 Establishing criteria

The establishment of critical limits for CCPs is one of the most important decisions in developing a HACCP plan.[4] A critical limit is the maximum or minimum value at which a process must be controlled at a critical control point to reduce to an acceptable level the occurrence of an identified food safety hazard. A CCP may have one or more control measures to assure that the identified hazard is prevented, eliminated or reduced to an acceptable level. This means that a specific objective in relation to food safety must be achieved. For example, in the manufacture of pepperoni, the normal fermentation process cannot control the growth of *E. coli* O157:H7, which can be a problem in such products.[5] In consequence, a heating step has been recommended during manufacture.[6] To ensure that this pathogen is eliminated or reduced to an acceptable level, data is required on the levels of *E. coli* O157:H7 present in pepperoni, the heat resistance of the organism, the heating parameters of the product, such as thickness, ingredient composition and the temperature and humidity at which the product ferments and matures during drying. Together these factors will determine the critical limits for this product. When this kind of data is available, a quantitative or measurable association between the critical limits in the HACCP plan and the risk to the health of the consumer can be established. This allows decisions to be made on acceptable levels of consumer risk and food safety.[7]

With fresh meat processing, the critical limits at a CCP are related to safety hazards, which represent only a potential risk to consumer health. This risk cannot be quantified, because detailed information on specific pathogens on carcasses is not available. In consequence, the relationship between the safety of raw meat and risk to the consumer is qualitative only.[7] In fresh meat processing the control of hazards at CCPs is generally associated with different processing operations, such as chilling or hot water treatment of carcasses. The critical limits for these processes are available from the scientific literature. In the case of hot water treatment of beef carcasses the critical limits for the washing of carcasses are 85 °C for 10 seconds.[8] The application of this treatment results in a 1.5 $\log_{10}$cfu/cm^2 reduction in total counts and about 2.0 $\log_{10}$cfu/cm^2 reduction in *E. coli* counts. These values are the levels to which the hazard on beef carcasses has been reduced at this CCP.

9.3 Determination of critical limits

Critical limits are process values that are used to control identified food safety hazards at CCPs. A critical limit may be readings or observations, such as a temperature, a time, a product property such as water activity (a_w), or a chemical property such as available chlorine, salt concentration or pH. Critical limits are specific values and do not include ranges. In processed meat products, these values are related to process values that are used to ensure the production of a microbiologically safe product. A specific example of this has already been given for the production of pepperoni, which is safe with respect to contamination from *E. coli* O157:H7. For this product, the heating step required to achieve a 5 $\log_{10}$ reduction is shown in Fig. 9.2. This shows that when pepperoni, inoculated with *E. coli* O157:H7, was heated for 50 minutes to a temperature of 61 °C and held at this value for 18 minutes, a 5 $\log_{10}$ or 5D reduction in pathogen counts was obtained.[9] For this product, heating at 61 °C for 18 minutes was the critical limit to ensure product safety and this process did not have an adverse effect on product quality. It was also noted that the added cost of this step would not be an obstacle, since it could be incorporated as part of the drying cycle of the product.

For fresh meat, the setting of critical limits is more difficult. The critical limits used will generally be derived from measures taken to reduce contamination such as the use of heat, chilling or organic acids. For these interventions, the critical limits will be temperature or pH values. Visual appearance of carcasses may also be used as a critical limit, in relation to faecal contamination, which should be absent on carcasses after processing. Some examples of critical limits for beef carcasses are shown in Table 9.3.

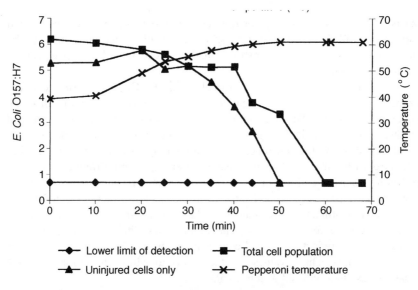

Fig. 9.2 Survivor curves for *E. coli* O157:H7 ($\log_{10}$cfu/g) in pepperoni heated to 61 °C post fermentation showing a 5 $\log_{10}$ reduction in counts.

9.4 Setting up monitoring systems

9.4.1 Function of monitoring

Monitoring is an essential process in the production of safe food, since it demonstrates whether a CCP is under control. This process shows whether there is a loss of control and whether the critical limits set for the control point have been exceeded. Monitoring is based on the use of observations or measurements to determine whether a CCP is under control. Where possible, monitoring should be continuous, but this is not essential. The data obtained during monitoring may be used in three ways:

1. It can be used to show whether there is a trend towards a loss of control, because the measurements or observations are showing a graded increase towards the critical limit for that process. In such circumstances corrective action can be taken before control is lost.
2. The data shows that there is a loss of control and corrective action needs to be taken to bring the process below the critical limit.
3. The information obtained during monitoring can be used as part of the HACCP verification process.

9.4.2 Monitoring procedures

As previously stated, monitoring may be based on visual observations or measurements.

Table 9.3 Critical limits for CCPs in beef slaughter operations and types of monitoring

Possible CCP for beef slaughter	Critical limits	Type of monitoring	Measurement[a]	References
Steam pasteurisation of beef carcasses at end of line	Steam, 100°C – carcass surfaces at 90–96°C	Continuous temperature	P	10, 11
Hot lactic acid washes after dehiding and/or evisceration	Hot (55°C) lactic acid, 1–2% (pH 2.2) – Mean carcass surface pH 2.8	Continuous/discontinuous temperature and pH	P + C	12, 13
Chilling	Chill temperature (°C) 4.0 Air speed (m/s) 2.0 Relative humidity (%) 94.0 Carcass weight (kg) 140.0 Fat class 3.4L Spacing in chill (cm) 6.0	Continuous temperature, air speed and relative humidity	P P	14, 15, 16

[a] P = physical, C = chemical

Visual observations

An example of using visual observations is the assessment of levels of faecal contamination on pig carcasses.[17] In the plant operating this system, trained factory personnel assessed the levels of faecal contamination on pig carcasses on line and determined the process stage at which contamination occurred. The online carcass monitoring system was implemented in June 1993. Carcass contamination rates were reviewed four times per day and a total daily rate was recorded, from which an average monthly carcass contamination rate was calculated (%) (Fig 9.3).

Initially, total carcass contamination rates decreased from 7.6% to 5.3%. However, two months later the rate had reverted to 6.7%. The online monitoring system identified the evisceration stage as being primarily responsible for this increase and an intensive training programme was implemented. As a result, total contamination rates decreased by approximately 3% after a further two months. Thereafter, the overall trend was downwards, reaching 1.8% by November 1995, some two and a half years after the system was first started. Despite a peak (3%) in February 1996, contamination rates continued to decrease, reaching an all-time low of 1.08% in October 1997.

Microbial data showing the total numbers of aerobic bacteria on carcasses were obtained for the first two and a half years of the project. Over this period, the levels of microbial contamination decreased consistently from an initial count of 4.8 to 2.0 $\log_{10}$cfu/in^2 (Fig. 9.3). Analysis of the data showed a good correlation ($R^2 = 0.88$) between carcass contamination rates and the total plate count, demonstrating the practical benefit of the online carcass monitoring system in improving microbial quality.

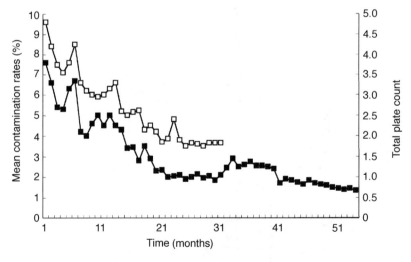

Fig. 9.3 Visual contamination rates (%) (■) and the corresponding total plate counts ($\log_{10}$cfu/square inch) (□) on pig carcasses after slaughter.

These data clearly demonstrate the effectiveness of online visible monitoring in achieving critical limits and helping to obtain significant reductions in contamination levels. While visible monitoring has proven successful for pigs, Biss and Hathaway[18] have stated that it should be used with caution for sheep carcasses.

Measurements
The most common forms of measurement are chemical, physical or biological. In the processing of fresh meat, chemical methods of monitoring would include the measurement of pH for decontamination washes with organic acids. The monitoring of temperature in the hot water washing of beef carcasses would be a physical measurement, where water is used at or above 80 °C.[19] Where acid decontamination washes are applied, the pH of the solution could be monitored (Table 9.3). Equally well, the surface pH of the carcass could be examined after treatment on a non-continuous basis, since there is a large reduction in the surface pH of beef carcasses after the application of a 2% lactic acid decontamination wash.[13] Monitoring pH on line can be accomplished using an instrument such as a Binar pH K2 pH-meter with an appropriate surface electrode, which is specifically designed for use with meat and has a facility which allows data transfer directly to a computer. Readings can be monitored by PC software to give an audible signal if the critical pH in the solution or on the carcass is exceeded.

There seems to be general agreement that carcass chilling is a CCP in generic HACCP plans for beef slaughter.[20–22] The reason for using chilling as a CCP for fresh meat processing is that it may prevent or reduce bacterial growth on carcasses. This may or may not be true, depending on the chilling conditions used, and to date there are no precise chilling data available which accurately define the conditions necessary to achieve this objective.

While chilling may prevent bacterial growth, it is unlikely to cause cell death. Prevention of growth will most likely occur through the inability of cells to grow at low temperature and a_w values. The a_w of the meat surface is controlled by the chilling conditions.[23] In the initial stages of chilling, the lower the temperature and the higher the air velocity, the higher the evaporative water losses from the meat surface. This loss of surface water leads to a decrease in the a_w of the meat. When the meat surface water content falls below 85%, the a_w is about 0.95, at which level bacterial growth ceases, because cells on the meat surface are deprived of sufficient moisture for growth. Later during chilling when the carcass surface and air temperature are in equilibrium, control of the surface a_w is dependent on the relative humidity of the chill and the air speed. Under properly defined conditions the carcass surface moisture continues to be kept under control and growth is prevented. Under these conditions of low temperature and a_w, cells are still viable but injured and stressed. Where chilling can have a lethal effect on bacteria is with cells that have been subjected to a previous stress. According to Presser et al.,[24] lactic acid and water activity synergistically affect the growth of *E. coli*. In experiments where beef carcasses were pasteurised with steam it was observed that the coliform and *E. coli* counts were reduced by about 2.0 $\log_{10}$cfu/cm^2, while the subsequent chilling reduced these counts by a further 1.0 $\log_{10}$cfu/cm^2. This may be an example of a synergistic effect between heating and cooling.[25]

Monitoring of the refrigeration process can be controlled by continuous recording of ambient and carcass temperatures, relative humidity and air speed, all vital indicators in the control of carcass chilling. Critical limits for these are shown in Table 9.3. These limits will differ depending on carcass fatness levels and weight, making proper segregation into chills an important procedure in achieving uniform chilling results. The data required to select carcasses of similar weight and fat classes is available online in all EU abattoirs from the carcass classification that takes place in all export slaughter facilities. As well as this, proper spacing of carcasses is essential to allow adequate air circulation, and for this purpose a gap of 6 cm has been recommended.[26] According to EU regulations, carcasses must be chilled to 7 °C in the deep round before transportation.[27] Using the data in Table 9.3, if carcasses of 140 kg and a fat level of 4 are chilled at an ambient temperature of 4 °C, an air speed of 2 m/s and a relative humidity of 94%, the time for the deep round to reach 7 °C will be 36 h, while the surface will be at this temperature in 9 h.[14–16]

9.4.3 Corrective actions

When CCPs exceed their critical limit, as indicated during monitoring, corrective action must be taken to bring them back into control. Corrective action must be planned in advance, in order to re-establish control as quickly as possible. Such planned corrective actions may not always be sufficient to control deviations in critical limits. Where this occurs, the product produced during the period of the deviation must be held and tested, and a review of the acceptability of the product for human consumption must be undertaken. This is essential to ensure that potentially contaminated product is not made available for human consumption. A reassessment of the HACCP plan is required to determine whether any modification is needed.

When a deviation occurs, positive action must be taken and where available, a suitable substitute process can be implemented. In relation to raw meat processing, if any of the CCPs identified in Table 9.3 exceed their critical limits, specific corrective actions must be taken. An example of this can be seen with the corrective action recommended if carcass chilling fails (Table 9.4).

In the Australian generic HACCP system[21] corrective actions are decided on the basis of a decision tree. This is a systematic approach to the problem of a failure during monitoring and is designed to cover factors such as the difficulty of the corrective action required and the amounts of product involved (Fig. 9.4). The severity of the failure is of great importance and emphasises the need to ensure that failure to control the relevant CCP must be avoided in the future, i.e. lessons have to be learned from all failures of non-compliance on critical limits.[21]

Table 9.4 Recommended corrective actions for carcass chilling in generic HACCP plans

Country		Reference
USA	Hold product and evaluate significance of deviation. Determine conditions of production and action required, i.e. reprocessing, cooking, or condemnation. Notify person in charge of monitoring. Identify cause and prevent recurrence. Notify maintenance to adjust refrigeration to comply with set critical limits. Adjust carcass spacing and retrain employees, if necessary.	20
UK	Reduce temperature of chill. If needed, remove carcasses to correctly functioning store. Investigate the cause of the high temperature and correct. Move carcasses to obtain correct spacing and retrain and/or discipline relevant staff, if necessary.	22

9.5 Verification of HACCP systems

9.5.1 Function of verification

According to the Codex Alimentarius, verification is the application of methods, procedures, tests and other evaluations, in addition to monitoring, to determine compliance with the HACCP plan.[28] As already stated, all the data generated during monitoring can be used in verification, including the direct data obtained from measurements taken and information concerned within corrective actions. Data from equipment certification, such as pH meters and thermometers, and data from the analysis of CCPs to verify that they are under control, are also relevant to verification.

In terms of fresh meat processing the deviations from critical limits need to be examined, in particular the product produced at the time of these deviations and how this was subsequently dealt with, i.e. cooked, sanitised, re-chilled or decontaminated, if appropriate.

During verification the complete working of the slaughter operations should be inspected with a view to verifying the HACCP plan to take account of any major or minor changes in the process that have been made since the initial writing of the plan. These would include changes in equipment, such as the hide puller in beef or lamb operations or in the chilling of carcasses as a result of the installation of new refrigeration plant. The influence of such changes on carcass contamination will be referred to later.

9.5.2 Verification systems

A further part of verification is to randomly check if the plant is producing product of an acceptable microbiological standard. In relation to the slaughter process, verification can be carried out by examining carcasses at the end of the

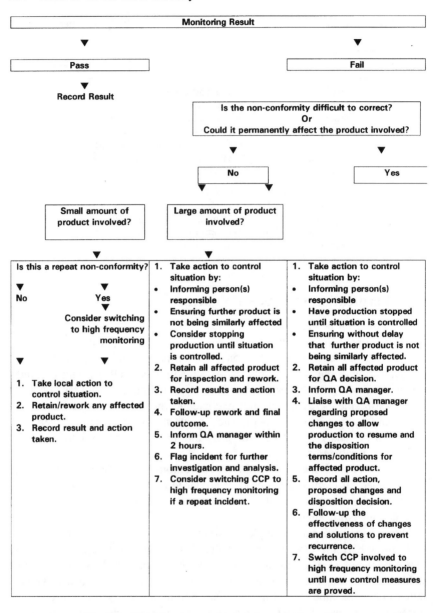

Fig. 9.4 Decision tree to determine corrective actions.

slaughter line or after chilling. A number of methods have been proposed in relation to verification based on indicator organisms which are used to show if the HACCP system is working satisfactorily. In general, the methods used rely on aerobic counts as indicators or generic *E. coli*. Aerobic counts have been suggested by Mackey and Roberts[26] as being the most appropriate organisms for this task. The objective in using these indicator systems is to identify target contamination levels, such that an advisory scale of plant performance can be identified. These advisory scales are not intended to be absolute in terms of acceptance or rejection of product but as indicators of control within the HACCP scheme and to act as general indicators of beef carcass hygiene.[29,30]

A similar approach has been adopted by the Ministry of Agriculture, Fisheries and Food (MAFF) in the United Kingdom. They have introduced a carcass evaluation scheme, the Hygiene Assessment System (HAS), which is based on a hygiene audit that assesses performance in five main categories: (1) ante-mortem, (2) slaughter and dressing, (3) personnel and practices, (4) maintenance and hygiene of premises and (5) general conditions and management. Each category is scored visually according to the standards observed and the scores are weighted according to the importance of each category in relation to carcass hygiene. An example of part of the assessment during slaughter and dressing is evisceration. The parameters used to assess this activity are as follows:

> Heads allowed to touch the floor or workstands at the removal point. Inadequate sealing of the oesophagus and anus allowing gut contents to escape. Opening of guts within the slaughter hall. Offals allowed to touch the floor and stands at removal point. Scabbard used for contaminated and sterilised knives. Pluck allowed to touch the floor or workstands at removal point. Gall bladder opened in slaughter hall to harvest bile.

The inspection of evisceration is based on the above, and depending on how well the process is carried out the risk of contamination is scored as a minimum risk 'a' to maximum 'd'. A number is then assigned to each of these as follows for evisceration:

$$a = 24 \quad b = 16 \quad c = 8 \quad d = 0$$

The score for each category of the slaughter and dressing process is multiplied by a weighted number, in the case of evisceration 0.37, and a total for the entire process, including all categories, is calculated. The weightings are highest for the categories that potentially contribute most to carcass hygiene. The higher the score the better the hygiene level of the plant.[31]

Recently this system has been examined to determine its relevance to microbial contamination of beef carcasses.[32] This showed a significant correlation ($R = 0.84$) between the HAS score and the total viable count, assessed at five carcass sites, and the unweighted scores for the most relevant categories of (1) slaughter and dressing and (2) personnel and practices. This indicated that the HAS score was a useful and simple method of predicting the capability of abattoirs to produce clean carcasses and of good or poor production practices. Based on this data the authors proposed an advisory graded scale of performance similar to that suggested by Mackey and Roberts[26] (Table 9.5). They stated that such a scale could be used to evaluate the efficiency of a HACCP system. The advisory scale shows a relationship between bacterial

Table 9.5 Graded advisory scale for use in verification of beef HACCP

Mean total viable count ($\log_{10}$cfu/cm^2) on carcasses	Advisory scale
< 2.0	Excellent
2.0–2.9	Good
3.0–3.4	Fair
3.4–4.5	Poor
< 4.5	Bad

numbers on carcasses and categories of hygiene which could be used to verify HACCP.

Although HAS is not used by MAFF in relation to the verification of HACCP, it does have considerable potential in this regard. It is a continuous assessment of the hygienic status of the slaughter process and the data obtained can be broadly related to microbial contamination. It has the added advantage of being a visual system and for the majority of slaughter facilities in Ireland or the UK would require the assessment of only 2–3 carcasses per day.[31]

In terms of HACCP verification systems, the most noteworthy is that presently being implemented by the Food Safety Inspection Service (FSIS) in the United States. All US meat processors are obliged to have a HACCP system in place and the verification procedure is based on the presence of generic *E. coli* on carcasses. These minimum performance criteria are directly related to determining whether plants are preventing or reducing the incidence of faecal contamination on carcasses. The criteria to be met have been determined from work carried out previously in beef plants throughout the country.[33] This study established the baseline for a range of organisms on beef carcasses, including *E. coli* and *Salmonella*. The basis of this approach is that all beef plants in the United States are subject to the same microbiological criteria. Furthermore the use of generic *E. coli* as an indicator organism is considered preferable to the use of aerobic counts, since it is directly related to faecal contamination. As such it is a good indicator of safety and its use in this way has been endorsed by others,[34,35] although some reservations have been expressed about the use of *E. coli* as an indicator organism on carcasses.[26,29] This is because they are usually present in low numbers making statistical treatment of results difficult.

The FSIS plan recognises three classes of results: acceptable, marginal and unacceptable. In this regard it has certain similarities to the graded advisory scale for aerobic counts mentioned previously. The values for m and M have been determined from the baseline study already referred to. The m and M values have been established from the 80th and 98th percentile values, rounded up to the nearest value of 10, i.e. 10, 100 or 1000. This means that 80% of the carcasses tested were at or below the minimum detectable level (5 cfu/cm^2) and are considered as negative, and 98% were at or below 100 cfu/cm^2.

FSIS have determined statistically that the number, n, of carcasses required to operate the performance criteria is 13, which constitutes a test window. If a plant is operating at the acceptable performance level, m, with an 80% probability, then it will have three or fewer results above m. This means that a plant may have three marginal results in 13 carcasses tested. It is necessary to have a number, c, of allowable marginal results to provide for variations in performance, hence $c = 3$. Equally however, if the number exceeds this level it raises serious concerns about slaughter performance and requires an investigation of the plant's HACCP plan. Finally in relation to this verification system there is the frequency of sampling required. This is largely dependent on volume or throughput of carcasses. For beef, the number of samples required is one in 300 carcasses. In high volume plants, 1000/day plus, a test window can be completed each day. With a lower volume the number for a test window would have to be accumulated over time, probably several days.

The use of a three-class attribute sampling plan for verification or validation purposes may not represent the most effective use of the data obtained from carcasses to determine the effectiveness of HACCP.[28,36] With an attributes plan for beef carcasses, bacterial counts are used to determine the group in which the sample will be placed, i.e. the sample will be assigned to one of two or three groups, depending on whether a two- or three-attributes sampling plan is used. This means that the data from carcasses is used only to assign them to groups that are either acceptable or unacceptable. The data obtained is not used in any way to determine the statistical variability between carcasses, and no measure of the overall performance of the HACCP plan or the effectiveness of a CCP in controlling contamination can be made.

An alternative to attribute sampling for HACCP plans is variables sampling. This involves the statistical use of the data from samples to determine whether the controls being introduced at CCPs are effective in reducing bacterial counts on carcass surfaces. This approach offers the possibility of quantifying the effectiveness of the controls and therefore of determining the most successful interventions to be used at different CCPs. Clearly numerical data can be used in a much more objective way to determine the safety of carcasses being produced.[36]

As well as the assessment of plant performance using *E. coli*, FSIS contend that HACCP should also be related to pathogen reduction on carcasses. The pathogen to be controlled is *Salmonella* and the standard for this is determined using the same criteria as for *E. coli* but using a two-class sampling plan. These tests are carried out by FSIS inspectors, in addition to the *E. coli* tests carried out by the factory. As with *E. coli,* the performance standards for *Salmonella* are derived from the beef baseline study and a set of performance standards have been set. These show that in a test sample of $n = 82$ steer and heifer carcasses no more than one may be positive. In a test sample of 58 cow and bull carcasses no more than two may be positive. This difference in contamination levels reflects the greater occurrence of *Salmonella* on older animals.[20] It should be noted that while *E. coli* is used as a plant performance indicator in these tests, *Salmonella* is a safety standard that the plant must aspire to.

Apart from the inappropriate use of attribute sampling in HACCP verification, the use of *Salmonella* as an indicator to verify the effectiveness of HACCP in terms of pathogen reduction, would seem to be untenable. This is because the presence of this and other pathogens, such as *E. coli* O157:H7, is low and sporadic on beef carcasses.[14,37] The sporadic contamination of carcasses is reflected in the seasonal appearance of this pathogen on carcasses. These factors make it impossible to use *Salmonella* statistically as an indicator of carcass safety.[26,38]

Given the shortcomings of using *Salmonella* as an indicator of pathogen reduction, the American Meat Science Association[38] have recommended the use of *E. coli* and total viable counts as indicators of process control. The higher numbers of these organisms on beef carcasses make them more suitable for use in validation and verification procedures.

9.5.3 Use of rapid methods in verification

Total viable counts
As indicated above, total viable counts (TVCs) may be used in verifying HACCP. The microbial testing required during the HACCP verification process may be time consuming and costly. In recent years a variety of methods for the rapid isolation and determination of TVCs and pathogenic bacteria from meat and meat products have been developed. Among these the direct epifluorescent filter technique (DEFT) is well documented.[39] Food samples are digested using specific enzymes for foods and the cells in the foods are captured on a membrane after filtration, where they are stained with acridine orange. Fluorescent cells are counted using an epifluorescent microscope. This technique has been adapted for meat by the use of specific enzyme and detergents.[40,41] The adapted system, the acridine orange direct count (AODC), has been successfully used to predict the TVCs on sheep carcasses.[42] The AODC method is capable of accurately predicting the standard plate count from carcass swabs as indicated by a correlation value (R^2) of 0.87 (Fig. 9.5).

The development of rapid methods for processed meats is a difficult problem. This is because live and dead cells are present in these products as a result of the use of heat, cold or salt during manufacture. Recently stains have been developed that are capable of accurately distinguishing between live and dead bacterial cells, which acridine orange cannot do. One of these, Baclight, has been used to develop a rapid method for processed meats, such as cooked ham, bacon rashers and frozen beefburgers. The data in Table 9.6 shows the relationship between membrane microscopic counts, using either Baclight or acridine orange, and the total plate counts. The R^2 values for the Baclight stain clearly indicate the improved relationship to the TVCs, compared to acridine orange.[43] The Baclight method could be used to verify the microbiological integrity of processed meat products or cleaned factory surfaces containing injured cells. These problems also arise in relation to the disinfection of plant surfaces where cells may be alive or dead.

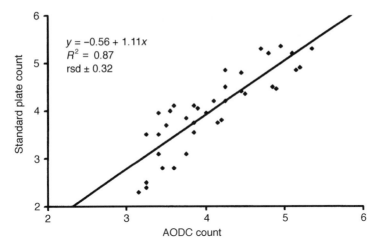

Fig. 9.5 Relationship between standard plate counts ($\log_{10}$cfu/cm^2) and acridine orange direct counts (AODC) ($\log_{10}$cfu/cm^2) on lamb carcasses.

Cells that are in an active metabolic state contain adenosine triphosphate (ATP). The levels of ATP in cells can be measured using the luciferase enzyme complex, which in the presence of ATP produces light. The amount of light produced is proportional to the concentrations of ATP present and this can be related to the numbers of cells present. Such a technique has been used to measure microbial contamination levels on beef and pork carcasses.[44] This

Table 9.6 Relationship between the acridine orange and Baclight direct count techniques and the standard plate count for fresh and processed meat samples

Product	Intercept	Slope	R^2 (correlation coefficient)
Minced beef	($n = 40$)		
AODC[a]	0.84	0.33	0.90
BLDC[b]	1.13	0.67	0.93
Cooked ham	($n = 28$)		
AODC	2.11	0.73	0.62
BLDC	0.39	1.03	0.87
Bacon rashers	($n = 36$)		
AODC	0.55	0.89	0.70
BLDC	0.32	0.92	0.92
Frozen burgers	($n = 20$)		
AODC	0.46	1.13	0.72
BLDC	0.26	0.92	0.88

[a] AODC, acridine orange direct count ($\log_{10}$cfu/g)
[b] BLDC, Baclight direct count ($\log_{10}$cfu/g)

technique separates microbial from somatic ATP and then quantifies the microbial ATP only (mATP). This separation and quantification is carried out using the same device. Using this system R^2 values of 0.85 and 0.86 were obtained between relative light units (RLUs) and the total plate counts for beef and pork, respectively.

While both the techniques described above could be used in the rapid determination of total counts on carcasses, there are major differences between them. The AODC takes about 15 minutes to complete a test, while the mATP requires only 5 minutes. The mATP test is fully automated, while the AODC is a microscopic system in which cells must be manually counted. An automatic version of the AODC does exist, the Cobra, in which counting is automatically carried out.[45] A simple AODC count is much cheaper than the equivalent mATP test. While both these systems have potential in respect to verification of microbial counts on carcasses, since they have short application times, the mATP with a test time of only 5 minutes may also be suitable for monitoring critical limits at CCPs.

Pathogens
Rapid methods for use in beef HACCP plans would mainly be concerned with tests for the presence of *Salmonella* or *Listeria*. As outlined above, *Salmonella* is used to verify that the HACCP plan is working and effectively controlling the presence of this organism on beef carcasses. As already stated, one of the prerequisite programmes that are required in conjunction with fresh meat HACCP plans is the introduction of a written plan to address control of sanitation operations. In the FSIS programme this is the Sanitation Standard Operating Procedures (SSOPs), which describe the procedures to be used to give effective sanitation in the plant environment both during and after production.[20]

The use of *Listeria* spp. as an indicator of environmental sanitation is now well established, and can be used to verify the efficiency of cleaning in meat plants. *Listeria* has been found on beef and sheep carcasses in Ireland and is a frequent contaminant on fresh beef and a number of meat products.[42,46–48]

Rapid methods for *Salmonella* and *Listeria* are readily available, for example the Lister Test for *Listeria* or immunomagnetic separation combined with electrochemistry,[49] along with many other tests which also exist for these pathogens.[50] Recently a simple new method has been developed using surface adhesion of cells to polycarbonate membranes, in conjunction with immuno-fluorescent microscopy, the SAIF technique.

Using this technique the target cells are absorbed from a meat enrichment broth onto the surface of a polycarbonate membrane. The *Listeria* or *Salmonella* cells isolated on the membrane are visualised using a fluorescent-labelled antibody specific from the pathogen to be identified (Fig. 9.6).[51] With this system the minimum level of detection for *Listeria* is about 3.11 $\log_{10}$cfu/ml and for *Salmonella* it is 3.5 $\log_{10}$cfu/ml. The test is capable of detecting one injured cell in a 25 g sample after overnight incubation for 18–20 h. In experiments with

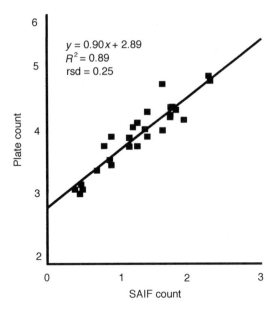

Fig. 9.6 Relationship between *Listeria* plate counts ($\log_{10}$cfu/ml) and surface adhesion immunofluorescent (SAIF) counts ($\log_{10}$cfu/mm^2) on minced beef samples.

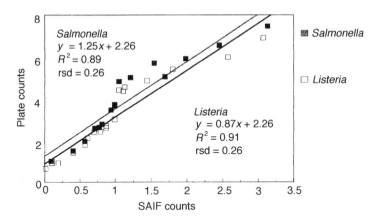

Fig. 9.7 Relationship between *Salmonella* and *Listeria* counts ($\log_{10}$cfu/ml^{-1}) and surface adhesion immunofluoresent (SAIF) counts ($\log_{10}$cfu/mm^2).

minced beef it was shown that both these pathogens could be isolated and detected using a simultaneous enrichment step (Fig. 9.7).[52] In further experiments it was shown that detection of *Listeria* cells on the membrane could be carried out using the polymerase chain reaction (PCR).[53]

9.6 Validation of the HACCP plan

9.6.1 Function of validation

Validation is part of the verification process that is concerned with evaluating scientific data and information to determine whether the HACCP plan, when properly implemented, can control the identified hazards at different CCPs. This means that validation is an overall evaluation of HACCP to determine whether it really works. In particular it is intended initially to provide scientific data that the critical limits identified at CCPs are capable of eliminating or reducing hazards that result in the production of food that is safe for human consumption.[1] It also involves a reassessment of the HACCP plan, usually on a yearly basis, to determine whether it is continuing to function properly.

9.6.2 Validation procedures

The information necessary to validate a HACCP plan initially can be obtained from two main sources: (1) scientific data on critical limits relevant to a CCP; and (2) practical in-plant data showing that the control being introduced is capable of achieving the desired control in terms of reductions in bacterial counts or growth prevention.

1. If a process, such as steam vacuuming, is to be introduced into a plant, all relevant scientific literature, experimental results or other data may be used to support its introduction into the HACCP plan. The parameters to be used with the system, such as application temperature, time and vacuum pressure, may be obtained in this way and in particular its effectiveness in controlling pathogens such as *E. coli* O157:H7 or *Salmonella* on carcasses.

2. While the technology to be introduced may be well documented in the scientific literature, the plant must be able to demonstrate its capability to use it effectively. An example of this is the introduction of a new technology, such as steam vacuuming for use in beef carcass decontamination. This would require a series of tests to show that it can remove faecal contamination from the surfaces of carcasses and that the reductions in some indicator organisms, such as *E. coli* or the Enterobacteriaceae are comparable to those observed in the scientific literature (Table 9.7). This shows a clear demonstration that the equipment is being used as intended and is achieving the desired reduction in the levels of contamination.[54] While the initial validation should include studies with pathogens inoculated on carcasses, in reality this is not feasible. It is not desirable to introduce pathogens into the slaughter environment for any reason so such studies, if undertaken, have to use indicator organisms.

Table 9.7 Mean total viable count (TVC), *E. coli* and Enterobacteriaceae numbers ($\log_{10}$cfu/cm^2), before and after Vac-San treatment on beef carcasses inoculated with 0.5 g of fresh bovine faeces (standard deviations shown in parantheses)

	TVC	E. coli	Enterobacteriaceae
Before treatment	5.18 (0.42)	4.22 (1.22)	4.32 (1.26)
After treatment	1.26 (1.13)	0.12 (0.49)	0.23 (0.56)

9.6.3 Reassessment on revalidation

An important part of validation is process reassessment, whereby the HACCP plan is reviewed. A review should take place at least every year and the plan should be modified if necessary. This is particularly important if any part of the slaughter process has been changed. Such changes include the sourcing of animals, the installation of new equipment, changes in the numbers of animals slaughtered each day or changes in personnel. For example, changes in line speed may require more than one operative at a CCP and a change in personnel will demand retraining.

9.7 Identifying problem areas

9.7.1 Establishing critical control points

Despite the fact that there are decision trees for the establishment of critical control points there is still confusion in this area. Reference to a number of HACCP documents demonstrates this confusion in the determination of CCPs (Table 9.8). The data show that with four processing steps common to many beef slaughter plants the criteria used to establish CCPs may differ, depending on the country. The matter is further complicated where specific interventions, i.e. decontamination procedures, are allowed, as in the United States.

Table 9.8 Critical control points in generic HACCP plans from different countries

Beef slaughter process steps	United Kingdom[22]	Australia[21]	United States[20]
Hide removal	CCP	Not CCP	Not CCP
Evisceration	CCP	Not CCP[a]	Not CCP[b]
Carcass washing	Not CCP	Not considered	Not CCP

[a] Not a CCP but may be in some HACCP plans.
[b] May be a CCP if microbiological interventions to reduce contamination are not in place or are in place only at a later point in the process

9.7.2 Online monitoring and the use of microbiological data

According to Lee and Hathaway[7] there is limited availability of online monitoring parameters to detect microbial contamination. As previously discussed, one of the few microbiological systems currently available that has potential in this regard is the use of microbial adenosine triphosphate (mATP).[44] The advantage of this system is that a test of the microbial status of the carcass can be carried out in about 5 minutes. Such a test would be successful, since it is sufficiently rapid to allow corrective action to be taken should a critical limit be exceeded. It could also be used to demonstrate that control had been re-established. Measurement of chill temperature is another area where continuous monitoring is carried out and the data stored. In this case both the temperature of selected carcasses and the ambient chill temperatures can be recorded.[26]

9.7.3 Implementing corrective actions for slaughter lines

Corrective actions at critical control points should be identified as part of the HACCP plan. This involves deciding the fate of carcasses produced during the period of the deviation at the CCP. This is a problem, since it demands identifying the corrective action in terms of the product involved at the time of the observed deviation. If carcasses are not sufficiently chilled as a result of the malfunction of the refrigeration plant, are the carcasses simply chilled to the correct temperature, condemned or decontaminated? If the latter options are to be used, how would such decisions be made, since contamination levels cannot be measured in sufficient time? In such circumstances, the use of the mATP or other rapid tests is invaluable and allows objective assessment of carcass disposal to be made.

9.8 Feedback and improvement

Since a HACCP plan is intended to be constantly evolving and changing, the use of records in this process is of vital importance. One of the main requirements of any HACCP plan is accurate recordkeeping. This applies to monitoring, corrective actions, validation and certification. All of these steps require accurate records which can be used to assess the efficient working of the system but can also be used to determine important decisions such as frequency of monitoring or verification requirements. These have important cost implications for the plant in operating the HACCP system. These records can also be used to identify changes that could improve the HACCP plan.

9.9 Future trends

Choosing CCPs is a difficult procedure in fresh meat slaughter. While CCPs are usually recognised on the basis that controls can be introduced that reduce contamination, potentially the most important CCP, the hide is seldom referred to. There is little doubt that the hide of cattle and the fleece of sheep are the principal sources of carcass contamination. Despite this there are few measures to decontaminate the hide. Recently it has become mandatory to bring clean cattle to the abattoir in Ireland, a procedure that has been in operation in other countries for some time.[55,56] While this policy can have beneficial effects in terms of reducing contamination on the carcass, the practical reductions, though significant, are small.[2,56] For the future there is an urgent need to develop effective decontamination systems for the hide of cattle and the fleece of sheep. A successful decontamination step at this point would have a potentially large impact on the safety of the meat produced.

As HACCP becomes more widely used in the meat industry the need for rapid microbiological methods will increase. These will have application only if they are rapid, i.e. if a result is available in 1–3 minutes. This is particularly important for monitoring where the necessity for corrective actions is required in the event of a loss of control at a control point. Ideally, the tests available should be relevant to safety and should therefore be capable of measuring some faecal indicator, e.g. the Enterobacteriaceae or *E. coli*. Because these organisms generally occur only at low levels, the rapid tests would have to have a very low sensitivity level, since any form of amplification of the target cells is not possible in the time available for the test. Fulfilling these criteria of speed and sensitivity will be very difficult.

The influence of such factors as line speed and changes in slaughter technology needs to be considered in relation to any fresh meat HACCP plan. Increases in line speed are generally considered to contribute to carcass contamination, yet the limited information available on this subject is confusing.[57] Of particular significance is the influence of changes in slaughter

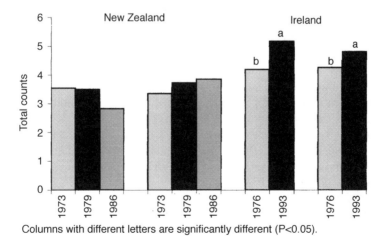

Fig. 9.8 Influence of changes in slaughter technology on total counts ($\log_{10}$cfu/cm^2) on lamb carcasses processed in different countries where plants had been upgraded.

technology. While technology changes are designed to increase productivity, the relationship to safety is less certain. An example of this is shown in Fig. 9.8 where changes in lamb slaughter in New Zealand and Ireland are compared. The data show that in some cases hygiene improvements did occur with technology changes but in others deterioration was evident. The reasons for these hygiene changes are not known but emphasise the need to establish the effect of the introduction of new technology. This is, of course, a stated aim of validation in HACCP plans outlined above and shows the necessity for such actions in the future.

9.10 References

1. SPERBER W H, Auditing and verification of food safety and HACCP, *Fd Cont*, 1998 **9** 157–62.
2. McEVOY J M, DOHERTY A M, FINNERTY M, SHERIDAN J J, McGUIRE L, BLAIR I S, McDOWELL D A and HARRINGTON D, The relationship between hide cleanliness and bacterial numbers on beef carcasses at a commercial abattoir, *Lett Appl Microbiol*, 2000 **30** 390–5.
3. ANON, International Commission on Microbiological Specifications for Foods, *Microorganisms in Foods 4. Application of hazard analysis and critical control point (HACCP) system to ensure microbiological safety and quality.* Boston, Blackwell Scientific Publications, 1989.
4. BUCHANAN R L, The role of microbiological criteria and risk assessment in HACCP, *Fd Microbiol*, 1995 **12** 421–4.
5. RIORDAN D C R, DUFFY G, SHERIDAN J J, EBLEN B S, WHITING R C, BLAIR I S and McDOWELL D A, Survival of *Escherichia coli* O157:H7 during the manufacture of pepperoni. *J Fd Prot*, 1998 **61**(2) 146–51.
6. REED C A, Approaches for ensuring safety of dry and semi-dry fermented sausage products. 1995 letter to plant managers, US Department of Agriculture and Food Safety Inspection Service, Washington, DC.
7. LEE J A and HATHAWAY S C, The challenge of designing valid HACCP plans for raw food commodities, *Food Control*, 1998 **9** 111–17.
8. GILL C O, BRYANT J and BEDARD D, The effects of hot water pasteurising treatments on the appearances and microbiological conditions of beef carcass sides, *Fd Microbiol*, 1999 **16** 281–9.
9. RIORDAN D C R, DUFFY G, SHERIDAN J J, WHITING R C, BLAIR I S and McDOWELL D A, The effect of acid adaptation, product pH and heating on the survival of *Escherichia coli* O157:H7 in pepperoni, *Appl Environ Microbiol*, 2000 **66** 1726–9.
10. PHEBUS R K, NUTSCH A L, SCHAFER D E, WILSON R C, RIEMANN M J, LEISING J D, KASTNER C L, WOLF J R and PRASAI R K, Comparison of steam pasteurization and other methods for reduction of pathogens on surfaces of freshly slaughtered beef, *J Fd Prot*, 1997 **60** 476–84.
11. NUTSCH A L, PHEBUS R K, RIEMANN M J, SCHAFER D E, BOYER Jr J E, WILSON

R C, LEISING J D and KASTNER C L, Evaluation of a steam pasteurization process in a commercial beef processing facility, *J Fd Prot*, 1997 **60** 485–92.

12. PRASAI R K, ACUFF G R, LUCIA L M, HALE D S, SAVELL J W and MORGAN J B, Microbiological effects of acid decontamination of beef carcasses at various locations in processing, *J Fd Prot*, 1991 **54** 868–72.
13. HARDIN M D, ACUFF G R, LUCIA L M, OMAN J S and SAVELL J W, Comparison of methods for decontamination from beef carcass surfaces, *J Fd Prot*, 1995 **58** 368–74.
14. SHERIDAN J J and LYNCH B, Effect of microbial contamination on the storage of beef carcasses in an Irish meat factory, *Ir J Fd Sci Technol*, 1979 **3** 43–52.
15. SHERIDAN J J and SHERINGTON J, The relationship of bloom to washing, bacterial numbers and animal type (cows, heifers, steers) in beef carcasses, *30th European Meeting of Meat Research Workers, Bristol*, 1984, 83–4.
16. BAILEY C and COX R P, The chilling of beef carcasses, *Inst Refrig Proc*, 1976 1–12.
17. BOLTON D J, OSER, A H, COCOMA, G J, PALUMBO, S A and MILLER A J, Integrating HACCP and TQM reduces pork carcass contamination, *Fd Technol*, 1999 **53** 40–3.
18. BISS M E and HATHAWAY S C, Microbiological and visible contamination of lamb carcasses according to pre-slaughter presentation status: Implications for HACCP, *J Fd Prot*, 1995 **8** 776–83.
19. DORSA W J, CUTTER C N and SIRAGUSA G R, Effectiveness of a steam-vacuum sanitiser for reducing *Escherichia coli* O157:H7 inoculated to beef carcass surface tissue, *Lett Appl Microbiol*, 1996 **23** 61–3.
20. ANON, Federal Register, Part 11. *Department of Agriculture, Food Safety and Inspection Service*, Pathogen reduction: hazard analysis and critical control point (HACCP) systems: final rule, 38806–38989, 1996.
21. ANON, *Meat Safety Quality Assurance System*. Canberra, Meat Inspection Division, Australian Quarantine and Inspection Service, 1998.
22. ANON, *HACCP Systems in Abattoirs and Meat Cutting Plants: Guide to Implementation*, Meat and Livestock Commission, Milton Keynes, UK, 1–81, 1999.
23. ROSSET R, in *Chilling, Freezing and Thawing*, ed BROWN M H, London, Applied Science Publishers, 1982.
24. PRESSER K A, ROSS T and RATHKOWSKY D A, Modeling the growth limits (growth/no growth interface) of *Escherichia coli* as a function of temperature, pH, lactic acid concentration, and water activity, *Appl Environ Microbiol*, 1998 **64** 1770–3.
25. GILL C O and BRYANT J, Decontamination of carcasses by vacuum-hot water cleaning and steam pasteurising during routine operations at a beef packing plant, *Meat Sci*, 1997 **47** 267–76.
26. MACKEY B M and ROBERTS T A, Improving slaughter hygiene using HACCP and monitoring, *Fleischwirtsch*, 1993 **73** 58–61.

27. ANON, Council Directive 64/433/EEC on health problems affecting intercommunity trade in fresh meat to extend it to the production and marketing of fresh meat, *Official Journal of the European Communities*, 1991, No. L, 268169.
28. ANON, *Principles for the establishment of microbiological criteria for foods*. Codex Alimentarius Commission, Joint FAO/WHO Food Standards Programme, Codex Committee on Food Hygiene, Supplement to Volume 1B-1997, CAC/GL 21.
29. JERICHO K W F, KOZUB G C, GANNON V P J, GOLSTEYN THOMAS E J, KING R K, BIGHAM R L, TANAKA E E, DIXON MACDOUGHALL J M, NISHIYAMI B J, KIRBYSON H and BRADLEY J A, Verification of the level of microbiological control for the slaughter and cooling processes of beef carcass production at a high-line speed abattoir, *J Fd Prot*, 1997 **60** 1509–14.
30. UNTERMANN F, STEPHAN R, DURA U, HOFER M and HEIMANN P, Reliability and practicability of bacteriological monitoring of beef carcass contamination and their rating within a hygiene quality control programme of abattoirs, *Int J of Fd Microbiol*, 1997 **34** 67–77.
31. SIMMONS A, MIDDLETON A and SOUL P, A hygiene assessment system for red meat and poultry meat slaughterhouses, *State Vet. J*, 1995 **5** 11–13.
32. HUDSON W R, MEAD G C and HINTON M H, Relevance of abattoir hygiene assessment to microbial contamination of British beef carcasses, *Vet Rec*, 1996 **139** 587–9.
33. ANON, Nationwide beef microbiological baseline data collection program – steers and heifers, *United States Department of Agriculture, Food Safety and Inspection Service, Science and Technology, Microbiology Division*, 1994, 1–24.
34. GILL C O, McGINNIS J C and BADONI M, Use of total or *Escherichia coli* counts to assess the hygienic characteristics of a beef carcass dressing process, *Int J Fd Microbiol*, 1996 **31** 181–96.
35. JERICHO K W F, KOZUB G C, BRADLEY J A, GANNON V P J, GOLSTEYN-THOMAS E J, GIERUS M, NISHIYAMA B J, KING R K, TANAKA E E, D'SOUZA S and DIXON-MACDOUGHALL J M, Microbiological verification of the control of the processes of dressing, cooling and processing of beef carcasses at a high line-speed abattoir, *Fd Microbiol*, 1996 **13** 291–301.
36. BROWN M H, GILL C O, HOLLINGSWORTH J, NICKELSON R, SEWARD S, SHERIDAN J J, STEVENSON T, SUMNER J L, THENO D M, USBORNE W R and ZINK D, The role of microbiological testing in systems for assuring the safety of beef, *Int J Fd Microbiol* (submitted).
37. McEVOY J M, DOHERTY A M, SHERIDAN J J and McGUIRE L, The incidence of *Escherichia coli* O157:H7 and *Salmonella* in faeces, rumen contents and on carcasses in a commercial Irish beef abattoir. *Handbook of Society for Applied Microbiology, Summer 1999 Conference, University of York*.
38. ANON, The role of microbiological testing in beef food safety programmes: the scientific perspective, *Consensus of the 1999 Symposium, American Meat Science Association*, 1–16.

39. RODRIGUES U M and KROLL R G, The direct epifluorescent filter technique (DEFT): increased selectivity, sensitivity and rapidity. *J Appl Bacteriol*, 1985, **59**, 493–9.
40. SHERIDAN J J, WALLS I and LEVETT P N, Development of a rapid method for enumeration of bacteria in pork mince, *Ir J Fd Sci Technol*, 1990 **14** 1–15.
41. DUFFY G, SHERIDAN J J, McDOWELL D A and HARRINGTON D, The use of alcalase 2.5L in the acridine orange direct count technique (AODC) for the rapid enumeration of bacteria in mince beef, *Lett Appl Microbiol*, 1991 **13** 198–201.
42. SIERRA M L, SHERIDAN J J and McGUIRE L, Microbial quality of lamb carcasses during processing and the acridine orange direct count technique (a modified DEFT) for rapid enumeration of total viable counts, *Int J Fd Microbiol*, 1997 **36** 61–7.
43. DUFFY G and SHERIDAN J J, Viability staining in a direct count rapid method for the determination of total viable counts on processed meats, *J Microbiol Meth*, 1998 **31** 167–74.
44. SIRAGUSA G R, CUTTER C N, DORSA W J and KOOHMARAIE M, Use of a rapid microbial ATP bioluminescence assay to detect contamination on beef and pork carcasses, *J Fd Prot*, 1995 **58** 770–5.
45. SHARPE A N, *Development and Evaluation of Membrane Filtration Techniques in Microbial Analysis. Rapid Analysis Techniques in Food Microbiology*, Glasgow, P. Patel, Chapman and Hall, 1994.
46. KOHN B A, COSTELLO K and BROOKINS PHILLIPS A, HACCP verification procedures made easier by quantitative *Listeria* testing. *Dairy, Fd Environ Sanit*, 1997 **17** 76–80.
47. SHERIDAN J J, DUFFY G, McDOWELL D A, BLAIR I S and HARRINGTON D, The occurrence and initial numbers of *Listeria* in Irish meat and fish products and the recovery of injured cells from frozen products, *Int J Fd Microbiol*, 1994 **22** 105–13.
48. McEVOY J M, DOHERTY A M, SHERIDAN J J and McGUIRE L, The incidence of *Listeria* spp. and *Escherichia coli* O157:H7 on beef carcasses, *44th ICoMST, Hygiene, Spoilage and Safety*, Barcelona, Spain, A43, 346–7, 1998.
49. GEHRING A G, CRAWFORD C G and MAZENKO R S, Enzyme-linked immuno-magnetic electrochemical detection of *Salmonella typhimurium*. *J Immunol Meth*, 1996 **195** 15–25.
50. FORSYTHE S J and HAYES P R, *Food Hygiene, Microbiology and HACCP*, 3rd edn, Gaithersbury, MD, Aspen Publishers, 1998.
51. SHERIDAN J J, DUFFY G, McDOWELL D A and BLAIR I S, Development of a surface adhesion immunofluorescent microscopy technique for the rapid detection of *Listeria monocytogenes* and *Listeria innocua* from raw meat, *J Appl Microbiol*, 1997 **82** 225–33.
52. CLOAK O M, DUFFY G, SHERIDAN J J, BLAIR I S and McDOWELL D A, Isolation and detection of *Listeria* spp., *Salmonella* spp. and *Yersinia* spp. using a simultaneous enrichment step followed by a surface adhesion immuno-

fluorescent technique, *J Microbiol Meth*, 1999 **39** 33–43.
53. DUFFY G, CLOAK O M, SHERIDAN J J, BLAIR I S and McDOWELL D A, The development of a combined surface adhesion and polymerase chain reaction (PCR) technique for the rapid detection of *Listeria monocytogenes* in meat and poultry, *Int J Fd Microbiol*, 1999 **49** 151–9.
54. McEVOY J M, DOHERTY A M and SHERIDAN J J, Decontamination of beef carcass surfaces using steam vacuuming (unpublished data).
55. ANON, *Clean cattle policy*, Department of Agriculture, Food and Forestry, Dublin, Ireland, 1997.
56. RIDELL J and KORKEALA H, Special treatment during slaughtering in Finland of cattle carrying an excessive load of dung: meat hygienic aspects, *Meat Sci*, 1993 **35** 223–8.
57. SHERIDAN J J, Sources of contamination during slaughter and measures for control, *J Food Safety*, 1998 **18** 321–39.

10

Validation and verification of HACCP plans

M. H. Brown, Unilever Research, Sharnbrook

10.1 Introduction

Most companies producing meat and meat products have HACCP plans. Validation and verification are integral parts of managing them and maintaining their effectiveness. The aim of this chapter is to show the different nature of the two activities. Validation checks that the HACCP plan is valid and appropriate for the company and its products. Verification checks that the actions are being carried out, achieving their objectives and being reviewed. At a practical level the chapter explains when and how this can be done and who does it.

In the meat industry of most countries the management of HACCP is shared between the producers and regulators, especially veterinarians (European Community, 1993). Because there are split and shared responsibilities, it is important that limits and responsibilities for the assessment of HACCP plans via validation and verification are established and accepted by all those involved. Currently, HACCP is limited to safety matters and therefore includes the microbiological, chemical and physical hazards likely to be carried into the food chain by meat and meat products. Validation is an essential part of the implementation and review of HACCP plans and verification can be integrated into many factory management systems such as 'continuous improvement' and ISO 9000.

Codex Alimentarius provides principles for validation and verification. Principle 6 of its guide to the management of HACCP (Codex Committee, 1997; Codex Alimentarius Commission, 1997a,b) asks that 'procedures are established for verification, to confirm that the HACCP system is working correctly'. And this in turn relies on Principle 7, which asks food producers to 'establish documentation concerning all procedures and records appropriate to those principles and their application'. Apart from these principles there is very little formal regulatory guidance on the presentation and content of HACCP plans, their degree of detail or the means and extent of implementation. This chapter explains how the principles and requirements for validation and verification can be effectively applied to plans using the Codex approach.

Meat quality and safety depend on the integrated application of effective management and control measures *from farm to fork*. Safety is a shared responsibility of the many individuals and organisations that comprise the supply chain. They establish and provide controls to prevent food safety problems reaching the consumer, and include farmers, breeders, feed manufacturers, livestock-market operators, transporters, processing-plant operators, retailers, regulatory authorities and veterinarians.

Currently, inspection systems providing microbiological food safety and based on sampling and analysis at economically realistic levels are not capable of detecting harmful levels of meat-borne food-poisoning bacteria. End-product sampling and testing cannot provide sufficient information for reliable decision making on safety. Testing is not reliable unless the food animal, or bird, shows clinical or pathological signs or the product is grossly contaminated. Over the years there has been an increase in human disease associated with the consumption of meat, even though pathogens are seldom detected by routine inspection. Hence there is a continuing demand for more effective means of preventing, or at least minimising, the risks of meat acting as a carrier of human pathogens.

The HACCP technique applies scientific knowledge to the identification and control of hazards during food manufacture. It has been advocated as the best means of achieving product safety as it focuses control and monitoring measures at the process steps where they can have the most beneficial effect on food safety, i.e. at the CCPs (Moy *et al.*, 1994; Gardner, 1995; Hathaway, 1995; Jayasuriya, 1995; Whitehead and Orris, 1995; Motarjemi *et al.*, 1996). Because HACCP focuses attention on a few of the many activities in a meat plant, it is essential to check that this focus can provide safe products, and that opportunities to control hazards have not been missed. Confirmation of this is the function of validation and verification. Typically in a meat plant items for attention will include the raw materials, packaging, hygiene procedures and standards, process and storage times and temperatures, prevention of cross-contamination, temperature control in chill rooms, finished product handling and storage. If products are preserved then additional stages relevant to safety and quality can include control of heat or drying processes, addition of preserving agents, prevention of post-process contamination, cooling rates and hygiene of specific areas, such as high care areas (Watson, 1994). If products are canned, then rigorous control of heat processing, primary packaging quality, container sealing, cooling water quality and post-process hygiene should also be included among the CCPs.

Many plants undertaking HACCP studies will have existing QA systems which have ensured the consistent manufacture of safe product over a number of years. HACCP should do no more than reinforce these systems and produce a workable plan based on them. If there are serious differences or omissions, for example unit operations identified by the study as CCPs have not previously been controlled by the QA system, then any differences must be explained and resolved prior to implementing the HACCP plan. An essential part of validation prior to implementation is ensuring that all the contributors to the supply chain (e.g. buyers, production and distribution staff) support the HACCP plan, not only QA. Where this has not been achieved the plan and its implementation must be reviewed with those involved in the argument, so that a technically justifiable solution is found to protect product safety.

Lee and Hathaway (1998) have suggested that validation of HACCP plans be done in conjunction with established food safety objectives (FSOs), that are defined as statements of the maximum level of a microbiological hazard considered acceptable for consumer protection. They may be process linked (such as below 12 °C or above 75 °C) or performance linked (such as the absence in 25g, or maximum numbers of specified microorganisms). Where they include targets, these should refer to particular segments of the food chain. In addition, FSOs should recognise that particular microorganisms may be important for the plant or line under examination or may be related to external or national standards. For some products additional or differentiated control targets (especially to do with the prevalence of pathogens) will be appropriate for products or in-process material with specialist end-uses (such as raw consumption or fermentation). Lee and Hathaway (1998) have also developed a decision tree which analyses the significance of identified hazards or hazard groups in terms of food safety objectives.

10.1.1 The principles of Codex Alimentarius
Codex has three Codes of Practice dealing with the hygienic preparation and handling of meat for human consumption (e.g. Code of Hygienic Practice for Fresh Meat: FAO, Rome, 1993). These Codes explain principles for applying the science of meat hygiene to the production of safe and wholesome fresh meat from slaughtered animals throughout the food chain, starting at the farm of origin. The Codes ask that any processing and hygiene requirements and procedures are based on scientific knowledge and their relevance to any particular product or process judged using risk assessment based on accepted scientific methodology (see below). Validation is required to ensure that requirements are based on current scientific knowledge. Indeed the World Trade Organisation (WTO), through the General Agreement on Tariffs and Trade (GATT) negotiations on food trade, advocates that sound scientific principles and harmonised standards based on them, are used to protect human health and prevent barriers to trade (Garrett *et al.*, 1998).

The US Food Safety Inspection Service is taking measures to reduce the occurrence and numbers of pathogenic microorganisms on meat and poultry products and modernise its meat and poultry inspection system. It has provided industry with specific HACCP-based objectives (WHO, 1996) that require all meat and poultry establishments to:

1. develop and implement written sanitation standard operating procedures;
2. have regular microbial testing by slaughter establishments to verify the adequacy of their process controls for the prevention and removal of faecal contamination and associated bacteria;
3. establish pathogen reduction performance standards for *Salmonella* that slaughter establishments and establishments producing raw ground products must meet;
4. develop and implement a HACCP system to improve the safety of their products.

In Europe, any HACCP plan for raw meat should embody requirements similar to these, except for the specific item 3, and in a plant that is ready to implement HACCP they would be covered under GMP, rather than being specified by CCPs. The United States National Advisory Committee on Microbiological Criteria for Foods (NACMCF) (1993) proposed a generic HACCP model for slaughter operations, which focused principally on the slaughter and processing areas, but also included an overview of live animal management and of hygiene during transport and retailing of beef products (see also Corlett, 1991). In 1994, the US (FSIS and FDA) submitted a view on the role of governments in developing a HACCP-based food safety system to Codex (FAO, 1995) and in it the functions of verification and validation were not clearly separated. Among other topics, the role of regulatory agencies was given as:

- Verify that HACCP plans are working as intended
- Establish verification inspection schedules based on risk
- Review the HACCP plan
- Review the CCP records
- Review the deviation and disposition records
- Visual inspection to observe whether the CCPs are under control
- Review the firm's verification audits
- Determine whether plan revalidation has occurred
- Review modification of the HACCP plan.

They proposed that the verification inspection report included coverage of the scientific basis of the plan, but did not specify the nature of comment required other than its presence. In 1998 NACMCF retained the seven principles of HACCP and clarified their meaning by providing definitions for validation and verification. The document included new sections on prerequisite programmes, education and training, and implementation and maintenance of the HACCP plan. It provided verification procedures for HACCP, including methods, procedures and tests used to determine whether the implementation and operation of a HACCP system is in compliance with the HACCP plan (NACMCF, 1997). Indeed Mortimore and Smith (1998) have proposed that the effectiveness of HACCP and related training are also included under verification.

Codex advocates that industry and the controlling authority should share the responsibility for the production of safe and wholesome meat. Industry personnel should be involved as widely as possible in voluntary quality assurance systems and the monitoring of meat hygiene, with supervision and audit by the controlling authority to ensure compliance. Although wherever possible after a HACCP study, existing quality assurance procedures should be utilised; training and education programmes involving both industry and the controlling authority may be necessary to implement a HACCP plan. Afterwards verification of the suitability and effectiveness of the procedures and systems should become a joint responsibility and can be done without the involvement of the original HACCP team.

10.1.2 EU legislation
The need for the verification of HACCP plans is mentioned in EU food hygiene legislation. The Directive (93/43) requires that 'competent authorities', e.g. veterinary inspectors or environmental health officers:

- propose or identify the potential food safety hazards associated with businesses;
- check that the necessary controls are in place, monitored and their performance verified within the HACCP plan;
- may have the option to use sampling to confirm the correct performance of the HACCP plan.

For meat production, microbiological monitoring may also be used to determine the prevalence of pathogens in pasture, housing and agricultural equipment, and in the case of poultry within the hatchery, the herd or flock, processing equipment or buildings or finished products. Such data is a key contributor to realistic hazard identification that constitutes the basis for risk assessment, because it describes the hazards requiring control for that plant and product (Notermans, 1995).

The legal requirement for hazard identification by the regulatory authority shifts responsibility at least partly from the HACCP team, as it means that the regulators, not the HACCP team, become responsible for hazard identification. This is not in accordance with the Codex principles of HACCP, where hazard identification is a starting point of a HACCP study and should take account of local circumstances. Under this Directive 'competent authorities' need to propose hazards and then judge whether they have been identified by businesses and are controlled by the HACCP plan (verification). Therefore the skills needed by 'competent authorities' under the current EU Hygiene Directive (European Community, 1993) are greater than merely checking correct implementation and operation of the HACCP plan; they extend to the expert activity of hazard identification and therefore contribute to validation of the plan.

In the Meat Products Directive (77/99) there are also more general requirements for the 'competent authority' specifically to verify cleanliness and staff hygiene and general compliance with the Directive. In practice the role of the 'competent authority' is most effective when they act as advisors ensuring that the minimum set of hazards have been addressed.

The relevant paragraphs from these two directives are reproduced below.

Council Directive 93/43/EEC of 14 June 1993 on the hygiene of foodstuffs

Article 1
1. This Directive lays down the general rules of hygiene for foodstuffs and the procedures for verification of compliance with these rules.

Article 8
2. Inspections by competent authorities shall include a general assessment of the potential food safety hazards associated with the business. Competent authorities shall pay particular attention to critical control points identified by food businesses to assess whether the necessary monitoring and verification controls are being operated.

Council Directive 77/99/EEC of 21 December 1976 on health problems affecting intra-Community trade in meat products

22. Constant supervision by the competent authority shall include the following:
    – inspection of the entry and exit register for fresh meat and meat products,
    – sanitary inspection of fresh meat intended for the manufacture of meat products for intra-Community trade and, in the case referred to in paragraph 3 (b) of Article 3 (1), of meat products,
    – inspection of meat products on dispatch from the establishment,
    – filling in and issuing the health certificate provided for in 34,
    – verification of the cleanliness of the premises, facilities and instruments and of staff hygiene as provided for in Chapter II,
    – taking of any samples required for laboratory tests,
    – any supervision measures considered necessary by the competent authority to ensure compliance with this Directive. The results of such tests shall be recorded in a register.

Woods and Hart (1998) have considered HACCP in the meat industry in the UK, especially following the 1990 Food Safety Act. They focus on the importance of good manufacturing practice (GMP) and training of personnel in GMP prior to establishment of HACCP systems. In common with the US FSIS they advocate the establishment of formal systems of cleaning schedules and hygiene audits, whose effectiveness should then be the subject of verification.

10.1.3 HACCP and risk assessment

Risk analysis (see below and ILSI Europe, 1998) is the recommended method for governments to determine the level of consumer protection they consider necessary and achievable for a defined population and product. The science of risk analysis is in its infancy and techniques of comparative or qualitative risk assessment are used at present to translate a 'level of protection' (the tolerable incidence of disease in a population) into 'food safety objectives'. The objectives are specific targets to be achieved by Good Manufacturing Practice hygiene requirements, CCPs and specification limits, such as levels of microorganisms in particular products. Any producer of meat products should achieve these objectives by carrying out and implementing a HACCP study. Therefore risk assessment can propose acceptable control actions for HACCP plans and can be used as a tool for validation.

Within the Codex definition, risk analysis has three components: *assessment*, *management* and *communication* which provide a formal and structured approach to understanding and reducing risk. In contrast, HACCP is a practical risk management tool that should be developed by the HACCP team and can take account of the output of a risk assessment in identifying hazards and fixing targets and limits, but remains a separate study. Existing HACCP studies can make use of data from risk analysis for validation of their scientific basis.

In the context of microbiological safety, *risk assessment* identifies hazards (in common with HACCP), estimates risks and looks at the factors influencing them. It is based on formulation of the problem (e.g. food poisoning by infectious pathogens caused by the consumption of undercooked beefburgers), which is similar to the definition of the scope of a HACCP study (e.g. control of infectious pathogens in beefburgers at the point of consumption). Both processes have as a starting point identification and description of the realistic hazards (e.g. infectious pathogens) associated with particular raw materials or product (*hazard identification*). A HACCP study may rely on in-house data from QA records, whereas a risk assessment may be more heavily based on data from various outside sources and experts. Microbial hazards can also be identified using surveys on the microbial quality of related products in literature. Based on this, *exposure assessment* provides an evaluation of the likely intake of the hazard by the consumer and the safety of any processing or consumer use instructions. *Hazard characterisation* defines the nature of the illness likely to be associated with the food. Information on the prevalence of the hazardous agent in the raw materials and products and the use and consumption habits of consumers is vital for validation, as such information should form the basis of any targets and limits specified. The level of the hazard, or the risk, may change according to processing and out-of-home or consumer use, because food poisoning bacteria can grow, may survive or be killed by processing. The overall estimation of the risks associated with specified raw materials and processes, etc., including any uncertainties, is known as the *risk characterisation*. The contribution of a risk assessment to HACCP is limited by our uncertainty that measures may be adequate. This uncertainty arises from doubts concerning the effectiveness of specific process control and QA procedures, the outcome of interactions between the many different microbial species on meat, and the variable nature of human disease.

Although predictive microbiology is still in its developmental stages as a food-safety tool, it can be used in the development and maintenance of HACCP systems. Predictive models can be used for hazard identification based on the effects of processing, formulation and storage, to assess risks or determine the fate of a microbiological hazard (such as salmonella) in food. By using predictive models, critical limits, ranges and combinations of process parameters can be assigned to CCPs. This can provide equivalent processing options (e.g. interchange of times and temperatures or of salt level and pH) while maintaining a uniform level of protection against a hazard. Validation or challenge testing of the targets and limits at CCPs can be reduced if predictive models are available for similar food types. Since HACCP is a risk-reduction tool, then predictive microbiological models can aid decision making in risk assessment so that process parameters able to achieve an acceptable level of risk are found.

10.2 The background to validation and verification of HACCP

HACCP is a long-term quality management technique for the production of safe meat products. It provides a focus for control and monitoring and can only succeed if the plant operates to GMP standards. Successful implementation requires that the scope and limitations of each HACCP study, including any processing objectives and constraints, are made clear and understood by all those involved at the operational or factory floor level. Any control, monitoring and review as required by the plan must be done effectively, on a continuing basis, by both workforce and management (CCFRA, 1997).

Validation examines the scientific basis of the study, and verification finds out how well it has been implemented and whether it is being used correctly to prevent food safety problems. Both are essential activities, requiring different techniques and personnel. The HACCP plan for a particular line may be summarised in the format of Fig. 10.1, which follows the process stages in sequence and forms a starting point for validation and verification.

Process stage	Preventative measure – CCP	Targets & limits	Monitoring	Corrective action	Records & outputs

Fig. 10.1 HACCP plan format.

The value of HACCP for controlling food safety is widely acknowledged in the food processing industry, though in the part of the meat processing industry producing raw chilled or frozen joints, mince or diced meat, usually for further processing or food-service use, its benefits are not uniformly acknowledged. This is because the processing technology used (slaughter, butchery, packing, etc.) lacks any means that is practical or acceptable to the customer of controlling or eliminating the most important microbial hazards. These hazards include infectious pathogens such as salmonella, campylobacters and pathogenic *E. coli*. Butchery and packaging stages cannot reduce microbial numbers. At best they minimise any increase, but more usually these stages will spread the pathogens.

If the conventional definition of a CCP is accepted (a step at which control can be applied and is essential to prevent or eliminate a food safety hazard or reduce it to an acceptable level), then the preparatory stages giving raw meat cannot include CCPs. But Gill *et al.* (1997) have suggested that the overall hygienic quality of uncooked beef hamburger patties can be improved only if the hygienic quality beef used to manufacture them is improved. To minimise risks, preparation stages should all be managed to Good Manufacturing Practice levels of hygiene, with special emphasis on those steps controlling the transfer of pathogens from the animal onto meat. A HACCP study will identify which process steps (including inspection, sticking and bleeding, bunging, evisceration and chilling) contribute to controlling cross-contamination and minimising the load of pathogens on meat. In the USA, the NACMCF (1993) suggests seven critical control points for controlling hygiene and processing immediately post slaughter. These are skinning, any post-skinning wash or rinse, evisceration, the final wash or rinse, chilling, refrigerated storage and labelling. For each CCP they provide critical limits, monitoring procedures and frequencies, recommended corrective actions, record and verification requirements. Gill (1995) has discussed the importance of using the HACCP approach for optimising conditions and procedures in pre-slaughter processes, slaughter and dressing, cooling, carcass breaking, storage and transport, and retailing processes. Whilst validation of the effectiveness of any individual step at ensuring product safety is difficult, there is no doubt that overall plant hygiene performance should be verified against GMP recommendations, as this will aid the optimisation of the various process stages.

To verify that HACCP, quality management and production systems are operated effectively to control microbiological quality, a procedure for objectively assessing the hygiene of carcass dressing is essential (Gill *et al.*, 1998). Biss and Hathaway (1995) have suggested that attempts to use visible contamination to monitor or verify the microbiological quality of lamb carcasses must be undertaken with caution. They suggest that an approach based on observation and records may be more effective than on-line monitoring of the product itself, as sampling cannot provide a continuing picture of adherence to good hygienic practices. The hygiene and management of the processing chain may be judged by monitoring the pre-slaughter condition of the animals, along with assurance from records and observations that the correct processing procedures are being used. Jericho *et al.* (1997) have investigated microbiological methods for verifying the adequacy of processes for butchering and chilling of cattle carcasses on a high-speed line. They found that low isolation rates dictate a large sample size, and therefore pathogens (verocytoxigenic *E. coli* and *L. monocytogenes*) cannot be used to verify routinely the workings of HACCP systems for beef slaughter processes (in Alberta, Canada). Alternatively, aerobic bacterial counts could be used to measure cleanliness directly, or *E. coli* counts to indicate faecal contamination indirectly. Mead *et al.* (1994) have used a marker microorganism (a non-pathogenic strain of nalidixic acid-resistant *E. coli* K12) to investigate the degree of microbial cross-contamination that could occur during poultry processing and the effectiveness of possible control measures. The approach of such studies can be used as a basis for providing data for microbiological validation. Microbiological and sensory quality data must also be considered in relation to quality and shelf life, which is an important part of validation. Use of surface plating has been suggested as a useful tool for small-scale meat processing facilities to visualise changes in fresh meat caused by the growth of aerobic microorganisms so that realistic levels can be targeted by GMP and within the HACCP plan (Jacquet and Peyraud, 1995). At present there is insufficient information on the impact of various agricultural, transport, slaughterhouse and butchery practices on the transmission of pathogens to allow reliable validation and the setting of scientifically justifiable requirements for GMP or pre-requisite procedures and conditions (see Sperber *et al.*, 1998).

Although the initial stages of the supply chain can exert a critical effect on product safety, the microbiological safety of meat and its products depends largely on downstream industrial or customer processing (e.g. cooking). When processing includes steps intended to reduce microbial numbers or eliminate pathogens (e.g. pasteurising heat treatments or fermentation and drying), then the story is different and HACCP has a better-defined role because a study can provide readily identifiable CCPs which are usually concentrated at, and downstream of, any step reducing microbial numbers.

For some prepared meat products, safety usually relies on multiple factors, for example in the production of cooked ham. Important factors needing control and monitoring include the initial microbial load of the meat, salt and nitrite levels, heating and chilling, vacuum packaging and chill distribution. For these factors validated scientific information is available (e.g. on the effectiveness of heating at killing microorganisms or their response to salt level, pH or storage temperature). This allows the effectiveness of any particular process or product formulation to be judged. Incze *et al.* (1999) have investigated the choice of critical, indicator or reference microorganisms and enzymes for checking the pasteurisation of meat products. As examples they used cured and picnic hams, with 10–15% brine, in cans of various sizes and heat-treated them in water baths at 60, 70 and 80 °C. They monitored heat penetration curves and measured EPT values (equivalent pasteurisation time in minutes) which they interpreted for lethality using *z*-values of several microorganisms (e.g. *Streptococcus faecalis* D, *S. faecium* P1-A) and acid phosphatase. Their results indicated that use of the phosphatase assay will lead to over- or underestimation of EPT and microbial markers are likely to give a better validation of the heat process. Residual catalase activity in cooked beef or pork products can be used to indicate endpoint temperatures over the range 60 °C to at least 70 °C and thereby provide verification of the thermal input for HACCP programmes (Davis and Cyrus, 1998).

Validation should identify those CCPs controlled by a single variable (such as chill or cooking temperatures) and those where hazards are controlled by several linked variables (such as salt and nitrite levels, cooking or the gas composition of a modified atmosphere, pack integrity, meat pH and storage temperature). Limits and targets for all the relevant variables must be specified for each CCP and validation should consider whether interactions between variables at or near their respective limits might lead to the manufacture of non-conforming or unsafe product. Drawing conclusions on safe or unsafe interactions or combinations of conditions is part of validation and requires thorough scientific knowledge of the interactions between microorganisms and processing or preservation systems, which must be left to experts. It should not be attempted as part of verification.

10.3 How far along the supply chain should a HACCP study extend?

HACCP studies must cover the critical microbiological, chemical and physical determinants of product safety in the core process (i.e. all the activities directly concerned with making product). But the impact of the non-core and non-production activities (i.e. supporting activities, such as procuring ingredients or animal feed or cleaning and plant maintenance) in the supply chain must not be overlooked and should be examined as part of validation. Ideally the scope of a HACCP plan should extend upstream to include possible sources of infection or contamination, and hence identify any means of control. Defining important activities is an essential part of deciding the appropriate scope for the plan because it determines its ability to improve food safety. The scope of the plan should be made very clear to the verification team. For example, in the manufacture of products that are sold uncooked, such as beefburgers or bacon, the origin and preparation of meat from the carcass and process hygiene will exert overriding effects on the presence and survival of pathogens in the product. But product safety in the end relies on consumer cooking for the elimination of pathogens. In a HACCP study for this type of product, animal rearing, meat buying and slaughter may often be incorrectly counted as non-core activities, if the limit of the HACCP study is taken as the factory gate, because butchered, boxed meat is used as the raw material. The limitations of such an incomplete study in identifying hazards and ensuring product safety must be recognised.

10.4 The importance of Good Manufacturing Practice (GMP)

Plants changing to or using HACCP-based QA systems must have basic operational procedures and the means for production of safe food already in place. Good Manufacturing Practice (GMP) describes requirements for hygienic design and construction of equipment, manufacturing and storage areas, hygienic operation and cleaning and disinfection procedures and defines the boundaries. GMP (or pre-requisites) will include the items listed in Table 10.1.

Kohn *et al.* (1997) have investigated the isolation of *Listeria* spp. and *L. monocytogenes* from two types of processing lines producing sliced meat, and have used the information to identify sites of contamination and to validate effective management and disinfection of processing environments to control this contaminant. GMP may also specify the quality and condition of raw materials, the safe operation of each process step and the aims of operative training in food manufacture and handling. A HACCP plan will not succeed in ensuring the safety of meat products by controlling CCPs unless the plant operates within Good Manufacturing Practice and has basic hygiene and control measures in place. For example if the study identifies hand and utensil washing as CCPs, then the infrastructure for the successful implementation of HACCP does not exist in the plant. Kukay *et al.* (1996) have developed a HACCP training programme for personnel in small-scale and medium-scale meat processing factories. It covers chemical and microbiological contamination and highlights the three most hazardous areas for this type of operation – employee hygiene, cross-contamination and control of heating/storage temperature.

10.5 Decision making within a HACCP-based QA system

Verification is used to check day-to-day decision making within HACCP-based QA systems. Usually decisions to release product, or not, are based on comparison of control and monitoring data with target and limit values; this may provide a range of correct or incorrect outcomes as shown in Table 10.2. Wrong decisions either way increase risk to the business or its customers. A 'customer risk error' occurs when unsafe product is wrongly 'cleared' by a QA system, and

Table 10.1 Factors to be included in GMP

Manufacturing	Materials	Personnel
Layout	Suppliers	Training
Facilities	Specifications	Hygiene
Hygienic design	Control of chemical and foreign bodies	Health screening
Cleaning and sanitation		
Pest control	Traceability and incident management	
	Transport and storage	

Table 10.2 Outcomes of decisions making based on data from CCPs

	CCP under control	CCP out of control
Failure to detect non-compliance	Correct decision making	Customer risk error
Detection of non-compliance	Producer risk error	Correct decision making

conversely a 'producer risk error' occurs when sound product is incorrectly rejected by the QA system. Wrong decisions can result from incorrect data on control at a CCP or incorrect decision making based on correct data. Where incorrect decisions are made repeatedly, the HACCP plan should be revalidated.

10.6 Monitoring

Monitoring is not the same as verification. Monitoring is done on a day-to-day basis by the QA department, by regulators or by production staff to check the effectiveness of control systems, especially at CCPs. In contrast, verification is done by a special team from time to time and includes review of the performance of the overall system of CCPs and the underlying GMP requirements or pre-requisites.

Monitoring provides continuing independent corroboration of the effectiveness of control activities at each CCP during production. It uses a low level of data generated on a routine basis, often independent of production or routine QA data, to show correct and effective control at individual CCPs. When microbiological testing is used for monitoring performance at CCPs, results may not be available sufficiently rapidly for any loss of control to be detected and remedied, so that loss of product is prevented. Therefore off-line testing cannot always be used for effective control of products, but it can generate data for trend analysis.

Any sampling scheme used for verification should examine monitoring data and the accompanying control data from production. This data should be taken in sufficient quantity and from a long enough period of time for it to show if out-of-specification material or inaccurate data have been produced (see Section 10.5, Decision making within a HACCP-based QA system). For verification, information and data should be collected and examined according to a predetermined scheme by a specialist team from the producer, customer or regulatory authority. The extent of data collected and reviewed should be influenced by the findings of previous verifications and should show whether or not data generation and control are in accordance with the HACCP plan. If the HACCP plan has been recently developed and implementation (or GMP) was poor, so that remedial actions were proposed, there should be an extensive review to check their effectiveness. If implementation was successful and the line operates consistently near to the targets of the HACCP plan, then a less extensive review will be appropriate. The scale of a verification review will also be influenced by its frequency and whether there have been changes to production or formulation leading to different control requirements. As with HACCP, validation and verification must be done on a line/product basis, recognising the unique nature of each line.

10.7 Validation, microbiological and other hazards

A wide variety of production and processing procedures ensure the safety of meat, beginning on the farm with feed and water and including the environment in which the animals are reared. Hazards carried by the animal may be biological (microbiological – in the gastrointestinal tract or on the skin, viruses or parasites, such as tapeworm in pigs), physical or chemical, including artificial or naturally occurring. Practical methods of hazard identification and risk estimation have been used for many years. An output of validation is confirming that the hazards identified by the HACCP study and the corresponding control measures are still realistic and, therefore, so is the extent to which product safety can be achieved by the implemented system. To do this, proposed or identified hazards can be compared with historical data from quality systems or published information, on-line process control or monitoring, special microbiological sampling, customer complaints and audit results. The experts undertaking validation for any hazard should appreciate that changes reducing or increasing risk occur all along the supply chain and include feed, water, pasture and silage, effluents, transport and lairage, process hygiene and control, etc. Hazards and risks may also change as the supply chain develops or suppliers are changed, and activities in the supply chain may act as primary sources of hazards, or as factors increasing or decreasing risk.

Scientifically-based predictive modelling and risk assessment methods, which are in the early stages of development, have the potential to improve the accuracy of practical risk estimates and hence may be used to propose scientifically-based limits and targets for processing and hygiene. They are able to do this objectively because they can take account of the kinetics of survival and growth of microbial pathogens on meat, in soil, water or other media, in response to many factors. Hence the predictive modelling of microbial growth can improve the management of risk. In the UK, Food MicroModel (FMM), and in the US, the Pathogen Modelling Program, are computer-based predictive microbiology databases applicable to meat and its products. Predictive models can be used to support the design and validation of HACCP plans and microbial risk assessment:

- Panisello and Quantick (1998) have used FMM to make predictions on the growth of pathogens in response to variations in the pH and salt content of a product when the pH of pâté was lowered. They suggest that a HACCP

system (for pâté) could be designed or validated using the predicted outcomes from FMM.
- Cattle carcasses need to be cooled immediately after leaving the slaughter floor. Within a HACCP system, cooling rates need to be monitored by the plant and verified by the regulatory agency. Jericho *et al.* (1998) have recommended the usefulness of the temperature–function integration technique (TFIT) based on predictive modelling of the growth of *E. coli* for verification of the adequacy of cooling processes.
- Zwietering and Hasting (1997) have taken this concept a stage further and developed a modelling approach to predict the effects of processing on microbial growth during food production, storage and distribution. Their model was evaluated using a meat product line and a burger processing line and was based on mass and energy balances together with simple microbial growth and death kinetics.

Such models can evaluate processes and predict the contribution of each individual process stage to the microbial level in the product. They have the potential to be valuable tools for a HACCP analysis and validation, as they provide a more quantitative basis on which to base decisions and set critical limits. The problems encountered when using predictive microbiology models for validation have been pointed out (Bourgeois, 1997). These focus on the impact of Good Manufacturing Practice on plant hygiene, and hence the state of any microorganisms present with respect to injury, growth rate and survival (Whiting and Buchanan, 1994). The accuracy with which plant and equipment temperatures and their variation can be accounted for and incorporated into models is also an important consideration if predictive models are to be used to validate the safety of the conditions required by a HACCP plan. Similar arguments also apply to the use of predictive models in Quantitative Risk Assessment.

Validation is likely to show that the focus of safety systems for non-preserved or mildly preserved fresh, chilled and frozen meat products is on the infectious pathogens. The main hazards are *Salmonella* spp., *Campylobacter* spp., enteropathogenic *E. coli*, *Listeria monocytogenes, Clostridium perfringens, Brucella* spp. and *Yersinia enterocolitica*. Where products are processed or more heavily preserved for prolonged chill or ambient stability, the toxin producers, e.g. *Staphylococcus aureus, Clostridium botulinum* and the other toxigenic clostridia are of greater importance. Other biological agents – *Taenia* spp., *Toxoplasma gondii* – are also of concern in specific products according to their processing, usage or ingredients, as are mycotoxins and anti-microbials.

Meat is spoiled by bacteria when their metabolic activities produce distinctive and unwanted changes such as off-odours or tastes (acid, sweet, putrid, ammoniacal or rancid), slime, loss of colour or mouldy appearance in the product. Spoilage bacteria include *Pseudomonas, Acinetobacter, Lactobacillus, Brocothrix thermosphacta, Shewanella putrefaciens* and the cold-tolerant Enterobacteriaceae. Where spoilage leads to changes in any of the preservation parameters, especially if spoilage leads to an increase in pH, then the potential effect on safety should be checked as part of validation.

Physical hazards can include foreign bodies, such as bone, metal fragments originating from tools and equipment, wood and plastic splinters, or unwanted or unspecified materials from the animal. As physical hazards are likely to be accidental contaminants, verification should check the effectiveness of procedures for their detection and removal. Validation should check that the measures for prevention or control are accepted as being effective and could form the basis for a 'due diligence defence'. The day-to-day effectiveness of detection of metal contaminants in meat and packaging materials is commonly done by metal detection systems (Anon, 1995). On-line equipment can have problems in detecting metal fragments in foods of different conductivities (flour and meat) and with the detection of flat pieces of metal. The practical effects of the limitations of such systems should be considered by validation, based on data from consumer complaints concerning foreign bodies. As the contamination of products with bone can be controlled effectively by training of boners and butchers, this should also be checked.

Similarly, validation should check which chemicals and residues are relevant to the materials being processed and to the process (i.e. processing aids such as lubricants). Validation should extend back onto the farm to check the composition of feed and the legality of any veterinary chemicals used. It may include the requirement for periods without their administration to animals, so that they will be eliminated from meat. The enforcement and monitoring of procedures on the farm for preventing the unwanted entry of veterinary chemicals or residues (e.g. synthetic hormones and antibiotics) into the food chain and the adequacy of procedures to exclude meat that may contain veterinary residues should be checked for effectiveness. Effective verification checks will cover records of control and monitoring by the farmer or primary producer and could include examination of veterinary records or analysis of material at the incoming raw material stage.

In the plant, inadequate cleaning or contamination with cleaning chemicals can provide significant hazards. Cleaning chemicals may be left on cleaned equipment, or utensils, or may be spilled onto food materials or equipment during production. Therefore it is essential that validation of the HACCP study covers measures to prevent this, including correct training, procedures, control and monitoring, especially covering:

- handling and diluting of cleaning chemicals during production in areas where there is exposure to food, and
- corrective actions in the event of sub-standard cleaning or a chemical spillage.

Validation of the study should especially focus on identifying chemical contamination hazards in areas where meat is handled or stored without any protective wrapping.

Validation and verification of HACCP plans   247

For preserved products, according to their processing and ingredients, the HACCP study should demonstrate the following.

- Storage temperature, cooking and cooling can produce safe product, and appropriate records (e.g. records of oven or autoclave or chiller temperatures) are produced.
- Any curing salts (nitrite or nitrate) or other preservatives, such as benzoate (mainly against *Clostridium botulinum*), added to products do not exceed the legally required maximum levels and are effective at ensuring the microbiological safety of products. To do this, processing and formulation should ensure there is a minimum residual level of nitrite during and after processing. For these reasons it is essential that validation consider whether factory procedures may allow the accidental omission or addition of preservatives to products.
- Where meat is processed or sold pre-packaged, any films or packs used for primary packaging are food grade and do not contain toxic chemicals, which may migrate into the meat during processing, storage, handling or the conditions of use, for example microwaving. The composition and performance of primary packaging materials is normally strictly specified by legislation and also should provide the correct gas barrier properties if gas or vacuum packaging is used.

10.8   Introducing validation and verification

The Oxford English Dictionary (1971) definitions make the differences between validation and verification clear and explain how the activities systematically analyse the relevance and working of a HACCP plan for a particular product, process and supply chain.

- Validation: *to render or declare legally valid; to confirm the validity; to make valid or of good authority; to confirm or corroborate.*
- Verification: *the action of documenting or proving to be true or legitimate by means of evidence or testimony; demonstration of truth or correctness by facts or circumstances; the action of establishing the correctness of a fact by means of special investigation or comparison of data.*

The latest definitions from US NACMCF (1997) provide very similar intent:

- Validation is 'that element of verification focused on collecting and evaluating scientific and technical information to determine whether the HACCP plan when properly implemented will effectively control the hazards'.
- Verification is 'those activities other than monitoring that determine the validity of the HACCP plan and that the system is operating according to the plan'.

Validation aims to examine whether the requirements for compliance (i.e. safety targets and limits) and the test methods are correct. Verification checks whether or not, based on QA and process data, the plan is being carried out and correct decisions have been made so that the HACCP plan is achieving the required level of food safety. In practice both activities are needed because the risks of producing unsafe or low quality meat and meat products can be increased by:

1. doing the wrong thing intentionally (or by design), e.g. missing a hazard, choosing the wrong equipment, procedure, product composition, storage or cooking temperature (*checked by validation*);
2. not carrying out and taking account of the well-designed or safe conditions, measurements or procedures noted in the HACCP plan – operational errors, e.g. not adhering to specified targets or limits (*checked by verification*).

10.8.1 Information requirements

To allow validation and verification of a HACCP study the items in the following list must be available from the producer and/or the supply chain. Some or all must be analysed by the teams, depending on whether verification or validation is being done:

- The terms of reference and scope of the HACCP study
- If the HACCP study is a new one, an outline of the existing QA system and historical records of the existing controls and procedures. In the case of a 'running' HACCP system, the outcome of previous verifications and validations
- The competence and skills represented on the HACCP study team
- The accuracy and validity of the product description and its intended use, including pack labelling (if available)
- A summary of the supply chain in total, indicating the scope of the study
- The accuracy of the flow diagram and the on-site location of the line, with an indication of the extent of the process covered by the HACCP study
- A list of the identified hazards and their control measures linked to specific step(s) or to materials
- Information that identified CCPs have the correct targets and limits for each hazard and are controlled within their limits and monitored
- Correct procedures and definitions of tasks for operating personnel and/or any software at each CCP
- Information on any background GMP or pre-requisites, with adequate records of control and monitoring results
- Records to show that data from CCPs is generated, analysed, responded to and archived
- Records to show the performance of the HACCP system, e.g. levels of compliance, exceptions and changes.
- Records of deviations and corrective actions, which ensure that CCPs are brought back under control; records of the incidence of deviations

CONTROL
timing – continuous hours
responsibility – manufacturing

MONITORING
timing – hourly, daily, weekly
responsibility – QA, regulatory or customer

VERIFICATION
timing – weekly, annual
responsibility – teams – QA, regulatory or customer

VALIDATION
timing – ongoing, annual, new hazard
responsibility – HACCP team or experts

Fig. 10.2 The links between control, monitoring, verification and validation showing responsibilities and typical timings.

- Details of the monitoring system
- Details of product dispositions after a deviation
- Triggers for review (such as process or ingredient changes or identification of new hazards) and any procedures used
- The programme and records of training for implementation and support of the HACCP plan.

Although verification may be a snapshot, validation needs to be a continuing or planned assessment of the scientific and technical content of a HACCP plan, to determine whether it is still correct. A team of technical experts, preferably including members of the original HACCP team working prior to implementation, carries it out initially. After the HACCP study has been implemented it becomes part of the review procedure, but is not part of verification. The frequency of validation will depend on changes in the hazards controlled by the plan. Validation and verification may require the disclosure of commercially sensitive data. The extent of disclosure is a matter that can only be settled by negotiation between the interested parties, but the criterion is that the minimum disclosure must be sufficient for the effective running of the systems and product safety to be demonstrated. Figure 10.2 shows the links between the various processes.

10.9 Validation – is it the right plan?

10.9.1 What is validation?
Validation should be a continuing expert process, assessing and reviewing the scientific and technical content of a HACCP plan to ensure it is effective and complete. It may result in modification of a plan and consequently some or all of the targets, limits and procedures may require change to improve product safety (van Schothorst, 1998). Validation should check the following.

- The HACCP plan is correct in microbiological, statistical, practical and economic terms.
- The scope of the plan is sufficient to ensure food safety.
- The producer has identified all realistic hazards using only reliable and relevant sources of data, and any new information has been taken into account.
- Suitable control measures are recommended.
- There is the appropriate level of response/control for each hazard according to its severity.
- The correct data for control is generated and used for decision making, including hygiene and monitoring measures.
- Staff responsible for control or monitoring of each CCP have sufficient training, resources and authority to do their job, as defined in the HACCP plan.
- Suitable remedial actions are used when processes or products go outside their control limits.
- Any products made and the HACCP plan comply with regulatory requirements and the plan produces sufficient regulatory records for the demonstration of due diligence.

10.9.2 What skills are needed?

The validation team must have practical experience of HACCP (ideally they would have done the study) and broad scientific and technical knowledge of the raw materials, process and product considered by the HACCP study. It should also have access to scientific data on microbiology, chemistry and animal science and an appreciation of the circumstances in the factory. These skills need to be employed to give a critical and informed appraisal of realistic hazards, the process and product design, and specifications, procedures and processes. Because, as a last resort, analysis for pathogens may be included to demonstrate that a line produces safe product, skills in microbiological analysis may be required. As a matter of good practice, such analyses should not be conducted where there is any possibility of contaminating food materials – ideally an outside laboratory would do this type of testing. Pathogen testing within the factory site increases risks of cross-contamination and will involve having high numbers of the pathogenic species within the laboratory buildings. The validation team must be able to recognise and judge reliable process flows, layouts and operations, so that the representation of the line by the flow diagram and HACCP plan can be challenged. The regulatory requirements for the commodity, e.g. EU Directives, should also be known, so that legislative demands can be interpreted in the context of the plant. Therefore it is necessary for the team to ask questions and challenge the identified hazards and CCPs, to find out if any are obviously wrong or absent and whether targets and limits are based on values appropriate to the food or process. If serious problems are found then the HACCP study is not valid!

A major question is whether or not an outside expert can validate or alter a plan in the absence of the HACCP team who have detailed knowledge of the plant. Clearly the answer is no, unless the CCPs are obviously outside the generally accepted principles (e.g. times and temperatures) for processing that type of product or operating particular unit operations, such as cooking or cooling conditions.

10.9.3 Who should be responsible?

Validation should be done by a team of experts, usually working prior to implementation and also later as part of a regular review procedure or one initiated in response to changes in hazards. The person responsible for product safety in the plant should head the validation team and be responsible for the quality of its output and representation of the HACCP team. If the original team is no longer available then the validation team must ensure they understand the scope of the study and the hazards and controls covered. Even so, validation by an external assessor or auditor may not be of the same technical quality as that done by a company expert or by a competent HACCP team. Any external study should at least check that accepted principles of product formulation, processing, packaging, storage temperature and hygiene have been applied. If necessary a non-expert team should suggest that external expertise be used to referee the study and have the power to initiate a technical review, for example prior to implementation.

In some countries it is the responsibility of an expert from a regulatory authority or other external experts to propose hazards or to check that the information used for hazard identification was adequate. The role of regulatory agencies in the US meat industry is explained by NACMCF (1994) and focuses on the implementation of HACCP; plans are verified by USDA inspection. Within the EU it is the responsibility of the 'competent authority' to identify potential food safety hazards associated with a business.

10.9.4 When should it be done?

Validation may be needed for a new or proposed plan when

- there have been changes in the raw material source, product composition or packaging, process or processing conditions,
- data, e.g. laboratory tests or consumer complaints, indicates loss of control,
- new scientific, epidemiological or recall information is published (e.g. identification of a new meat-borne hazard), or
- the record, or other requirements, for regulatory compliance have changed.

As a result of validation, control, monitoring and verification procedures may need to be updated, for example after a change in processing.

10.9.5 What should be done?

The validation team needs to check that the team which produced the HACCP plan included the necessary skills, based the plan on the identification of realistic hazards and drew justifiable links between hazards and specific control points. The output should be critical limits ensuring customer safety. The effectiveness of limits may be checked by some practical assessment, such as reference to similar processes, controlled processing runs or intensive investigational sampling and analysis of production. Use of marker or indicator microorganisms for validating compliance with food processing procedures and hygiene requirements has been proposed. The choice of suitable or reliable markers is difficult and may be plant or line specific, especially if particular types or species are chosen from the range of different microorganisms present in the process or the food. Depending on the type of process, various markers may provide information of use in validation. *Escherichia coli* may indicate contamination with faeces or gut contents; the presence of coliforms, Enterobacteriaceae or Enterococcae may indicate lack of hygiene of processing conditions. Other microbiological methods can be used for verification, including direct epifluorescent microscopy, ATP bioluminescence and catalase activity (as a hygiene monitoring tool) (Griffiths, 1997). The use of Enterobacteriaceae to indicate the possible presence of salmonellas in processed foods has also been proposed (Mossel and Struijk, 1995), but there is little data to show a quantitative relationship. Bautista *et al.* (1997) used a Latin square design to investigate a poultry-processing line for microbial contamination. Samples were taken and analysed during the processing of several flocks ($n = 16$) over four separate days. They showed this design was able to show significant correlations ($p \leq 0.001$) between flocks and between processing days and indicated that microbiological examination can be useful for validation of HACCP programmes.

Many meat plants produced safe products prior to HACCP and corroborating data may be collected from existing QA, process control and customer complaint records. Therefore the validation follows the general steps of the HACCP study:

- Hazard identification
- Identification of critical control points (CCPs)
- Establishment of control criteria and critical limit values
- Monitoring of the CCPs
- Remedial actions
- data recording

The validation team checks that:

1. All realistic hazards and product uses have been identified, based on reliable and relevant sources of data. Control of the identified hazards must be addressed by control measures at specific process stages or groups of stages.
2. Whether any features of the process, formulation or usage that could increase the safety risks associated with the product have been approved by the HACCP plan. This could include
    - the specification of unhygienic equipment

- recommendation of working practices, targets, timings, controls or layouts which may lead to product contamination or deterioration
- unstable preservation, or distribution systems
- incorrect consumer use instructions
- unsafe remedial actions.

3. Reliable methods of control, monitoring and decision making are in place, and can distinguish between acceptable and unacceptable material and conditions and initiate appropriate control actions.
4. The means of dealing with process deviations or receipt of unsafe raw materials will prevent the release of unsafe product.
5. There are effective lines of communication within the supply chain.

Pre-work for the validation team includes assembly of relevant scientific data and documentation from the HACCP study, including the skills employed by the HACCP team. It should be supplemented with up-to-date information covering the same materials and processes and taking account of any new hazards. This is followed by an analysis of the content and output of the plan, including discussion with the HACCP team and agreement of any conclusions, recommendations or alterations. Lastly, an agreed validation report and action plan must be agreed with management, including recommendations for review based on changes in hazards or in the product, raw materials or process. The conditions for revalidation should be agreed, especially if there are frequent changes in the hazard landscape.

10.9.6 Sources of information for validation

Sources of information include publications such as books and journals, and web sites on the Internet (see Table 10.3); experts, including scientists working for government or local authorities, research institutes and universities; trade associations, consultants, and equipment and chemical suppliers. A list of useful Internet sources on food microbiological safety is given in Table 10.3.

10.10 Verification – are we doing it correctly? Is it working?

10.10.1 What is verification?

If correct process and product design principles were identified by the HACCP study, so that all realistic hazards are covered, unsafe products can still result if production or QA does not carry out the requirements of the HACCP plan. Verification should show this and also how consistently agreed specifications are met. In many ways this is like an ISO 9000 audit covering systems, procedures, practices and documentation. It should be a structured, ongoing check carried out by manufacturers and/or regulators at a certain frequency (e.g. 3–6 monthly) to assess the effectiveness of the plan in the plant and find out if the procedures and conditions currently in use are those documented in the plan.

254 HACCP in the meat industry

Table 10.3 Internet sources on food microbiological safety

Internet address	Topic
www.who.int/inf-fs/en	World Health Organization (WHO) home page
www.who.ch/wer/wer-home.htm	WHO Weekly Epidemiological Record
www.who.int/wer-home.htm	
www.foodsafety.gov	US Government 'gateway' site with useful links
www.cdc.gov/ncidod/diseases	National Center for Infectious Diseases United States (CDC home page)
www.arserrc.gov/mfs/pathogen.htm	USDA Pathogen Modeling Program
www.cfsaan.fda.gov/~mow/intro.hmtl	Bad Bug Book (US FDA)
www.cdc.gov/ncidod/eid/index.htm	Emerging Infectious Diseases
www.cdc.gov/od/oc/media/facts.htm	Facts sheets from Morbidity and Mortality Weekly Report (MMWR)
www.fsis.usda.gov	Food Safety Inspection Service
www.nfpa-food.org	National Food Processors' Association Web Site
www.nmaonline.org	National Meat Association
www.eurosurv.org/main.htm	Eurosurveillance Weekly
www.hc-sc.gc.ca/hpb/lcdc/biosafety/msd/index.htr	Health Canada Material Safety Data sheets
www.cfia-acia.agt.ca/english/corpaffr/publications/foodfacts/fofistac.hmtl	Canadian Food Inspection Agency
www.exnet.iastate.edu/pages/families/fs/inftox.hmtl	FSNET Foodborne Pathogens
www.easynet.co.uk/ifst/hottop.htm	Institute of Food Science and Technology (UK) Hot Topics
www.campden.co.uk/index.htm	Campden and Chorleywood Food Research Association
www.lfra.co.uk	Leatherhead Food Research Association
www.phls.co.uk/publications/cdr.htm	Communicable Disease Report Weekly
www.phls.co.uk/publications/cdph.htm	Communicable Disease and Public Health
www.phls.co.uk/facts/index.htm	PHLS (UK) Facts and Figures
www.phls.co.uk/facts/bac-inf.htm	PHLS (UK) information on specific pathogens
www.phls.co.uk/facts/campyinf.htm	
www.phls.co.uk/facts/clospinf.htm	
www.phls.co.uk/facts/ecoliinf.htm	
www.phls.co.uk/facts/list-inf.htm	
www.phls.co.uk/facts/salm-inf.htm	
www.phls.co.uk/facts/stap-inf.htm	
www.maff.gov.uk/foodinfsheet/index.htm	MAFF Food Surveillance Information Sheets and Bulletin
www.maff.gov.uk/food/bulletin/intro.htm	
www.beef.org/saf_libr/saf.library.htm	Beef Industry Food Safety Council
www.safefood.net.an/index.cfm	Food Safety Campaign – Australia

All data collection for verification should be based on a sampling plan clearly related to the severity of the hazard and the associated risks. This should cover the inspection of schedules and records of control, monitoring and decisions at CCPs, routine analytical or process data originating from the line and period under examination and any supplementary information. Its use will ensure that the verification process is consistent from one review to the next, any activities are repeatable and conclusions have known reliability.

Because in any meat-processing operation, manual operations play a major role in ensuring food safety, verification should examine whether line operators still adhere to the implemented procedures and objectives or not. If not, then the causes should be sought – whether it is that further training is required, operatives are becoming complacent and require motivation, or whether management pressures have prejudiced product safety. If a previous review has revealed that the plan is working reliably and no significant changes have been found, then there is little need for additional sampling, as safety is evidently 'built in' by the HACCP plan. Where the HACCP plan has not been implemented adequately or non-complying product is produced, day-to-day information from QA or regulatory monitoring and control may need to be supplemented with broader data. This should be targeted to identify the causes of the problems, so that the performance of the plan and the degree of confidence in the conclusions can be assessed. Such data may include:

- Information from personnel in the agricultural and supply chain (workers, line operatives, supervisors, managers and experts)
- Inspection of facilities for livestock rearing, transport and lairage
- QA data and data on consumer complaints.

10.10.2 When should it be done?

HACCP plans should be verified at a fixed frequency, or when there are changes in hazards, processes or materials. NACMCF (1998) advocate that companies establish verification schedules covering the activities included in the review, its frequency and who is responsible. There should be an additional independent review by an unbiased or independent authority. Topics to be covered include:

- The verification schedule giving frequencies and triggers
- The initial validation of the HACCP plan and information on any changes and their impact on the current HACCP plan
- Monitoring activities
- Evidence of compliance from monitoring and also initiation and management of corrective actions
- Examination of the HACCP system in the context of the plant QA system.

When the plant is running without problems, a minimum verification may be done by record review according to an agreed sampling plan, but it should preferably include an inspection of the line to assess hygiene, maintenance and housekeeping against GMP and HACCP plan requirements (Sperber, 1998). If a new type of product has been manufactured, or if procedures or a formulation have changed, so that there is some rationale for suspecting increased risks concerning a particular hazard, then additional or investigative microbiological sampling during production, or sanitation, may be needed to provide corroborative data. Where there have been substantial changes in the supply chain or in products, examination of the HACCP plan is outside the scope and skill of a verification team and revalidation should be recommended or initiated. In the absence of records, suspected unsatisfactory plant control or knowledge of potentially hazardous conditions in the plant, additional sampling will be essential. Most seriously, verification including this type of sampling will be done if food from the plant has been involved in a food-poisoning outbreak or routine data reveals severe contamination with indicator organisms. In these circumstances it is essential that data be collected according to a recognised and documented sampling plan so that it can be related to a specific production period or location on the line and food safety risks can be assessed.

10.10.3 What skills are needed?

Verification should involve a team of people with the skills to audit systems and plant layouts, because the focus of verification is on how well the workforce and management are complying with the requirements of the HACCP plan, and that the process is working within its specification. In the meat industry the team will usually include a representative of the regulatory authority, and possibly the trade customer, plus production line operatives and supervisors, plant technical staff (plant manager, plant engineer and Quality Assurance and development staff) and other important functions, such as the buyer or transport or lairage manager. It should have access to the original and revised HACCP plans and contact with the HACCP study team. Their skills should allow them to assess the performance of the designated line against the process flow diagram, the HACCP plan and other information provided by the factory management. They should be familiar with the technology used, but should not take on the role of technical experts. They should be sufficiently familiar with the principles of the technology to know the essential CCPs and their characteristics, so that they can recommend revalidation where they are not satisfied. Because their activities usually include examination and evaluation of a wide range of process, microbiological, analytical, consumer and other QA data, their skills must allow them to appreciate the scientific principles underlying the safe manufacture, handling, distribution and use of the product under examination. They should be able to check the line or supply chain by observation, sample analysis and inspection of process and analytical data generated by production and QA on a day-to-day basis.

10.10.4 Who should be responsible?

A multi-skilled team of trained 'systems' experts or process auditors best does verification activities, because of the range of data requiring examination and analysis. An auditor working alone will not normally have complete access to all the skills required for verification. Verification audits may be done by

- an internal team from in-house QA departments as part of assessing the performance of the QA system,
- personnel who are external to the plant, e.g. corporate auditors, or
- third parties, e.g. those inspecting suppliers on behalf of customers or government agencies.

The teams are responsible for reporting their findings to management and making any necessary recommendations for improvements and therefore must understand the unique circumstances of each operation or line.

10.10.5 What should be done?

Verification should be done after implementation of a HACCP plan and as part of its review procedure, according to a defined scheme. In any plant, verification may be a routine part of production and QA activities, but preferably it should be a periodic formal review of the HACCP plan and its implementation. This may be done in-house, or by external or independent bodies, and initiated by either producers or regulators. Internal verification audits will be the responsibility of the plant management and may be limited by resource or skill availability. External or corporate audits may be done at a lower frequency but consider more data. Regulatory or government audits may come under the broad heading of inspection and examine the complete system for controlling product safety along the supply chain to see if it meets mandatory requirements. These audits are not generic exercises, but must be done for each product/production line. Verification specifically examines the performance of a line making a product from defined raw materials, to determine whether its control delivers set limits, monitoring is effective, corrective actions are promptly taken and documentation and records are consistent. General guidance in assessing plant-specific plans can be obtained from generic HACCP plans that exist for many types of process:

- The slaughter and processing of raw broiler chickens (McNamara, 1997)
- Raw beef, including slaughter operations, slaughter and processing areas, live animal management and hygiene during transport and retailing of beef products (USDA-FSIS, 1993)
- Vacuum-packaged, sliced, cured meat product (Tompkin, 1994)
- Beef franks (including a flow diagram and process description from raw materials to palletised products and a verification schedule (Anon, 1997).

Information gathered on non-core activities needs to be associated with data on the core process and may include additional microbiological, chemical and physical analyses of raw materials, in-process material and products, shelf-life determination of products and analysis of consumer complaints. There are different views on the value of additional data or sampling for verification. The principle should be that any additional data should improve the confidence of the verification team in their assessment of how well the HACCP plan has been implemented and is being carried out within the process line under examination. Additional sampling should not encourage unofficial revalidation of the HACCP plan. Therefore the verification team should thoroughly analyse existing control and monitoring data for trends and non-conformances, and document and communicate any conclusions.

Verification will typically involve four stages. These are described in subsections 10.10.6–10.10.8 below.

10.10.6 Pre-work – assembly of information

The team should devise a plan to assemble and organise documents and data concerning the GMP or pre-requisites in the plant and control and monitoring systems at CCPs specified by the HACCP plan, so that compliance can be judged. Effective recordkeeping is essential to document the performance of the HACCP plan, and records outlining the producer's performance are the only starting point. The quality and accuracy of these records will be a good indication of how well the HACCP plan has been implemented, and it should be possible to cross-reference this data with other records (such as deliveries of hygiene materials) outside the QA system. The main documents needed are:

- The HACCP plan, scope, documentation and explanation, accompanied by a list of any changes since the last review. This will indicate the documentation and records required for review. Where there are mandatory or contractual requirements relevant to safety (e.g. product composition or preservation, process and storage temperatures) these should be noted for use by the team.
- The process flow diagram including its relation to the factory or department layout.
- Details of the allocation of tasks and responsibilities to plant personnel and an outline of the information flow in the plant, for example from operational and methods manuals, or indications of plant product safety performance on notice-boards, etc.
- Control and monitoring data from production or Quality Assurance can include the results of testing, line control, process control outputs (such as chiller temperatures), hygiene monitoring, end-product testing, consumer or trade complaints and marketplace conformity samples. Much of this information will be readily accessible if it is stored within the ISO 9000 format (ISO, 1994) and should show compliance with procedures and with GMP.
- Details of any other internal audits and their follow-up actions.
- Records of process deviations or delivery of out-of-specification raw materials and the product dispositions and actions taken to bring the process back under control. The effectiveness of these actions should be assessed as part of verification.
- Training records and details of personnel changes at CCPs.
- Records of control systems.

Validation and verification of HACCP plans 259

Supplementary information may include shelf-life studies on chilled products (e.g. perishable products such as packaged or minced meats) or investigational microbiological examination of lines, their surroundings or materials being processed. This type of sampling should aim to identify vulnerable areas of the plant that may provide sources of contamination (Silliker, 1995). Bautista *et al.* (1997) have suggested that good statistical planning, experimental design and appropriate measurement techniques are required to verify the control of microbial hazards in food processing systems.

10.10.7 Comparison and review, control and monitoring

The next stages cover comparison and review of work practices, control and monitoring data with the HACCP plan to judge compliance, identify problem areas, and if necessary establish the causes of any problems.

This can be done using an approach of audit and checklist and should cover the parts of the supply chain included in the HACCP plan (scope). It may conveniently be divided into two parts, one covering how well the systems and procedures are followed and the other showing how well these systems work, as demonstrated by records and actions. This stage should always include an audit visit to the line and observation of work practices when the product noted in the HACCP plan is being made. An integral activity is talking to the line operatives to find out if they are properly trained and appreciate the key safety aspects of their activities. It should identify

- poor or inadequate training and management procedures,
- insufficient time for the safe execution of tasks, poor hygiene, inadequate management or control of lairage or butchery,
- ineffective segregation of raw and cooked, or non-food, materials,
- faulty process control or packaging, e.g. uncontrolled cooking, chilling procedures, etc.

The team should check the accuracy of the flow diagram on the plant layout and confirm the operations and locations recorded in the flow diagram by observation, so that any changes in layout, practices or process times and temperatures can be identified by comparison with the original plan. These activities can also show whether the HACCP plan is really working in production areas. The auditors doing this should not be from the line or department concerned, as their familiarity with procedures may be a hindrance, not a benefit. Trend analysis data is a valuable source of information if it includes 'average' or summary values and can indicate variability.

10.10.8 Documentation and explanation of the HACCP plan

Typical questions which could be answered by this stage of the process of verification and information that should be formally recorded are listed under the following headings.

HACCP plan documentation

- Is GMP in place and was it considered by the current HACCP plan?
- Has the current plan been correctly implemented?
- Which hazards were considered, when was the plan last reviewed, and is the hazard list up-to-date?
- Was consumer use of the product considered?
- Is there reliable reference to the original HACCP plan and any revisions?
- Are current equipment, processes, products, distribution and storage accurately represented by the documentation?
- Are appropriate control measures for the type of product present?
- Is there a calibration record for critical measuring equipment?
- Were any process or product changes discussed with the HACCP team beforehand and their conclusions respected?
- Is there effective feedback and communication within the supply chain, e.g. with the supplier or transporter?

The flow diagram

- Are process stages indicated on the diagram still used?
- Is the factory location and layout still valid, especially for CCPs?
- Have there been equipment or personnel changes?
- Have there been modifications to equipment?

Raw materials, specifications and limits

- Is the scope of the HACCP plan sufficient to ensure product safety?
- Are suppliers audited and approved?
- Is there an animal/meat intake programme? How is it managed?
- Is there acceptance or intake monitoring with records?
- Are changes in suppliers recorded and acted on?
- Is there segregation and tracking of stock during transportation and in the lairage, with effective ante-mortem inspection and segregation of animals with features likely to adversely affect food safety? (There should be effective post-mortem inspection of carcasses and offals, exclusion of suspect material and record keeping.)

CCP records

- Are the specifications, limits and specified control measures documented, up-to-date and being followed?
- Are the control measures being followed and documented?
- Are there clear instructions for the CCPs?
- Are the same equipment and personnel involved at the CCPs?
- If there have been changes, what is their impact on product safety and was the HACCP study team consulted beforehand?

Production and hygiene information

- Are there specifications, limits and records covering key process steps (e.g. GMP) outside the CCPs? Have there been changes?
- Is there adequate training of staff, do they have adequate working instructions and time for their tasks?
- Is there trend analysis on product safety and GMP matters?
- Is there raw material and product rotation, e.g. FIFO?
- Is there product traceability? A reliable lot or batch tracking system should be managed, controlled and operated throughout the food chain. All information recorded should be available to the producer, factory management and regulatory authorities.
- Are non-core and supporting activities, such as cleaning, disinfection, maintenance and waste material removal, controlled?
- Are there documented cleaning procedures and recording of hygiene monitoring results?

Control system calibration

- Do control systems have specifications and a test or calibration frequency with documentation of results?
- Do control systems cover critical conditions?
- How are control sensors maintained?
- Changes?

Deviations and corrective actions

- Is there data on the frequency and extent of non-conformances affecting safety, the delivery of out-of-specification materials and the actions taken to restore control and minimise risk to customers?
- Is there compliance with procedures for dealing with product that has left the main product flow (for example trimmings, arisings or rework) or has arisen as a result of a line breakdown? Attention should be focused on procedures and records kept of storage or holding conditions and dispersal of the material.

10.11 Reporting conclusions and agreeing an action plan

This should include approval of HACCP plan operation or recommendations for updating or modifying it. The verification report is a formal assessment of the findings by the nominated team, reporting on the competence of the plant and demonstrating whether systems and procedures are being followed. Over time these reports should provide a continuing record of the performance of the plan and show whether there are any trends in product safety within the plant, for example increases or decreases in out-of-specification product or consumer complaints. The report should contain the data supporting any conclusions. The audience should understand

- whether or not implementation and use of the plan is in line with the output of the HACCP study and with legislation,
- whether any remedial actions or changes are necessary to improve product safety, their importance and time scale and a means of checking their effectiveness.

Management and the workforce must agree any recommendations, and if these involve changes to monitoring, control or hygiene procedures, these should be checked with those involved.

10.12 Specific additional requirements for the meat industry

In addition to the information normally collected for verification of manufacturing processes, in the meat industry additional supply chain stages should be considered as either GMP or CCPs during verification. These should be as follows:

- The effectiveness of checks carried out on the farm of origin to ensure that only healthy stock are sent for slaughter and records are kept, including disposal or treatment of affected stock. Farm records should cover stock records, veterinary treatments, feed records, in-feed medication and animal health problems
- Animal production on the farm of origin, including husbandry practices and the prevention and control of infectious agents
- Transport of the animals to the abattoir and their condition and cleanliness on arrival
- Holding and handling of animals awaiting slaughter (lairage)
- The physical structure of the abattoir and its equipment
- Operating requirements and practices during slaughter, hide removal and the evisceration/dressing of carcasses
- Carcass chilling
- Boning
- Chilling or freezing, packaging and storage of the boneless product for retail or further processing
- Temperature control during transport
- Requirements for inspection
- Casualty and emergency slaughter.

Each of the above specific stages should have a statement of the hazard(s) dealt with and the principles used. If possible the verification team should produce a description of the range of practices used by the various suppliers, or shifts, contributing to the line and the product, the basis for the process control and monitoring systems, including targets and limits. For example, a provision

Validation and verification of HACCP plans 263

dealing with cleanliness of an abattoir which requires that 'a cleaning and sanitation programme should be established by the manager of the abattoir that ensures ...' should contain enough detail to allow the objectives and procedures to be identified and judgement of suitable results made.

10.13 Involvement of plant management in validation and verification

10.13.1 Validation

Although plant management may initiate validation of a HACCP study, they should not be involved in the validation process. Usually they will not be expert in the technical and scientific aspects that are covered by a HACCP study. One of the difficult tasks for the team is to judge whether the HACCP team was technically up to the task of identification and control of hazards. Where the experts undertaking validation recommend revision of a HACCP study, they should be willing to work with the HACCP team to ensure the correct revisions are agreed and made. Common areas for disagreement are:

- Differentiation between CCPs and important aspects of the process accepted as GMP
- Relation of targets and limits to the severity of the hazard and capability of the process
- Correct and safe handling of process deviations, as this can involve disposal of product.

10.13.2 Verification

Plant management and regulatory authorities should be frequently and actively involved in verification. Within the proposed scheme, they can contribute to and learn from an impartial comparison of plant data with the agreed requirements of the HACCP plan. Management will usually have set the scene for implementation and daily use of the HACCP plan and may have set up the onward reporting system of summaries and trend analysis. These should be assessed to see how well plant performance is represented. In large plants, management may not be familiar with changes in departmental layout or procedures, and the reporting back of the HACCP study verification should highlight any changes and indicate whether review or revalidation of the plan and its implementation is necessary.

10.14 Involvement of the HACCP team in validation and verification

If possible the HACCP team should be fully involved in both activities. With the initial validation, soon after the study, this may be easy to arrange, but for subsequent reviews it may be less easy; therefore the team should leave sufficient detail in the study documentation to allow its evaluation by those responsible for validation. If possible the team should be consulted when verification shows that procedures are in need of modification, as they will have been involved in implementation and will know what is required and what is to be avoided.

10.15 How to validate a new HACCP study

A team should validate a new study. In a large company this may include head office technical auditors, but in a small- or medium-sized enterprise an external consultant or auditors may be used. It is important that the best available level of technical expertise is used to prevent increased levels of consumer or producer risk. Regulatory authorities may be involved in validation in all sizes of company. Typical activities will correlate any proposals and activities with existing QA systems and relate the product and process design to the specifications in the plan. At a later stage they should reconcile the plant layout to the process flow diagram and check (new) control procedures, equipment, training, responsibilities and communication. This is done to provide evidence of the effectiveness of CCPs, the monitoring methods and the safety of corrective actions including disposal of any non-conforming product, as indicated above.

10.16 How to validate an implemented HACCP plan

Loss of control leads to unknown risks, therefore an implemented plan which has been running for some time should be examined by auditing which covers not only the line, process and QA data, but also the competence, training, motivation, and authority of operatives and supervisors. Core and non-core parts of the supply chain should be covered. The plant layout, including comparison of the current flow diagram and layout with the original HACCP plan, should be included to validate that forward flow, access and separation are as originally specified. The condition, suitability, performance and maintenance of equipment and work practices should be checked, with the emphasis on inspection schedules at CCPs; frequency should be related to risk and the severity of the hazard controlled by the CCP. The scientific relevance of the plan should be checked to ensure that any new hazards or consumer requirements are covered. Finally, information flow and effective communication are key aspects of the continuing reliable performance of a HACCP plan, and the development of short-cut systems, which exclude either data or particular functions in the organisation, should be assessed.

10.17 Sampling plans for validation

Sampling plans for validation should concentrate on hazards and their means of control at CCPs. The plan should take the hazards currently addressed and check that the means of control used are still valid; this is especially important as there is continuous pressure on manufacturing costs and consumer pressure for milder processing or preservation. Conversely the hazards considered realistic for the raw materials used should be reviewed to ensure that all those currently problematic are covered. Over the past few years the range of hazards that must be considered realistic for meat and meat products has grown from the salmonella and *Staphylococcus aureus* traditionally considered, to now include *Listeria monocytogenes*, campylobacters and enteropathogenic *E. coli* (see White *et al.*, 1997).

10.18 Sampling plans for verification

The examination of routine, and any additional, data to verify implementation and operation of a HACCP plan should be done according to a sampling plan, so that any conclusions have a known basis and reliability as illustrated by Table 10.4. The verification team should use sampling plans to demonstrate the reliability of the control and QA systems at CCPs. For verifying that a HACCP system is achieving its objectives, sampling plans should also be able to detect any out-of-specification material or loss of control at CCPs with a high degree of confidence.

Sampling plan terms can be used to describe the data collection and handling needed to do this, e.g. sampling period and frequency, the amount of data taken and the analytical methods used, with estimates of the proportions within and outside limits. Confidence in a correct assessment of the performance of the sampling plan can be increased by increasing the amount of data inspected and this should always be done as the severity of the hazard being controlled or the variability of the process and its materials increases. Process and analytical data from process stages should be sampled on a random basis, unless there are reasons for focusing on particular times or activities. The amount of data examined and hence the stringency of this review should be based on the hazard to the consumer from the materials and the operation of the process.

There should be review and analysis of existing microbiological data and process control records from CCPs rather than point sampling and analysis at the time of the review. Preferably the data should extend beyond the CCPs and consider 'Good Manufacturing Practice' within the establishment and its supply chain over the same period of time. The task of verification is made easier if data is organised so that trends and out-of-control occurrences can be easily distinguished. Additional analysis for pathogens will not normally be included in an audit, because of the low probability of detection. Indicator organisms may be used to show the effectiveness of hygiene or other procedures, but they must be

Table 10.4 Probability of detecting defective material

Composition of data		Number of results examined and probability of finding a defective result (%)							
Percentage of all results within limits	Percentage outside limits	3	5	10	15	20	30	60	100
98	2	6	10	18	26	33	45	70	87
95	5	14	23	40	54	66	79	95	99
90	10	27	41	65	79	88	96	>95	>99

chosen with care to prevent misleading conclusions being drawn. The reliability and relevance of QA and process data may also be judged by comparison with consumer complaints.

Depending on the number of samples of data examined (out of the total available for a particular time period) there will be a fixed probability of detecting out-of-specification material, as indicated in Table 10.4. Sampling plan components similar to those given by ICMSF (1978) for defining microbiological raw material or end-product sampling plans are relevant, because they have predictable chances of detecting 'defective units', at a fixed sampling frequency (see Table 10.4).

Ideally the intensity of sampling should detect very low levels of defectives (i.e. out-of-control conditions) with a high level of confidence, 90% certainty. Typical features of a sampling plan for verification include:

- Clear specification of the activity or data to be considered
- The methodology used to produce analytical or process data
- The volume of data and the time frame to be examined, including a specified amount of control and monitoring data from CCPs.
- The criteria to be used to assess the effectiveness of implementation of the HACCP plan (e.g. distribution of results at each CCP with respect to targets and limits) and the confidence in the conclusions
- Assessment of the incidence and outcome of any deviations
- The reporting method. For convenience, control and monitoring results may be banded into the performance categories based on those used by attribute sampling plans
  - acceptable control (i.e. within the target values),
  - marginally acceptable (i.e. between the target and the limit), and
  - outside the limit values
- If, or when, the process or materials have values outside the limits, or the process has become out-of-control, any remedial actions taken and the disposition of the resulting defective product must also be examined and conclusions drawn on the safety of the actions.

10.19 Output from validation and verification

Each time a process is verified, a written report must be issued to the interested parties and copies kept. Prior to issue, the findings should be discussed by the verification team and with the original HACCP team, if possible. The report should:

- be presented to management in understandable terms, so that its impact on their business and any necessary actions can be proposed,
- conclude whether or not the HACCP plan is being followed,
- indicate whether or not the correct hazards are covered, and whether targets, limits, monitoring procedures and corrective actions are appropriate for the severity of the hazards,
- identify any discrepancies and their causes, accompanied by the information used by the team to conclude this,
- summarise discussions with the plant management and the other personnel involved in discrepancies,
- provide agreed recommendations for correction, or improvement, with a clear timetable,
- identify the resources needed for improvement and those responsible, and
- propose an interval between verifications.

The success of the plant at completing the actions should be examined by a subsequent verification and the report should be considered as continually increasing the plant's knowledge of its products and processes and providing evidence of its progress.

A validation report is a more specialist document and should contain

- scientific information on the current relevance of the HACCP plan,
- whether the right hazards are controlled and the degree of control is adequate,
- whether the objectives or risks to the consumer (or business) have changed from a food safety point of view.

In summary, it should indicate whether or not the HACCP plan, if properly implemented, would ensure the production of safe food by controlling realistic hazards. If this is not the case then it should make recommendations for changes or the collection of more information. Examples are given in Fig. 10.3.

10.20 Conclusions

Validation and verification are essential tools for the management of HACCP and the assurance of product safety. There are poor and confused definitions of these activities in the legislation, even though the meaning of the terms is clear. Structured storage of data makes the task of verification easier, but validation remains an expert activity. Developments in the use of validation and verification as HACCP plans become more widely used must take account of

268 HACCP in the meat industry

(a)

	Validation		Verification
Target microorganism	*Clostridium botulinum*	Target microorganism	(*Clostridium botulinum*)
data used	Literature or legislation	data used	Literature or legislation
control response	> 3 minutes × 121 degC in product	control response	Process software & management procedures
Process design	scheduled process and 'matching & segregation' procedures	Process design	scheduled process and process records
Equipment requirements	high reliability, independent control & monitoring – MTI	Equipment performance	defect deviation & maintenance records
calibration	National standard for MTI, < annual	MII & control calibration	records
HACCP study team	Thermal process expert, Dev. QA, Engineer, Production, Maintenance	Training	records & interviews
Operational responsibility	Cook room manager	Operational responsibility	records & interviews

(b)

	Validation		Verification
Target microorganism	*Listeria monocytogenes*	Target microorganism	*Listeria monocytogenes*
data used	Literature or factory data	data used	Factory QA records
control response	management procedures, disinfection & monitoring	control response	management procedures & interviews, hygiene records
Process design	prevent re-contamination + environmental + personnel hygiene	Process design	in-line with hygiene requirements in published GMPs
Equipment performance	easy to clean, weight control + vacuum, minimum temp. rise	Equipment performance	hygiene procedures & records *sampling or swabs*
calibration	environmental temps.	control & calibration	records
HACCP study team	Production, hygiene crew, QA, engineers	Training	records & interviews
Operational responsibility	high care area manager & operatives	Operational responsibility	records & interviews

(c)

	Validation		Verification
Target microorganism	*Salmonella*	Target microorganism	*Salmonella*
data used	Literature or preventive	data used	preventive
control response	water hygiene & change, evisceration, chilling	control response	management procedures & interviews, hygiene records
Process design	limit cross-contamination, rapid chilling + environmental hygiene	Process design	in-line with hygiene requirements in published GMPs & Directives
Equipment performance	easy to clean, good heat exchange, water management	Equipment performance	hygiene procedures & temperature & suspended solids records
calibration	water temps.	control	temperature & Cl_2 records
HACCP study team	Production, hygiene crew, QA, engineers	Training	records & interviews
Operational responsibility	manager & operatives	Operational responsibility	records & interviews

Fig. 10.3 Examples of (a) food sterilisation (retort step) CCP, (b) sliced meat packaging CCP, (c) poultry chilling CCP.

the competence and capability of the personnel available, so that false expectations of their reliability are not created.

10.21 References

ANON (1995) Metal detection, ISO 9000 and HACCP, *International Food Hygiene*, **6** 5–6.

ANON (1997) Beef franks, *Dairy, Food and Environmental Sanitation*, **17** 417–26.

BAUTISTA D A, SPRUNG D W, BARBUT S and GRIFFITHS M W (1997) A sampling regime based on an ATP bioluminescence assay to assess the quality of poultry carcasses at critical control points during processing, *Food Research International*, **30** 803–9.

BISS M E and HATHAWAY S C (1995) Microbiological and visible contamination of lamb carcasses according to preslaughter presentation status, implications for HACCP, *J Fd Prot*, **58** 776–83.

BOURGEOIS C M (1997) Microbiological quality and safety assurance, *European Food and Drink Review*, Summer, 65–7.

CAMPDEN AND CHORLEYWOOD FOOD RESEARCH ASSOCIATION (1997) *HACCP, A Practical Guide*, 2nd edn Technical Manual No. 38.

CODEX ALIMENTARIUS COMMISSION (1997a) *Report of the twenty-eighth session of the Codex Committee on Food Hygiene*, FAO/WHO Food Standards Programme, ALINORM 97/13A.

CODEX ALIMENTARIUS COMMISSION (1997b) *Report of the twenty-ninth session of the Codex Committee on Food Hygiene*, FAO/WHO Food Standards Programme, ALINORM 97/13.

CODEX COMMITTEE ON FOOD HYGIENE (1997) *HACCP System and Guidelines for its Application*, Annex to CAC/RCP 1 – 1969, Rev 3 in Food Hygiene Basic Texts. Secretariat of the Joint FAO/WHO Food Standards Programme, Food and Agriculture Organization of the United Nations. Viale delle Terme di Caracalla, 00100 Rome, Italy, ISBN 92-5-1040121-4.

CORLETT D A JR (1991) Regulatory verification of industrial HACCP systems. *Food Technology*, **45** 144–6.

DAVIS C E and CYRUS S (1998) Evaluation of a rapid method for measurement of catalase activity in cooked beef and sausage, *J Fd Prot*, **61** 253–6.

EUROPEAN COMMUNITY (1993) Council Directive 93/43/EEC (14 June) on the Hygiene of Foodstuffs, *Official Journal of the European Communities*, 19 July, No. L 75/I.

FAO (1995) The use of hazard analysis and critical control point principles in food control. HACCEXP 94/2 The role of government in developing a HACCP based food safety system. FAO Food and Nutrition Paper 58. FAO, Rome.

GARDNER S (1995) Food safety, an overview of international regulatory programs, *European Food Law Review*, **6** 123–49.

GARRETT E S, JAHNCKE M L and COLE E A (1998) Effects of Codex and GATT, *Food Control*, **9** 177–82.
GILL C O (1995) Current and emerging approaches to assuring the hygienic condition of red meats, *Can. J Animal Sci*, **75** 1–13.
GILL C O, RAHN K, SLOAN K and McMULLEN L M (1997) Assessment of the hygienic performances of hamburger patty production processes, *Int J Fd Microbiol*, **36** 171–8.
GILL C O, McGINNIS J C and BRYANT J (1998) Microbial contamination of meat during the skinning of beef carcass hindquarters at three slaughtering plants, *Int J Food Microbiol*, **42** 175–84.
GRIFFITHS M W (1997) Rapid microbiological methods with hazard analysis critical control point, *Journal of AOAC International*, **80** 1143–50.
HATHAWAY S (1995) Harmonization of international requirements under HACCP-based food control systems, *Food Control*, **6** 267–76.
INTERNATIONAL COMMISSION FOR MICROBIOLOGICAL SPECIFICATIONS FOR FOOD (ICMSF) (1978) *Microorganisms in Foods. 2. Sampling for Microbiological Analysis, Principles and Specific Applications*, University of Toronto Press, ISBN 0-8020-2143-3.
ILSI EUROPE (1998) *Food Safety Management Tools*, eds JOUVE J L, STRINGER M F and BAIRD-PARKER A C, ILSI Europe Task Force, 83 Avenue E. Mounier, B-1200 Brussels, Belgium.
INCZE K, KORMENDY L, KORMENDY I and ZSARNOCZAY G (1999) Considerations of critical microorganisms and indicator enzymes in connection with the pasteurization of meat products, *Meat Science*, **51** 115–21.
INTERNATIONAL STANDARDS ORGANIZATION (ISO) (1994) ISO 9000 series, *Quality Management and Quality Assurance Standards, Guidelines for selection and Use*.
ISO 9001, *Quality Systems – Model for Quality Assurance in Design/ Development, Production, Installation and Servicing*.
ISO 9002, *Quality Systems – Model for Quality Assurance in Production and Installation*.
ISO 9003, *Quality Systems – Model for Quality Assurance in Final Inspection and Test*.
JACQUET B and PEYRAUD D (1995) Microbiological control of meat-based processed products, *Viandes et Produits Carnes*, **16**(6), 203–6.
JAYASURIYA D C (1995) The regulation of the trade in food in Asia and the Pacific, *Food Control*, **6** 307–10.
JERICHO K W F, KOZUB G C, GANNON V P J, THOMAS E J G, KING R K, BIGHAM R L, TANAKA E E, DIXON-MACDOUGALL J M, NISHIYAMA B J, KIRBYSON H and BRADLEY J A (1997) Verification of the level of microbiological control for the slaughter and cooling processes of beef carcass production at a high-line-speed abattoir, *Journal of Food Protection*, **60** 1509–14.
JERICHO K W F, O'LANEY G and KOZUB G C (1998) Verification of the hygienic adequacy of beef carcass cooling processes by microbiological culture and the temperature-function integration technique, *J Fd Prot*, **61** 1347–51.

KUKAY C C, HOLCOMB L H, SOFOS J N, MORGAN J B, TATUM J D, CLAYTON R P and SMITH G C (1996) Application of HACCP by small-scale and medium-scale meat processors, *Dairy, Food and Environmental Sanitation*, **16** 74–80.
LEE J A and HATHAWAY S C (1998) The challenge of designing valid HACCP plans for raw food commodities, *Food Control*, **9** 111–17.
KOHN B A, COSTELLO K, BROOKINS-PHILLIPS A (1997) HACCP verification procedures made easier by quantitative Listeria testing, *Dairy, Food and Environmental Sanitation*, **17** 76–80.
McNAMARA A M (1997) Generic HACCP application in broiler slaughter and processing, *Journal of Food Protection*, **60**(5), 579–604.
MEAD G C, HUDSON W R and HINTON M H (1994) Use of a marker organism in processing to identify sites of cross-contamination and evaluate possible control measures, *British Poultry Science*, **35** 345–54.
MORTIMORE S E and SMITH R A (1998) Standardized HACCP training: assurance for food authorities, *Food Control*, **9** 141–5.
MOSSEL D A A and STRUIJK C B (1995) Advantages and limitations of *Escherichia coli*, other Enterobacteriaceae and additional indicators as markers for the microbiological integrity of foods, *Microbiologia*, **11** 75–90.
MOTARJEMI Y, KAFERSTEIN F, MOY G, MIYAGAWA S and MIYAGISHIMA K (1996) Importance of HACCP for public health and development. The role of the World Health Organization, *Food Control*, **7** 77–85.
MOY G, KAFERSTEIN F and MOTARJEMI Y (1994) Application of HACCP to food manufacturing: some considerations on harmonization through training, *Food Control*, **5** 131–9.
NATIONAL ADVISORY COMMITTEE ON MICROBIOLOGICAL CRITERIA FOR FOODS (NACMCF) (1993) Generic HACCP for raw beef. *Food Microbiology*, **10** 449–88. United States of America, Department of Agriculture.
NATIONAL ADVISORY COMMITTEE ON MICROBIOLOGICAL CRITERIA FOR FOODS (NACMCF) (1994) The role of regulatory agencies and industry in HACCP. *International Journal of Food Microbiology*, **21** 187–95. United States of America, Department of Agriculture.
NATIONAL ADVISORY COMMITTEE ON MICROBIOLOGICAL CRITERIA FOR FOODS (NACMCF) (1997) Hazard analysis and critical control point principles and application guidelines. *Journal of Food Protection*, **61** 762–75. United States of America, Department of Agriculture.
NOTERMANS S *et al.* (1995) The HACCP concept: specification of criteria using quantitative risk assessment, *Food Microbiology*, **12** 81–90.
PANISELLO P J and QUANTICK P C (1998) Application of food MicroModel predictive software in the development of Hazard Analysis Critical Control Point (HACCP) systems, *Food Microbiology*, **15**(4) 425–39.
SCHOTHORST M VAN (1998) Introduction to auditing, certification and inspection, *Food Control*, **9** 127–8.
SILLIKER J H (1995) Microbiological testing and HACCP programs, *Dairy, Food and Environmental Sanitation*, **15**(10) 606–10.

SPERBER W H (1998) Auditing and verification of food safety and HACCP, *Food Control*, **9** 157–62.
SPERBER W H, STEVENSON K E, BERNARD D T, DEIBEL K E, MOBERG L J, HONTZ L R and SCOTT V N (1998) The role of prerequisite programs in managing a HACCP system. *Dairy, Food and Environmental Sanitation*, **18** 418–23.
TOMKIN R B (1994) HACCP in the meat and poultry industry, *Food Control*, **5**(3) 153–61.
UNITED STATES DEPARTMENT OF AGRICULTURE (USDA-FSIS) (1993) *Generic HACCP Model for Beef Slaughter*. USDA, Washington DC.
WATSON B (1994) UK's chilled food guidelines lead the world, the 2nd edition of the UK Chilled Food Association (CFA) Guidelines, *Food Industries*, **47** 35–42.
WHITE P L, BAKER A R and JAMES W O (1997) Strategies to control Salmonella and Campylobacter in raw poultry products, *Revue Scientifique et Technique Office International des Epizooties*, **16** 525–41.
WHITEHEAD A J and ORRIS G (1995) Food safety through HACCP. The FAO approach, *Food, Nutrition and Agriculture*, No. 15, 25–8.
WHITING R C and BUCHANAN R L (1994) Microbial modelling, *Food Technology*, **48** 113–20.
WOODS K and HART B (1998) HACCP action in the meat industry. *Food Manufacture*, **73**(5) 36–7.
WORLD HEALTH ORGANIZATION (WHO) (1996) Pathogen reduction: hazard analysis and critical control point (HACCP) systems (Title 9, Part 304), *International Digest of Health Legislation*, **47**(4) 500–1.
ZWIETERING M H and HASTING A P M (1997) Modelling the hygienic processing of foods – a global process overview, *Food and Bioproducts Processing*, **75**(C3) 159–67.

11
Auditing HACCP-based QA systems

N. Khandke, Unilever Research, Sharnbrook

11.1 Introduction

The food industry today is increasingly under pressure from the outside world. Food legislation is becoming more comprehensive internationally, standards are tightening and inspection authorities are better trained and have a greater understanding of the hazards and their means of control. Similarly, consumers are more aware, have higher expectations and are concerned about food safety and quality, and the media is quick to pick up food-related stories. In order to remain competitive in the marketplace, meat product companies are changing their approach to product safety and quality. They are moving away from systems based on checking the finished product, to a system of assuring safety and quality through design and control of manufacturing and supply chain operations.

In order to facilitate this change in the trading environment, food producers are adopting standardised systems, or frameworks, within which quality systems can be developed and demonstrated to customers and regulatory authorities. The two major systems currently utilised to manage quality systems are Hazard Analysis Critical Control Point (HACCP) and the ISO 9000 series of quality standards. The ISO 9000 system describes 20 elements required to build a quality system (not all elements are required for each of the different standards) (see Table 11.1). The basic premise of the system is that the producer defines systems and procedures, developing a quality system for his whole operation, documents these procedures and demonstrates compliance with his own internal standards. Because of its structured nature the ISO 9000 system offers the added benefit that certification can be gained from third party certifying bodies, to demonstrate to customers that you have a documented quality system in place.

Table 11.1 The 20 elements comprising the ISO 9000 standard

Element	ISO 9001	ISO 9002	ISO 9003
Management Responsibility	✓	✓	✓
Quality System	✓	✓	✓
Contract Review	✓	✓	
Design Control	✓		
Document Control	✓	✓	✓
Purchasing	✓	✓	
Purchaser Supplied Product	✓	✓	
Product Identification and Traceability	✓	✓	✓
Process Control	✓	✓	
Inspection and Testing	✓	✓	✓
Inspection and Test Measuring Equipment	✓	✓	✓
Inspection and Test Status	✓	✓	✓
Control of Non-Conforming Product	✓	✓	✓
Corrective Action	✓	✓	
Handling, Storage, Packaging and Delivery	✓	✓	✓
Quality Records	✓	✓	✓
Internal Quality Audits	✓	✓	
Training	✓	✓	✓
Servicing	✓		
Statistical Techniques	✓	✓	✓

HACCP is a different tool for identifying and controlling product safety hazards, and unlike ISO 9000 is specific to a line and product. HACCP is internationally accepted and is mandatory in many countries. External normalisation companies and agencies are beginning to offer certification services for HACCP, but this is still in its early stages. However, it is likely, through pressures from customers, that HACCP certification will become more of an issue in the future.

A fundamental process within any quality system is auditing. Auditing is not a new concept, but in the past may often have been viewed as a tool for 'checking up' on a company, or policing the company's systems. Auditing is in fact the main tool for driving continuous improvement, by identifying weaknesses in a quality system and recommended changes for improvement. The two major types of audit applicable to safety and quality are the technical audit and the system audit and these will be addressed in this chapter. A common misconception is that anybody can turn up to a company with a blank sheet of paper and audit the company. This is definitely not the case. Audits must be carefully structured and planned, and must be carried out by trained personnel. The major areas of auditing discussed in this chapter are:

- Scope
- Standards
- Preparation
- Format
- Assessment and scoring

- Follow-up
- Frequency

By carefully addressing each of the above areas, a company can develop comprehensive, effective auditing systems for both internal auditing of their own quality systems, and external auditing of suppliers and third party producers.

11.2 HACCP and quality systems

The majority of processors in the meat industry now accept the fact that the traditional approach of testing a product to detect defects, post production, is statistically unsound, gives no assurance that defective or hazardous product is not released onto the market and provides no opportunity for remedial action.[1,2] As a result, many processors have moved away from this traditional 'quality control' approach to more preventative systems based on design and operational control. In order to facilitate this change producers are adopting standard quality systems, such as the ISO 9000 series,[3-7] HACCP[8] and Total Quality Management (TQM)[9] to name but a few. All the above quality systems share a common element, in that they do not provide a company with a ready-made quality system, but define a framework upon which a company can build quality management systems of the required complexity and focus to enable the consistent manufacture of products of a defined quality. The ISO 9000 quality management series offers the additional facility in that the systems and procedures making up the system are formally recorded so that they can be assessed externally and accreditation/certification given if the system meets the requirements of the standards.

There is extensive information in the literature on the quality systems mentioned above, and it would be futile to try to cover all the topics here. However, we should briefly consider the main systems currently favoured.

11.2.1 The ISO 9000 series

The ISO 9000 series of standards for quality systems[3-7] were published in 1987 and were based upon the British Standard BS5750[10] and a similar Canadian standard.[1] The ISO 9000 system is comprised of five separate standards. ISO 9000 'Quality Management and Quality Assurance Standards – Guidelines for Selection and Use', and ISO 9004 'Quality Management and Quality System Elements – Guidelines', offer advice and guidance on selecting the appropriate standard and implementing the guidelines. The standards themselves are encompassed in ISO 9001–9003. ISO 9003 covers the quality system for final inspection and test and is not normally applicable to food processors. ISO 9002 covers the quality system for production and installation and is the standard most commonly sought in the food industry, and ISO 9001 is the quality system for

276 HACCP in the meat industry

design/development, production, installation and servicing, and is the most comprehensive of the three standards.

The ISO 9000 standard is composed of 20 requirements (see Table 11.1) which guide a company into the areas which need to be contained within the quality system. Not all the requirements are relevant for all the standards (ISO 9001 uses more than ISO 9003). The standard itself does not define specific criteria for any of the 20 requirements, but the standards do give guidance on what is required in each. It is up to the company to define the specific criteria required in each section.

There are a number of key features which need to be mentioned with regard to ISO 9000.

1. The ISO 9000 system, as a quality system, normally specifies a quality system for the whole company, covering all quality-related activities.
2. The ISO 9000 quality system is based on the contracts and relationships between customers and suppliers.[1]
3. The ISO 9000 system requires companies to define their own standards, systems and procedures, which they believe will result in the production of product of a consistent quality.
4. In order to gain certification in ISO 9001–9003, the company only needs to define their own standards, to document these standards and associated systems and procedures, and to demonstrate to the assessor that they adhere to these internal systems. There is therefore always the chance with ISO 9000 that a company will not have covered all critical elements for product quality or safety within their internal standards, but nevertheless may achieve certification by demonstrating compliance with those set.

The method by which ISO certification is achieved varies, depending on which certifying body is used, but in general the certification processes involves:

- Selection of the appropriate standard and the development of internal standards, systems and procedures covering the 20 elements
- Pre-review of documentation by the third party certification body to identify any early non-conformances, and subsequent remedial action. In a labour-intensive industry this may result in considerable training
- Formal assessment, in house, by the third party certifying agency
- Correction of any non-compliance
- Certification
- Maintenance and reassessment (normally six-monthly maintenance visits and a full review every three years). (This may vary depending on the certifying body used.)

11.2.2 HACCP

Although it is probably fair to say that HACCP predates ISO 9000 (and the BS 5750 series before this), it was not until the publication of HACCP in its current form, based on the seven principles, in the late 1990s[8] that HACCP has come to the fore as a key safety system utilised in the food industry.

The various features of the HACCP system have already been discussed in this book. However, we can draw a number of comparisons with the ISO 9000 system. The first point to mention is that HACCP is a quality management system, and is similar to ISO 9000 in that it provides a framework on which a system can be built. HACCP does not come 'ready made' and a company implementing HACCP will establish criteria to control hazards, based around the requirements defined in the standards.[8] The HACCP system, however, does have a number of important features, distinct from the ISO 9000 system.

1. HACCP as a quality system focuses on product safety, and is targeted at individual production lines and products. This is unlike ISO 9000 which specifies a quality system for the whole company.
2. Although HACCP provides an empty framework, the safety hazards, limits and in many cases the controls for many of the food processes are very often universally accepted and quantified. This makes it easier for a company to gain information on the hazards and controls relevant to a particular food process. It also has the effect of making it easier for an inspector to assess the completeness and technical accuracy of a HACCP plan.

11.2.3 Total Quality Management (TQM)

TQM is unlike HACCP and ISO 9000 in that it does not provide a rigid framework within which to build up a system. TQM focuses on continuous improvement, through the participation of employees in identifying and implementing improvements, and focuses on 'delighting the customer'. TQM therefore provides a philosophy, culture and discipline within which quality systems such as HACCP and ISO 9000 can be built and operated.[11]

11.3 Establishing benchmarks for auditing

Auditing is a fundamental part of a food safety or quality system, whether it be auditing to certify a supplier or a quality system (such as seen in the ISO 9000 system), or internal auditing to assess compliance to Good Manufacturing Practice (GMP), to verify a HACCP plan or to monitor internal compliance to quality systems and procedures. An audit can be defined as a 'systematic evaluation of a system against a set of defined criteria'. Audits are often viewed as being surreptitious checks on companies' systems, with the auditors being viewed as policemen. This should not be the case, and if an audit is perceived in this way it is not being carried out correctly. An audit is a quality tool which allows an auditor to assess performance against a set of criteria. The main purpose of an audit is to drive continuous improvement by identifying areas of weakness which may pose a risk to product quality or safety (and hence a business). There are essentially two types of audit, each of which can be further subdivided into a number of types of audit. At the broadest level audits can be defined as:

- Technical audits
- Systems audits.

Technical audits are generally of a limited scope and performed by technical experts in a specific field, such as microbiological safety, hygienic design, or thermal processing. This type of audit will examine a particular process in detail to assess its technical performance against set criteria. In most cases the criteria set for such audits will be defined externally, in either national legislation or industry codes of practice. The technical audit is more often used to assure the manufacturer that the products manufactured, and the processes or unit operations employed, meet a minimum set of requirements to ensure the safety of the end product.

Systems audits are more commonly applied in the food industry and are not necessarily carried out by technical experts. A systems audit is examining compliance with a set of systems or procedures which make up a company's quality system. The systems or procedures covering supply or production procedures may be internationally or nationally defined, but in most cases will be developed internally by the company. The most commonly recognised systems audits in the food industry are those of the ISO 9000 certification system. The key issue with regard to systems audits is that, where the systems are developed internally, they do not necessarily ensure the quality or safety of the product or process. The absence of a 'judgemental element' can be a problem with the ISO 9000 system where the approach of 'say what you do, do what you say, show that you have done it' can get a company certified as ISO 9000 without the company addressing the critical safety or quality issues within the product or process design.

Within the two audit types above, companies will be carrying out, or receiving, audits of different types, the main being:

- Internal audits
- External audits
- Regulatory audits
- Certification audits.

These types of audit will be discussed later in this chapter.

11.3.1 Establishing the ground rules for an audit

Irrespective of the type of audit that will be carried out, there are a number of ground rules which must be followed to ensure that the output of the audit can be used for reporting and improvement. No matter how experienced the auditor, auditing is not simply a case of turning up to the company or department to be audited with a pen and paper to see what you can find. When this approach is used, it inevitably leads to omissions and inconsistencies in the audit process and

Table 11.2 Main elements required in setting up a successful audit system

Element	Rationale
1. Scope	Defines the type and limit of the audit
2. Standards	Define the depth of the audit
3. Preparation	Allows the auditor to develop an understanding of the product, process and standards
4. Format	Determines the method of the audit, e.g. using check lists, questionnaires
5. Assessment and scoring	Describes the method by which the audit will be evaluated
6. Follow-up	Checks progress against an agreed action plan resulting from an audit
7. Frequency	Defines how often audits will take place

assessment. The key elements of an audit, which must be considered, are shown in Table 11.2.

Scope

The scope of an audit is determined by a number of factors, the two most important being the type of audit being undertaken, and the resources available to carry out the audit. The scope of the audit will be made up of a number of different elements. The first element is whether the audit is a technical or systems audit, together with the type of audit (internal, external, etc.). This level will immediately determine the type of auditor required to carry out the audit, as a technical audit will require specialist expertise in the subject area being audited.

The second element should define what the audit will cover. This is always an important question and is more often than not determined by the resources available. HACCP audits will, by the nature of the HACCP study, be product and process line specific. ISO 9000 audits focus on the company's quality system as a whole. The common trap is to focus on in-house operations during the audit, which may result in critical elements which are important for product quality and safety, but which lie outside the core manufacturing process (upstream or downstream from the processing establishment), being missed. As a minimum, a company's quality audit system should include upstream audits as far as the raw material supplier or primary producer (e.g. farmer). These audits should cover how the supplier manages their own upstream and downstream supply chain, but it is often impracticable actually to audit these elements yourself, and downstream audits extend as far as the end user of the product (for retail goods this would normally be down to the retail outlet) or in the case of a further processing plant the inwards goods reception.

Standards

All audits should be carried out against defined standards. Without standards there is no benchmark or frame of reference, and the auditor's personal belief becomes important in defining what is acceptable and unacceptable. Audits of this nature are rarely satisfactory and can lead to disagreements between the auditor and the company or department being audited over the action points raised. Another consequence of not setting fixed or published auditing standards is that it becomes almost impossible to draw conclusions when trying to evaluate the results of different audits, especially where different auditors are used, because the audits will have been carried out to different standards.

The standards used will depend on the type of audit. For any audit, local legislative requirements which may be agreed with the local veterinary service will be important, but in many cases a company's internal standards may well be stricter than the local legislation. For internal audits, internal procedures and specifications form the basis of the standards against which the audit is carried out. These internal standards should include any published GMPs, and should cover the control, monitoring and corrective actions defined in the HACCP plan. For external audits, e.g. supplier or third party producers, it is more normal to use external standards or industry guidelines. Two good examples are the 'General Principles of Food Hygiene' produced by the Codex Alimentarius[12] or the 'Food and Drink Good Manufacturing Practice Guidelines' produced by the Institute of Food Science and Technology.[13] These are two of many such guidelines which can be useful. When carrying out an external audit, it is important that the auditor takes note of any internal standards being applied by the third party, specifically those defined within the HACCP plan, to assess how well the company is adhering to their own standards.

In all cases the standards to which the audit is being carried out, and its scope, should be mutually agreed in advance of the audit (it is not the objective of the audit to 'catch people out').

Preparation

The key to any successful audit is preparation, whether it is an internal ISO 9000 audit of a department, a HACCP audit of a line or a complex audit of an external supplier. Auditors should familiarise themselves with the scope of the audit and the applicable standards well in advance of the audit. For internal audits they will need to familiarise themselves with the process, products, systems and procedures being audited (it is not good practice to allow auditors to audit within their own department of the plant, and it is good practice to rotate auditors within a company to avoid auditors becoming over familiar with any area or department).[14]

For external audits, the auditor may not be familiar with the product or process in operation because in the meat industry processing covers the scope of processes from slaughter and butchering right through to the preparation of cooked, sliced meats. It is therefore important that the auditor is pre-armed with knowledge of the following:

- The typical hazards associated with such processes and materials
- The controls which should be in place

- The limits within which the process should be capable of working
- The minimum CCPs which should be included in the HACCP plan and GMP requirements
- Product usage
- The process stages and personnel involved, etc.

Format

The audit format determines the method of the audit. There are many different approaches to auditing, each having their own benefits and shortcomings. Whatever approach is used, it should be designed to aid the auditor in covering all the areas defined in the scope of the audit. Some of the more common approaches are:

- Experience based
- Check sheets
- Questionnaires.

Audits based only on experience should generally be avoided, due to possible inconsistencies and the difficulty in interpreting their results. This type of expert audit is more suited to technical audits which are carried out by technical experts and have a very narrow scope. The outcome of this type of audit will be a technical evaluation of a line or process.

Check sheets are the simplest form of 'organised' audit. They normally consist of a series of simple questions designed to cover specific elements of a process or quality system, together with a set of check boxes for each question which can indicate 'Yes' or 'No' at the simplest level to an indication of 'fully compliant', 'partially compliant' or 'non-compliant' in more complex cases. Check sheets often have scores allocated to the individual questions to allow an overall score to be calculated. Scoring is discussed in more detail later in this chapter. Check sheets can be very useful for internal auditing, especially hygiene and GMP auditing, and their relative simplicity enables them to be used by less experienced auditors. The nature of a check sheet is that it is very regimented and guides the auditor in specific directions. This type of audit is less likely to look at areas outside the checklist which in certain situations may provide relevant data for the audit. For example, a check sheet may look at the temperature of a meat slicing operation, it may check that the slicer is clean and that the records of cleaning and disinfection are adequate. However, an auditor using a check sheet is unlikely to pick up whether the slicer is hygienically designed or being operated correctly. Check sheets are therefore more suited to operations where the technical evaluation of suitability has already been performed and the auditor is required to check that systems in place are being adequately performed (verification). Check sheets are therefore particularly suited to the regular auditing of a defined set of specified activities, such as internal hygiene auditing, verification of HACCP systems and ISO 9000 type internal audits.

Check sheets are not well suited to auditing unfamiliar premises (third parties) as their scope is too limited. However, it is often very useful to develop standard check sheets which can be sent ahead of the audit, with the request that they are completed and returned to the auditor before the audit. These can then be very useful in making an initial assessment as they can often identify areas where attention needs to be focused during the audit.

Audit questionnaires come in many different guises and are widely used for auditing. Audit questionnaires differ from check sheets in that they ask open-ended questions which are a prompt for the auditor to cover a specific (subject) area of the processes or systems in a plant, rather than the specific yes/no type of questions used in a check sheet. Effective use of the open-ended style of audit questionnaires require that auditors are experienced in the topic of the audit and must understand the requirements set out in any standards that are available. A good auditor will use each question in the questionnaire as a starting point for a discussion in a particular subject area with the personnel involved, and will not move on to the next question until they have assured themselves that the personnel involved understand their role in processing and that the company being audited is, or is not, complying with the requirements. When preparing a questionnaire, care must be taken that the questions guide the auditor into all relevant areas, but also that they give the auditor enough freedom to fully investigate issues in sufficient depth. This is illustrated below where we ask the auditor to look at the same subject, traceability (i.e. the ability to trace a particular material from its origin to the retail trade or consumer), but in different ways.

1. Does the company (plant) have a lot traceability system in place?
2. To what extent can a company (plant) trace products in the marketplace?
3. Lot identification on packs, bins or product is an essential tool for product recall and helps effective stock rotation. Each container (primary pack) of food should be permanently marked to identify the producer and lot.[15]

Question 1 is very restrictive and more suited to a check sheet. It leads the auditor to make a yes/no assessment and relies on the experience of the auditor to actually go beyond the simple issue of whether a traceability system is present to look at its suitability and extent.

Question 2 is more balanced and asks the auditor to look into traceability to determine whether a system exists and whether or not it is suitable. This question requires that the auditor knows what the applicable standard or internal requirement for traceability is, and is able to judge the level of compliance.

Question 3 is not in fact a question but a quote from the standard on which the audit is being based. This serves two purposes. It firstly tells the auditor to look into traceability during the audit. However, because the question is a quote from the standard, it also tells the auditor what is required. It is important to note that this does not mean that the auditor need not prepare, or be familiar with, the standards. It does, however, provide a convenient *aide mémoire* for the auditor to use during the audit.

Question 1 is not suitable for use in an audit questionnaire, and it is advised to use the approach given in question 2 or 3 above when developing audit questionnaires.

Assessment and scoring

The information collected by all audits needs to be evaluated. The methods by which the evaluation is done are very dependent on the type of audit carried out. The audit process will generate data which informs an auditor how well the activity in question complies with the given criteria defined in the standards. Criteria have been mentioned several times, but it is at this stage that they become very important. When making recommendations, based on non-compliance to a standard or criterion, these must be based on non-compliance with the agreed criteria, such as a temperature, stock rotation regime or hygiene standard. It is not good practice for the auditor to make recommendations based on personal belief, as these will be open to debate. A non-compliance based on an agreed standard, whether it be an internal standard such as a work procedure, or an internationally agreed standard, is much more likely to be agreed and accepted by the company or department being audited.

There are no fixed rules determining the amount of information handed over to the plant being audited at the end of the audit. For third party or supplier audits, it is common only to give an indication of 'Pass' or 'Fail', rather than a detailed written report. It should be remembered that one of the main purposes of auditing is to drive continuous improvement. The auditor should therefore leave an agreed list of recommendations with the Plant manager or QA manager, whether a third party or internal audit has been done, and if possible the auditor should give advice on how to solve any problems found.

At this stage we need to mention scoring. Many auditors or audit systems utilise a scoring system by which the findings of the audit are converted into a single score, expressed, for example, as percentage compliance or an approval grade (A, B, etc.). There are as many different scoring systems as there are audit methodologies, but each provides a means by which the results of the audit can be quickly and easily interpreted or compared by persons not involved in the audit. Scored audits also have the advantage in that, if the audits are carried out to the same standard, different audits can be compared quickly and easily, simply by using the score.

There are a number of points to remember when developing scoring systems for audits. The first is that, if not developed carefully, scoring systems can hide critical deficiencies. This can often happen if the scoring system allocates points for excellence or above standard. This immediately allows a company to overachieve in a section of the audit and to underachieve in another section, and when the results are averaged at the end they come out with a standard score. For this reason it is not advised to develop scoring systems which increase the score by overachieving; this should be rewarded in other ways.

There are several ways of ensuring that critical issues are accounted for in the overall score of the audit. The first is to have a weighting system, where the score for each question is multiplied by a weighting factor to give the final score for the question. The weight given to each question should reflect its contribution to product safety or quality. Thus, personnel wearing hair covers and overalls, whilst important, would not be weighted the same as having a calibrated cooking process and strict raw/cooked segregation in an area preparing cooked meats. Where trained auditors are used, it is possible to develop scoring systems where individual questions are not scored, but sections of the audit are scored. The score given to each section represents how well the company or department complies with the given standards, taking into account any critical areas covered in the section. The auditor is therefore looking at the overall picture, placing emphasis on critical issues when giving a score. This can be a very effective system but is obviously more subjective than the method mentioned above. It relies on having well-trained, experienced auditors, good standards and a well-developed audit questionnaire. This approach cannot be used with a check sheet. Where more than one auditor is used to carry out audits of this nature, it is also useful to set up a referee system, either by exchanging reports for discussion between auditors or by having the audit reports refereed by an experienced auditor to ensure consistency between auditors.

Follow-up
The food industry is ever changing. At the external level, new legislation and standards are introduced, new hazards, microbiological or chemical, are discovered which affect the way we work and the risks to our customers, and new process technologies become available. Within a business, new procedures are written, to take account of internal and external pressures, new processes are introduced and new products are manufactured. For this reason auditing cannot be 'one off'. For both internal and external auditing, regular audits are required in order to ensure that the systems and procedures keep pace with the external pressures on the business, and that internally, new procedures are implemented and effective.

Where an audit is part of an audit programme, follow-up is a vital part of ensuring that any actions resulting from a previous audit are being put into place.

Frequency
The frequency at which audits take place is dependent on the nature of the operation being audited. Major suppliers or suppliers of high risk ingredients (i.e. those which may carry pathogens or chemical contaminants) or finished packed product for direct sale will need to be audited more frequently than suppliers of minor ingredients.

11.3.2 Auditing HACCP systems

The principles described above are applicable for all types of audit. In the same way, auditing HACCP systems is no different from auditing other quality assurance systems such as ISO 9000. However, there are a number of points which should be considered. ISO 9000 as a system concentrates on the contractual relationship between supplier and customer, and the conformity to customer specifications.[1] The systems and procedures developed under the ISO 9000 system are therefore derived internally and specify the quality system for the whole company. HACCP differs from ISO 9000 in that it defines the hazards and controls related to a specific product or process, and a plant will have several different HACCP plans in place, one for each line/product, covering the total manufacturing operation. When auditing HACCP systems, therefore, the scope of the audit is likely to be very different from an ISO 9000-type audit.

Before auditing a HACCP system, it is important that the objective of the audit is very clear. HACCP audits make no check on the technical accuracy of the HACCP plan. This activity is part of the validation process which is discussed elsewhere in this book. A HACCP system audit is used to establish whether or not the controls, monitoring procedures and corrective actions defined in the HACCP plan are being applied correctly, and whether or not they are effective. It is a common misconception that HACCP audits will indicate whether a HACCP system is 'safe' and covers all applicable hazards. This is definitely not the case.

A HACCP systems audit would generally cover the following elements:

1. Have the HACCP studies been carried out according to the seven principles described by the Codex Alimentarius,[7] or an equivalent system?
2. Has a team approach been used to generate the HACCP plan, and what technical expertise has been available to the team?
3. Does that HACCP plan cover all the expected CCPs, together with targets, limits, monitoring systems and corrective actions? (This would normally be a part of validation, and would not be covered in an internal audit.)
4. Is there evidence that the HACCP plan has been validated?
5. Has the HACCP plan been discussed with operators, and do operators have access to work procedures based on HACCP? Have they been sufficiently trained and do they have sufficient tools and authority to carry out their responsibilities?
6. Are monitoring procedures being carried out and recorded on the factory floor? Is there any indication that the control procedures are not effective?
7. Are there clear priorities for action in the event of a process deviation?
8. Has the process changed since the study was carried out?
9. What verification data is available to demonstrate the effectiveness of the HACCP plan?
10. When was the HACCP plan last reviewed?

The above is not an exhaustive list but covers the main elements normally associated with a HACCP audit.

Internal auditing of a HACCP plan

In general there is very little difference in auditing a HACCP system in your own plant and in that of a third party. Both audits will require that the auditor assess the elements described above. However, in an audit 'in house' elements 1–4 above will be assessed initially and then left out of the regular audit system which would focus on elements 5–9. The key to auditing HACCP is not to spend a great deal of time examining the HACCP plan to check its accuracy – this will have been done when the plan was validated – but to focus on the operational side of HACCP. What we mean here is that the HACCP plan will define a number of controls and monitoring systems associated with each CCP. The aim of the audit is to check that working procedures are available which adequately cover the requirements at the CCP, that the operators have, and understand, these procedures and that any required data collected is being recorded and action taken if the process or material is outside the critical limits.

An important part of the HACCP audit is not only to check that the HACCP plan is implemented and the procedures are in place, but also to check that there have been no changes on the line, to working procedures (e.g. times, temperatures or hygiene) or to product formulation (e.g. preservation system or packaging) which may affect the effectiveness of the HACCP system. Although this is normally associated with the formal review of the HACCP system, it is normally not sensible to leave this type of check for the yearly review but to keep on top of the changes in this more frequent audit system.

External auditing of a HACCP plan

Auditing a third party HACCP plan follows the same principles as defined above. However, although it is not normally necessary to check the content and accuracy of your own HACCP plan, the auditor will need to make a judgement on the content and accuracy of the third party plan, to check its suitability for ensuring the safety of the supplied product. It is very difficult to assess another team's HACCP study, especially if you are not familiar with the product or the processes used by the third party. The way to tackle this problem is to identify the minimum CCPs that you would expect to find for the type of process being audited. This information can often be found in industry guides, or in generic HACCP plans which are produced for different sectors of the food industry. A note of caution here is that by their very nature these guides are generic and can be superficial. However, they should be of use in identifying the minimum number and location of CCPs which you should be able to find in the HACCP plan of the third party. If these minimum CCPs are not present this immediately warns the auditor that this HACCP plan is not likely to be effective at controlling the hazards in the process. As an aside, the Internet is a source of significant information with regard to HACCP, and many food companies post their HACCP plans on the Internet. These can be a useful source of information, but again they must be used carefully as these are individual company plans and have not undergone a peer group review, unlike the industry guides available. The USDA/FSIS provide a number of generic HACCP studies at http://www.inppaz.org.ar/MENUPAL/Bvirtual/FOS/haccp/usda/haccpmod.htm.

If the HACCP plan is acceptable, the auditor will then proceed to determine if the plan has been implemented in the factory and is working as intended.

11.4 What the auditor should look for

In any audit, time plays a crucial factor. The auditor never has sufficient time to cover all the elements they would like to, and good time management is critical to the success of an audit. As an auditor it is therefore important to remember that you will never be able to check everything, and should not try to do so. As a general guide the auditor should carry out following procedures.

- Start with a brief tour of the factory, starting with the raw materials and finishing where the finished goods leave. This tour is not a fact-finding exercise but is intended to give the auditor a general feel of the operation being audited. It will also provide an insight into the management attitude of the company with regard to quality and safety. A clean, tidy, well-organised factory with hand washing, clean operators with suitable protective clothing, notice-boards and signs instructing operators in good practice is always a good indication that the management are committed to quality and safety. On the other hand, an untidy, dirty and haphazardly organised factory gives a clear indication of a general disregard of the management for quality and safety. First impressions are significant, and although it is important that the auditor does not jump to too many conclusions from the initial visit, an experienced auditor will normally be able to tell what the outcome of the audit will be from this visit.
- The auditor should now check whether or not the required systems and procedures are in place to cover the required elements of the HACCP system, and whether they contain the necessary depth of information. The use of a well-designed check sheet or questionnaire is a vital aid to ensuring that all the relevant systems are covered during the audit. Remember that the auditor here is assessing against standards and not making a personal judgement.
- The existence of a well-written procedure is not an indication that the system is implemented in the company. It is the role of the auditor to check that what is written on paper is actually working and is effective. Although the auditor should check whether or not all the required procedures exist, they will not be able to verify that all procedures are in place and working. Therefore he or she should select a number of key elements to check. Selection of the elements to check should not be a random process and the auditor should always check a number of the CCPs defined in the HACCP plan, to assess whether what is described in the HACCP plan is in fact happening on the factory floor. This therefore involves checking that the work instructions for operators cover the work practices and any control measures and that the targets and limits are clearly specified to enable the operator to judge whether the CCP is in control. Monitoring procedures should be available on the line or in a laboratory and records should be meaningful, available and up to date.

- The auditor should also check a selection of other procedures so as to ensure the quality of implementation of the HACCP plan and the background of GMP. It is often useful, during the initial factory visit, to note any activities that do not appear to be in line with a given standard. If a procedure exists which covers the activity observed and which is not in accordance with the standard, clearly there is a problem with implementation.
- It is extremely important to talk to people, especially the operators on the production floor. It is possible to find out more about the current state of implementation of the company's quality system by talking to the operators than in any other way. (Do they know what a CCP is? Have they been told about HACCP?) Ask to see work procedures and line check sheets used for monitoring CCPs and other quality parameters. If the operator does not have the relevant procedure readily available, it is more than likely that the procedure is not being followed.

11.5 Future trends

Food quality and safety is continuously evolving and the foods industry needs to keep abreast of these changes to remain competitive and meet customer requirements. HACCP is here for the immediate future, and future trends in HACCP are discussed earlier in this book. However, one point to note is that many major customers now see HACCP as a key requirement from their suppliers, whereas in the past ISO 9000 was seen as the key requirement. As such, HACCP certification may become a more important feature of the HACCP system. Already many third party accreditation companies are offering HACCP certification services, either as a stand alone, or combined with existing ISO 9000 certification. International standards for assessing and certifying HACCP are being developed[16] with the aim of standardising the certification process. Currently, HACCP certification looks at the approach taken and the standards used for developing the HACCP plan and subsequent implementation of the plan. Technical accuracy of the HACCP plan will not usually be assessed and this may become a weakness of the certification process.

ISO 9000 was the dominant quality system in the early 1990s and is currently under revision (the so-called ISO 9000:2000 standards). This standard will retain the original 9001–9004 standards, but has changed the structure of the elements making up the standards. The ISO 9000:2000 standard has five elements, each with a number of sub-components:

- Quality Management System Requirements (one sub-component)
- Management Responsibility (six sub-components)
- Resource Management (three sub-components)
- Management of Processes (seven sub-components)
- Measurement, Analysis and Improvement (two sub-components).

Many of the sub-components are further subdivided. The 20 elements of the existing ISO 9000 system are covered within the five elements in the new system. However, the new system places more emphasis on validation and will hopefully address the issues associated with the current standard, whereby it is not inconceivable for a company to miss key activities within their internal system but be able to gain accreditation by demonstrating compliance with incorrect or incomplete standards defined internally.

For many companies it is difficult to find the resources, and the necessary skills, within their company for auditing third parties. In addition, many companies are faced with an increasing number of customer audits, which takes valuable resources from the day-to-day activities of the company. Third parties are picking up on these facts and offering third party auditing services, and even accredited auditing services. One such which is operational in the UK is the European Food Safety Inspection Service (EFSIS).[17] The system audits a plant against 35 set criteria in quality, safety and hygiene, and if the audit is acceptable will grant accreditation. Accreditation is a continual process and the frequency of re-accreditation will be determined depending on the type of process and the previous audit score. The rationale behind the EFSIS scheme is that it will reduce the number of third party or customer audits by providing third party auditors who will assess suppliers, and it will allow companies to show they have reached a set of fixed standards defined by EFSIS. Third party auditing and accreditation schemes are becoming seen as a good means of reducing the resource requirements in a company with regard to auditing, and offer independent assessment of a company's safety and quality system. Such systems are dependent upon the skills and professionalism of the auditors who carry out the assessments, but are likely to become more important in the food industry.

Current quality systems, and many of the associated auditing systems, focus on whether or not a system exists, and check that the system is actively implemented within the company. Very few systems require that the subsequent results of the implemented system are evaluated. Within Europe, the European Foundation for Quality Management (EFQM) has developed a model for quality excellence.[18,19] In common with ISO 9000 and HACCP, the EFQM system provides a framework for achieving excellence. This framework is built up of the following nine elements:

1. Leadership
2. People
3. Policy and strategy
4. Partnerships and resources
5. Processes
6. People results
7. Customer results
8. Society results
9. Key performance results.

However, unlike other systems, the EFQM divides these elements into 'enablers' and 'results', elements 1–5 being defined as enablers and elements 6–9 as results. Enablers are those criteria which define what the organisation does, and would be focused on internal policy, systems and procedures making up a quality system. The results are intended to cover what the organisation achieves, the premise being that there cannot be results without enablers. The EFQM website (http://www.efqm.org) describes the system in detail. The EFQM system is not the only system which focuses on results but the future lies with such systems, which look outside the organisation to ensure that what is defined internally has the desired results both internally and externally.

11.6 References

1. HARRIGAN W F, The ISO 9000 series and its implications for HACCP, *Food Control*, 1993 **4**(2) 105–11.
2. MAYES T, The application of management systems to food safety and quality, *Trends in Food Science & Technology*, 1993 **4**(7) 216–19.
3. ISO 9000, *Quality Management and Quality Assurance Standards – Guidelines for Selection and Use*, 1987.
4. ISO 9001, *Quality Systems – Model for Quality Assurance in Design/ Development, Production, Installation and Servicing*, 1987.
5. ISO 9002, *Quality Systems – Model for Quality Assurance in Production and Installation*, 1987.
6. ISO 9003, *Quality Systems – Model for Quality Assurance in Final Inspection and Test*, 1987.
7. ISO 9004, *Quality Management and Quality System Elements – Guidelines*, 1987.
8. Codex Alimentarius, *Hazards Analysis and Critical Control Point System and Guidelines for its Application*, Alinorm 97/13A, Codex Alimentarius Commission, Rome, 1997.
9. OAKLAND J S, *Total Quality Management. The route to improving performance*, 2nd edn, Oxford, Butterworth Heinemann, 1995.
10. British Standard Quality Systems (BS 5750), British Standards Institution.
11. JOUVE J L, STRINGER M F and BAIRD-PARKER A C, Food safety management tools, *Food Science and Technology Today*, 1999 **13**(2) 82–91.
12. *Recommended International Code of Practice – General Principles of Food Hygiene*, CAC/RCP–1 (1969), rev.3 (1997).
13. Institute of Food Science and Technology, *Food and Drink Good Manufacturing Practice Guidelines* (a guide to its responsible management), 4th edn, 1998.
14. CHESWORTH N, Implementing a factory auditing programme. *Int. Food Hygiene*, 1993 **4**(4) 11–13.
15. Codex Alimentarius, *Food Hygiene, Basic Texts*. Joint FAO/WHO Food Standards Programme, FAO Rome, ISBN 92-5-104021-4, 1997.

16. Criteria for testing an operational HACCP system, Central Board of HACCP Experts, The Hague, PO Box 93093, 2509 AB, The Hague, 1996.
17. RICHARDSON D, Audit after audit, is there an alternative? *Food Manufacture*, 1998 **70**(4) 20.
18. ROGERS V, EFQM, a model for management excellence, *Food Manufacture*, 1998 **73**(12) 20.
19. The EFQM Excellence Model, http://www.efqm.org

12

Moving on from HACCP

J.-L. Jouve, Ecole Nationale Veterinaire de Nantes

12.1 Introduction

Throughout the world, food manufacturing, distribution, retailing and food service is becoming a highly complex business. Raw materials are sourced on a global scale and an increasing number of processing technologies are used to provide a wide variety of products to the consumer. In addition, consumers' expectations are changing, with a desire for convenient foods with less processed, fresher and more natural characteristics.

Against this background of change, there has been a permanent improvement in the performance of the food industry at large, associated with unprecedented efforts by private and public organisations to create an environment that fosters better prevention and control of foodborne hazards. In this context, the Hazard Analysis Critical Control Point (HACCP) system is now widely accepted as the most effective means of ensuring that a high standard of food safety is maintained.

However, common experience shows that food systems are still vulnerable to disturbance when challenged by internal or external factors such as equipment breakdowns or new hazards. Public and private organisations have been thrown into disarray when faced with recent food scares, such as *Escherichia coli* O157:H7, bovine spongiform encephalopathy (BSE) or the crisis related to the presence of dioxin in animal feeds that affected several countries in the European Union in 1999. The debates and controversies raised by endocrine-disrupting chemicals or genetically modified organisms (GMOs) have also shown the vulnerability to disturbance or failure that has many consequences. Failure of food systems may result in a serious threat to public health, but also has economic or legal consequences and the potential for financial loss to businesses and countries. The major outcome of these failures is the scepticism of the public at large about the security of the systems responsible for producing, manufacturing and safeguarding food safety. This is regardless of the improved level of safety actually achieved.

Several factors may contribute to this perceived vulnerability.[1] All are derived from the central position that food plays in everyday life. The vulnerability of the food supply affects the welfare of citizens, consumer choice and food price, the income of farmers, manufacturers and others and the strength and international competitiveness of national agro-food economies. Any disturbance in consumer confidence feeds back to all links in the chain.

Food may contain many different hazards. Food-borne microbial pathogens are highly changeable and adaptable; foodstuffs can also carry a number of chemical/toxicological risks, which lead to new areas of concern, such as endocrine perturbation or allergenicity.

Risks or perceived health risks linked to food are becoming increasingly unacceptable to society. As, in reality, food becomes safer, the public at large becomes less tolerant of the remaining and occasional risks. This trend is enhanced because the general public feels more and more alien to the preventive or control activities of experts, where decisions appear to be made in isolation by technological (e.g. the food industry) or administrative (e.g. the public agencies having jurisdiction) organisations. This results in a new perception of food safety risks as an 'outrage'. An outrage is seen where the level of risk is controlled by others who may be coerced by industry and linked to other untrustworthy sources. The public wants outrage risks to be taken more seriously.[2] Despite substantial progress in food science and food control, the current technological or administrative approaches pay little attention to the perception of outrage, thus magnifying any residual risk.

Other forms of vulnerability are associated with the production, manufacturing or distribution of food. These sourcing systems are characterised by the following features.

- Their interdependence (e.g. high integration, specialisation, global susceptibility to one single adverse event, series of connected responsibilities)
- Their fragility (e.g. existence of several weak points in commercial and business processes)
- Structural obsolescence
- The possibility of drift in specifications and the application of control and quality assurance schemes
- The existence of weak segments whose failure may adversely impact on the activity of all other segments
- Their 'black-box' nature (e.g. lack of a global and transparent presentation of food safety assurance and management; this is often a feature of the systems used or may happen by accident)
- Managerial shortcomings (e.g. uneven allocation of resources, complexity of technical regulations, poor internal or external communication, lack of

consistent, comprehensive and flexible food safety programmes and structures).

Over the past decades, industry and regulators have focused mainly on scientific and technological advances aimed at preventing or controlling hazards in food. Today, the analysis of major food safety problems teaches us that many of these problems have their origins in organisational deficiencies. Increasingly complex food systems are vulnerable because a global organisation for approaching food safety issues is not growing at the same rate as the food business is changing or consumer concerns are growing.

The principal lesson to be drawn from this overview is that the challenges of providing food safety have changed. The main challenges for the total food chain are to maintain the highest standards of safety, to meet new challenges and reduce the vulnerability of food systems in order to restore and develop public trust. Meeting this challenge basically requires more effective use of HACCP, as discussed by the previous chapters of this book. Running alongside is the need to develop a broader approach to food safety management extending beyond HACCP. This chapter outlines a framework for this approach and considers its rationale, its components and some of the tools that may be used. Although the primary focus will be on microbiological issues, the principles are equally applicable to control of chemical or physical contaminants.

12.2 Future trends

It is widely recognised that HACCP has the potential to provide enhanced assurance of product safety by focusing resources on the control of raw materials and other key steps in the supply chain. However, as a rationale or for prediction of future trends, it is important to understand what HACCP can actually do for food safety improvement and to recognise that the approach has some limitations that need to be overcome.

12.2.1 Strength and limitations of HACCP

HACCP originated in the food industry. It is a system owned by food producers, presently widely accepted and utilised across their industry. Its principles can be applied at all stages of the food chain, although some difficulties may require specific adjustments to the system (e.g. in the primary production, at the slaughterhouse, or even in the home). It has also become a cornerstone of many national regulations and international recommendations related to food control. It can never be overemphasised that HACCP was primarily designed as a tool to establish or improve product/process control activities and provide assurance that operators focused control of their processes where product/process sensitivity and/or food safety requirements were greatest. HACCP was derived from the armoury of reliability tools, in particular from HAZOP (Hazard

Analysis and Operability) type studies and FMEA (Failure Mode and Effect Assay). Both are derived from engineering quality management systems that look at a product, its components and manufacturing and ask what can go wrong within the total system. In this respect, the HACCP study is designed to identify hazards and find potentially hazardous conditions that may exist in a product or process (contamination, development or persistence of hazards) in order to eliminate or control them and their causes. This leads to all the process stages being controlled in the most effective way.[3]

There is no need here to describe HACCP in detail, as this information is available in the preceding chapters of this book. Suffice it to say that HACCP requires users to foresee where problems may occur and to take steps to reasonably ensure that they will not. Under the HACCP concept, potential problems are identified and steps are taken to analyse likely causes and to develop and implement preventative measures at the stages involved. Appropriate evidence has to be produced. According to its principles and rules, HACCP focuses on hazard management, through identifying hazards and hazardous conditions during the HACCP study, leading to their assessment and control by its implementation. It can minimise the chances of sporadic problems, while establishing more reliable control. It is a systematic and very effective approach for reducing the probability of unsatisfactory supply chain performance, thus making the process and products safer.

However, whilst industry and government bodies have published their expectations of what HACCP will achieve and there is genuine commitment to its use as a key food safety tool, there are a number of inherent limitations in the approach. In particular, although HACCP is a systematic and practical approach to hazard control in food manufacturing, it has not been designed for, nor is able to guide, managerial decisions about:

- the nature of hazards whose elimination, or reduction to an acceptable level, is considered essential to the production of a safe food and to maintenance of consumer confidence, i.e. the identification of hazards and assessment of their public health significance,
- the definition of an acceptable level of hazard(s) in a food after processing, relative to the level desirable for consumer health protection (or a level of risk that a society considers as acceptable or tolerable). There is no means for the HACCP system to check whether its outcome or objectives (design or adjustment of products, processes and control measures) are appropriate to the needs of society in public health terms. It does not consider these versus other considerations, such as technical feasibility or the cost of achieving control. The hazard analysis stage is very weak.

Clearly, such managerial decisions are 'outside' the HACCP process and the necessary linkage of HACCP to public health requirements has been generally overlooked.[4] To be meaningful and effective, HACCP needs to be driven by an understanding of the relationship between the reduction of risk in a food process, and the level of food safety required by consumers. In other words,

HACCP needs to be directed by consumer requirements for food safety. This is the only means for appropriate identification of the hazards that concern consumers, for the allocation of resources and determination of extent and stringency of the HACCP plan and its control measures. Determination of the outcome in this way would give consistency in evaluating what is acceptable and what is not, allowing for the validation of the components of the HACCP plan, its Critical Control Points (CCPs), their critical limits, controls and monitoring procedures.

Determined in such a way, the outcome may also provide a reference for comparing the objectives of different HACCP plans. To that point, it has to be borne in mind that HACCP originated in the NASA space programme and was originally applied by various parts of the food industry (e.g. canning) and for the control of specific types of hazard (e.g. foreign bodies) to give zero defects. For these applications the objective was chosen prior to development of the plan. For microbial pathogens, a zero defect or zero tolerance level may or may not be appropriate. To design food safety into products and processes and to provide appropriate assurance along the supply chain that food safety requirements are effectively met, implemented HACCP systems should evolve so that they operate with reference to the broader framework of food safety and public health.

12.2.2 The way forward

The future challenge is to improve the management of food safety by providing a clear link between control and public health benefits. If HACCP remains the chosen system then the challenge is to improve its efficacy whilst still keeping its practical nature.[5] There might be several approaches to achieving this, but we firmly believe that there are three crucial components.

The first is government-led and corresponds to the development of a risk-based food safety strategy, with a public health perspective. This approach would include

- determination of the requirements for food safety,
- development and use of specific procedures for risk analysis, extending from scientific understanding and characterisation of the actual risk to the appraisal of managerial options,
- monitoring of the implementation of any measures,
- assessment of their effectiveness,
- review of hazards and measures.

Mechanisms would need to be provided to ensure the involvement and participation of interested and affected parties (the stakeholders) at all stages. The practical implementation of this approach would rely on two basic principles, the limitation of exposure and optimisation by review.

The second is the development of effective food safety management programmes and systems by food business operators. This radical approach should help businesses take account of their role in the food chain and ensure that product safety is given the highest priority. Food safety programmes should help to determine the food safety requirements of each business, identify where improvements are necessary and how any organisational and technical issues can be tackled and the improvements reviewed in the light of changing food safety issues. The systems should incorporate HACCP as a key tool for implementation.

The third is to realise the full potential of HACCP for ensuring that food safety requirements will be met. This involves using (quantitative) risk assessment techniques at the Hazard Analysis stage of the HACCP study. The aim of this is to link the probability of failure (in the process) with the severity of the consequences for public health and to use these findings to apportion safety management resources throughout the different stages and elements of the process.

These three components are considered in turn in the following sections.

12.3 Development of a risk-based food safety strategy

Food safety results from the successful interaction of government agencies, business, private organisations, consumers, and other supporting players. Government agencies are in the best position to influence how the other partners work together. Until very recently, these agencies had a mainly reactive approach to food safety problems and focused their interventions on specific contaminants posing immediate hazards and on preventing poor hygiene in the food chain. As a consequence, food safety problems were dealt with mostly in a pragmatic way. Now because of the complexity of the food safety problem, there needs to be a move from this hazard-based approach towards a more comprehensive, risk-based approach taking a public health perspective. This requires the development of national food safety plans[6] that should encompass the following elements:

- Formulation of a food safety policy and objectives (see the newly introduced concept of 'Food Safety Objectives', in Section 12.3.2)
- Identification of systems and means for ensuring that the objectives are achieved
- Development of supporting food control activities
- Evaluation of the effectiveness of activities undertaken
- A mechanism for review against current needs.[7]

Hence a national plan would provide a framework and infrastructure for the development of horizontal functions. These should include targeted research, data collection and analysis; surveillance and monitoring; management of emergency situations and the building of adequate resources, including personnel; strategies for the reduction and containment of identified risks; and building the confidence of consumers. The international debate until now has focused attention on risk analysis as the foundation for decision making in the process of designing and implementing a food safety strategy.

12.3.1 Risk analysis

Risk analysis is becoming a cornerstone for the development of food safety plans. The technique can legitimise and communicate decisions regarding programme priorities, allocation of resources, levels of protection appropriate to populations, preventive interventions and research. Risk analysis has been described as a process consisting of three components: risk assessment, risk management and risk communication.[8] Although it is oriented towards decision making, which is a managerial activity, risk management always needs to be supported by risk assessment and risk communication and cannot function well in isolation.[9]

An emerging consensus suggests that the risk management process should encompass four elements.

- Risk evaluation, which includes
  — identification of a food safety problem,
  — establishment of a risk profile,
  — ranking of the hazards for risk assessment and risk management priority,
  — establishment of risk assessment policy for conducting the risk assessment,
  — commissioning of a risk assessment,
  — consideration of the results of the risk assessment.
- Risk management option assessment, consisting of
  — identification of available management options,
  — selection of preferred management option,
  — final management decision.
- Management decision on implementation.
- Monitoring and review, including
  — assessment of effectiveness of measures taken,
  — review of risk assessment and/or risk management as necessary.[10]

Risk assessment provides essential factual support for the risk management decision. It is a scientific process aimed at understanding known, or potential, adverse health effects resulting from human exposure to foodborne hazards, how likely they are to occur, and their consequences.[11] It covers documentation and analysis of the scientific evidence to measure the risk and to identify factors that influence it. Risk assessment is the domain of the sciences and uses concepts and information from many fields that is structured and passed to risk managers to assist them in making informed decisions. In this context, it is the duty of the managers and decision makers to create an environment safeguarding the scientific independence and integrity of the risk assessment, while at the same time ensuring that it is documented, structured, transparent, reliable and credible.[12] To do this it is essential that risk assessment is functionally separated from risk management. But to ensure a realistic outcome, a balance has to be found between ensuring the scientific independence, integrity and transparency of the risk assessment and interaction with risk management.

Risk communication is understood as an exchange of information and opinions between the risk managers, the risk assessors and the other concerned or affected parties (the stakeholders). It must extend over the whole process and is crucial for bringing together the necessary information, bridging gaps in understanding, values and perceptions, for ensuring that public values are considered and, finally, for generating better accepted and more readily implemented decisions.[13] How, and how much, stakeholders are involved will depend on the nature of the study, but however it is done should not compromise the independence and integrity of the risk assessment. It should not hide any responsibilities for risk management.

Providing an outline of developing risk analysis concepts and practices goes far beyond the scope of this chapter and additional information can be found in the literature. The elements of risk analysis in one form or another have been utilised by many government agencies to deal with food safety problems. However, critical evaluations[14,15] have stressed that there is now a need for a more systematic and comprehensive application of risk analysis to food safety strategy and programmes, by both governmental authorities and food companies (see Section 12.4).

12.3.2 Principles for implementation

The food safety risk associated with microbial contamination of foodstuffs can be virtually eliminated by eliminating pathogens. An array of cheap technologies to do this is available, or may soon be. However, a strong school of thought recognises that it might be unrealistic to require all microbiological contaminants to be eliminated from all foods. If we share this view, then the objective becomes risk reduction or minimisation to the level of 'acceptable' or 'tolerable' risk. This should be guided by two principles: limitation of exposure(s) and optimisation, using risk analysis as an essential supporting tool.

The *principle of limitation* suggests that the exposure of individuals to a hazard should be limited, so that no one is exposed to an 'unacceptable' extent. Between unacceptable and negligible exposure, a band of concern over levels or concentrations ('brightlines') can be made based on risk.[13] This concept is intended to convey the idea that there is not an exact boundary between safe and unsafe, but equally it is not intended to indicate that the risk, provided it is low, is acceptable. Rather it serves as a source of information about the level of health protection necessary for any hazard.

There are several examples of its use and usefulness outside the food industry, in sectors such as radiological protection. It is amazing that the scientific and public health community in the field of microbiological food safety[16] has not debated its usefulness. Health-based 'brightlines' should be established during risk analysis, by merging the scientific and analytical process of risk assessment with an analysis of the different options from a societal perspective, taking into account the wishes of stakeholders. Differences between a technical appraisal of risk and the risks concerning stakeholders contribute to the breadth of the brightlines. Thus communication can help decision making on risk reduction and process optimisation. In practice, risk-based brightlines could be expressed as pathogen distributions, levels or frequencies. This is the new concept of Food Safety Objectives (FSOs) proposed for the management of microbiological hazards by the Codex Committee on Food Hygiene; several groups are currently considering procedures for the application of these objectives.[17–19] Multiple brightlines may be established when there is a need to protect sensitive sub-populations, or there may be a range from the upper boundary of 'acceptable' risk down to lower levels where technology and other considerations allow virtually 'zero risk' to be achieved.

The *principle of optimisation* is the essential basis for risk management, because it requires that within the brightlines, exposures should be 'as low as reasonably achievable'. Risk assessment plays a pivotal role in determining actual levels or frequencies in question. Optimisation starts with a scientific characterisation of the actual situation (unrestricted risk assessment), identifying and ranking the most important factors contributing to the risk. This process may eliminate some scenarios and support the development of particular mitigation measures. A restricted risk assessment will allow comparison of mitigation measures for robustness and effectiveness, while providing an essential basis for further considerations, such as cost-effectiveness.[20]

These principles may be implemented by government agencies when considering a food safety problem from production to consumption and establishing the basis for intervention. They may also be applied by food business operators for the management of a specific segment of the supply chain or a process, through a comprehensive food safety programme.

12.4 The Food Safety Programme

Within a food company, food safety is typically cross-functional, including activities that draw on many functions and departments, using different skills and levels in the organisation. It also extends outside an organisation, to suppliers and customers. Because food safety interacts with all aspects of food production and distribution, such as quality, productivity or costs, 'trade-offs' or compromises between areas may create difficult problems.

Expectations of HACCP should not extend beyond what it can realistically achieve. HACCP is only a powerful tool for reliability improvement. To contribute to public health and food safety within a food company, HACCP has to be operated within a longer-term managerial strategy concerning food safety. This is the rationale for integrating the Food Safety Programme (FSP) with other managerial initiatives of a company.

The intent of this section is to provide a 'road-map' for the development of an FSP and it utilises many of the same concepts which underpin other quality tools, such as the Total Quality Management (TQM) approach, the series of standards related to quality management (ISO 9000 series of standards) or environmental management systems (ISO 14000). Detailed guidance may be found in the related literature,[21-26] standards and guidelines for application.

12.4.1 FSP – a managed programme

The Food Safety Programme (FSP) must address the needs and expectations of customers in relation to food safety and be compatible with a company's capability. Total commitment of the senior company management is crucial to successful implementation. The FSP should cover both organisational and technical issues and should be focused where improvements are likely to be necessary. In line with TQM, it is based on the concept of continuous improvement and the participation of all members of the organisation.

Preparation for the programme needs a consistent and disciplined approach. The starting point is the development of a food safety policy and a review of food safety issues. The policy is a means to guide and inspire the development of food safety activities within a company. It is a statement by senior management establishing the overarching goals with regard to food safety performance and the direction for action to maintain and improve its performance. For example, a food safety policy may include core values and guiding principles of food safety, their relationship with other policies and guidance on best management practices and best practicable technology. It should state the company's commitment to compliance with legislation and regulations and provide a framework for prevention and risk reduction. It needs to make provisions for evaluating performance, change and continual improvement. Policy should require the involvement, education and training of internal and external interested parties. Management should ensure that the food safety policy is fully communicated to, understood and supported by all employees.

Under the guidance of the policy, issues should be prioritised so that requirements and actions to achieve full implementation are identified. To meet its commitment, management needs to ensure that sufficient resources are available at the right time and that there are on-going activities to improve the food safety performance of the company. It is particularly important to involve all employees.[27] Communication is a key for success and management should ensure that the importance of food safety for consumers and the company is properly understood at all levels of the organisation. Motivation is of primary importance. To help this, cross-functional teams and teams addressing specific activities (e.g. the HACCP team or teams providing assistance or support to suppliers and customers) should be built to develop the opportunity for everyone to contribute. This implies a clear presentation of the food safety policy and its requirements with adequate supporting information. Within the company both awareness of the FSP and the linkage of company activities to food safety should be developed. Correct presentation should ensure support from everyone in the organisation.

Management should ensure that all employees develop a positive, proactive attitude towards food safety issues. Emphasis should be placed on prevention, and not only on control activities. To support this, the Food Safety Programme should provide evidence of reasonable care and regulatory compliance. Most importantly, it must provide a climate for continuous improvement. As a whole, the Food Safety Programme should contribute the development of a food safety-conscious culture within the company, providing consumer satisfaction and increasing the company's competitiveness.

12.4.2 Stages and components

The Food Safety Programme can be developed through a series of stages,[28] as illustrated in Fig. 12.1.

Preparation
Preparation is the initial stage. Its aim is to establish the company's position with regard to food safety and determine the needs and opportunities for establishing an FSP. This should include the following elements.

1. An inventory of constraints, including the level of protection required by public health authorities (legal and regulatory requirements) and other professional or contractual requirements.
2. Identification of food safety issues relevant to the company's products and processes.
3. Identification of the impact of products and processes on food safety, including a comparison of the company's performance and internal criteria with consumers' concerns and needs and external standards and regulations. The degree of practical control of food safety achieved by current safety practices and procedures should be considered, not only for normal or abnormal conditions, but also for potential 'emergency' conditions. Contractual and liability issues should also be considered.
4. A review of past problems or shortcomings and information gained from investigation of previous non-compliance or incidents.
5. Identification of opportunities for improvement or change.
6. Consideration and description of accepted risks (see below, product/process planning and Section 12.5).

The preparation phase continues with the identification of company or structural barriers and enablers. Barriers may include poor organisation, inadequate resources, cultural and technical factors and marketing constraints. Enablers may include a motivated workforce, and the identification of areas for improvement, and for programme implementation (e.g. organisation, resources, technology, motivation and training of the workforce).[27] It concludes with the

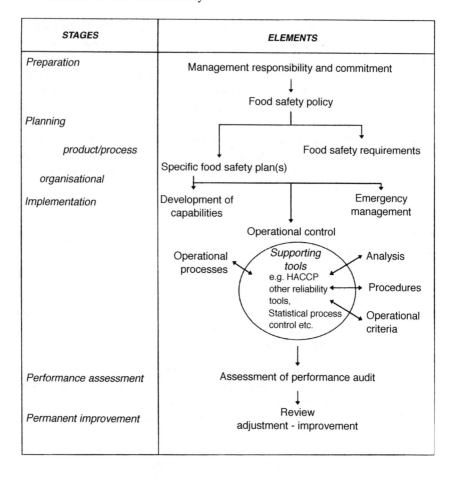

Fig. 12.1 The Food Safety Programme.

development and communication of a food safety policy and the commitment of resources for planning and implementation by management.

Planning
Planning is the activity that ensures a logical and structured approach to changes and improvements. It covers organisational planning and product/process planning.

Planning should make provision for the preparation of specific food safety plans or the improvement of existing ones, and establish the company's requirements for food safety, including new and existing products and processes. It should address the specific designs, practices, resources and sequence of activities relevant to the safety of products or processes; a HACCP plan is an example. It needs to assign responsibility, authority and resources, and identify those specific actions, procedures and lines of communication necessary to gain support and commitment from all personnel. Organisational planning should also include the development of mechanisms for assessing performance and continuous improvement. Requirements should be expressed as a set of measurable quantitative or qualitative requirements for the product and process. Food safety requirements should cover all significant aspects of food safety and be based on the company's preferred level of protection, to meet public health requirements and protect the company from the adverse consequences of unsatisfactory performance. The current capability and any improvements to the technological or control within the supply chain, plus envisaged product developments or new markets, must be covered and may lead to periodic review and revision.

Company food safety requirements may be expressed in several forms, such as rules to be followed, procedures to be implemented or attributes to be met (e.g. a target level for a certain contaminant in a food). Where they are not directly measurable, a company should consider developing measurable indicators of performance (e.g. rate of failure associated with a particular hazard) to allow for the assessment performance versus requirements. Food safety requirements may exist at several levels. At the highest level there are global requirements for a broad system within a multi-site company. Lower down there are site-specific requirements for a local business and below them requirements for a product or process design or operational requirements, covering a group of steps, or even a single step, in a production process (e.g. a known reduction or inhibition of specified microorganisms).

The consideration of risk plays a major role in product or process design and the determination of its food safety requirements by a company. A clear statement of requirements is necessary to provide a basis for decisions and to determine what is acceptable and what is not with regard to food safety performance. At a company level, consideration of risk includes two interacting dimensions. One is consumer oriented (the consumer risk) and corresponds to the unacceptable probability of illness resulting from consumption of an apparently safe food product, leading to public health problems and possible litigation. The other is company oriented (the producer risk) and refers to the unacceptable chance that a process step does not consistently produce foods meeting specified requirements for safety or may be falsely rejected by QA. Even if there is no increase in risk to the consumer this may result in claims, loss of confidence, loss of image, or loss of market.

The control or minimising of consumer safety risk should be the primary consideration of a business. It must provide the level of control of food safety risk that is 'accepted' or 'tolerated' by society at large, or to express this positively, to the level of consumer protection that should be guaranteed. The minimum level (LOP) is usually fixed by public authorities, and may be developed through a government-led risk analysis process (see Section 12.3) leading to the proposal of specific 'Food Safety Objectives (FSO)'. These identify the level, frequency or concentration of a hazard in a food that is tolerable to provide a specified level of protection, i.e. the level of control that should be achieved. Hence an FSO provides a basis for a company to determine its own minimum food safety requirements. However, a company may in practice wish to operate to another (lower) level of risk, perhaps defined by the rate of failure it will tolerate in the marketplace. This may take into consideration additional factors such as the reputation (or loss of image and position) of the brand or consumer perception. In any case, the level of risk accepted by a company should always be lower, never higher, than the level determined by reference to public health authorities in their Food Safety Objectives.

Food producers have traditionally considered and managed the risks of their products using an empirical, experience-based, qualitative approach. Today, there is a need to rationalise decisions about food safety requirements using a more formal, quantitative approach within the framework of risk analysis. Its purpose is to ensure that industrial risk management is linked to the consumers' perspective on public health risk and to facilitate communication of decisions regarding food safety requirements and their basis inside and outside the company, especially to consumers.

The framework identifies significant aspects of food safety and places them in a risk context, by assessing information on potential hazards, exposure assessment and hazard characterisation, followed by risk characterisation. Based on actual practices within the factory (unrestricted risk assessment), this analytical process allows appreciation of the risks actually taken by a company and how these match with, or differ from, the risks deemed 'tolerable' by public authorities, society or even the company. As such it supports the sound identification of food safety requirements in a prospective manner. Participation of staff in the technical and analytical stages of risk assessment and in decision making ensures understanding and communication throughout the food business.

For efficient development, risk assessment should be introduced according to a 'tiered' or 'phased' approach. At the preparation and planning stages of the food safety programme, risk assessment needs only to be developed on a general or global basis ('screening' risk assessment). This characterises the present situation, placing it in a comparative context for establishing priorities, and determining overall requirements. Later on, at the implementation stage, a more detailed, operational level of risk assessment should be developed to assess the impact of any deficiencies on a given process and develop preventive or corrective actions and limits necessary to meet food safety requirements. Risk assessments may be conducted both prior to the development of a specific food safety plan (e.g. the HACCP plan) and as one of its vital elements.

Implementation
Implementation of the food safety programme is done through specific activities using appropriate tools. Management should coordinate this and ensure consistency of results, decisions and actions with regard to food safety requirements. Implementation needs capabilities, procedures and systems to be developed, so that operational control can be established.

Developing capabilities within the factory requires the allocation of human, technical and financial resources. The workforce should gain an increased awareness of food safety requirements and develop an understanding of their importance. Good communication, recognition of work well done and encouragement to make suggestions will lead to motivation and improved food safety performance of staff. Identification of the knowledge and skills necessary will lead to the development of an ongoing training and education programme. It should be oriented towards helping the workforce understand their roles and responsibilities and perform tasks in an efficient and competent fashion. The development of technical capability might also include medium- or long-term changes in premises, equipment or control technologies within the financial capability of the company.

The establishment of operational control is crucial to ensuring that food safety performance is consistent with an organisation's policy and requirements. The impact of manufacturing operations and activities on food safety must be determined. And controls must be identified to ensure that these activities are carried out to ensure compliance with the food safety requirements. This implies a thorough understanding of all stages of the production cycle from raw material sourcing through to finished products and their use, including understanding of product–process interactions, key product or process parameters, monitoring and verification procedures, and where and which improvements are necessary to comply with requirements.

A detailed analysis may be necessary to identify which factors affect safety, which preventive or control measures should be implemented to ensure compliance with food safety requirements, and which operational criteria are necessary for control and monitoring. This is the same scope and functions that HACCP methodology has developed in the food sector.

The food safety programme provides a framework to create a favourable environment for the use of HACCP. The quality of HACCP plans will be improved by the use of risk assessment tools and techniques (see Section 12.5) during the hazard analysis stage.

Preparation for emergency situations is becoming more and more important; companies need to be able to respond in a consistent and timely manner to unexpected incidents or situations, to minimise their impacts on food safety. A company should develop an emergency plan covering organisation, responsibility and authority. It should outline any services required and their coordination, and contain information on hazardous materials, possible contaminants or hazardous situations with their impact and actions to be taken. This should be covered in a training plan with periodic testing for practicality and responsiveness. Risk assessment, used in a predictive manner, should play a crucial role in emergency planning. Firstly, it can be used to predict the robustness of the food manufacturing system (its stability or capability to function without disturbance) and its ability to run under emergency conditions without overwhelming the company and having an adverse impact on food safety. Secondly, it can be used to predict the likely impact of an incident on public health, allowing the company to provide a response that is proportionate to the risk. Much remains to be done to reduce the vulnerability of food systems. A good start would be an effective system for exchange of information within the food industry and between the latter and public authorities.

Performance assessment
Assessment of performance is a key activity that ensures a company is performing according to its intended food safety programme and plan(s). Management should ensure that performance is regularly monitored against food safety requirements. Performance indicators should show the reliability of data and test results (e.g. laboratory accreditation schemes, management of the hardware and software capability, etc.) and may include audits of the food safety programme to determine if it conforms to planned arrangements. Audits will show if it is properly implemented and maintained and will identify any weaknesses, causes of unsatisfactory performance or drifts that need appropriate correction.

Review, adjustment and improvement
These processes should ensure that changes in the food safety 'context' are taken into account in a timely manner. The review should cover new or emerging hazards, changing consumer expectations, changing regulations, changes in products or activities of the company, lessons learned from incidents, and advances in food science and technology, and should ensure that assurance of food safety is not compromised. In particular, the findings and conclusions of audits and reviews should be documented, with the necessary corrective actions and their completion identified.

12.4.3 Integration with other management programmes

There is a need for a company to develop an overall vision of its managerial goals. Food safety management is not a standalone activity; its management should be intimately integrated with other activities within the overall managerial strategy. Sub-systems that are independent of the overall management of an organisation, or not in line with it, will not operate successfully and may not even survive.[22] The integration of the food safety programme with other management systems in an overall managerial strategy is illustrated in Fig. 12.2.

As previously stated, the TQM approach provides a recognised road-map. Within its framework, a company should endeavour to manage risks in a consistent manner. Industrial risks may be categorised into four areas, apart from food safety:

- Quality (the risk of delivering a product that does not conform to implied or expressed customer specifications or is unsafe)
- Occupational safety (risk of workplace injury)

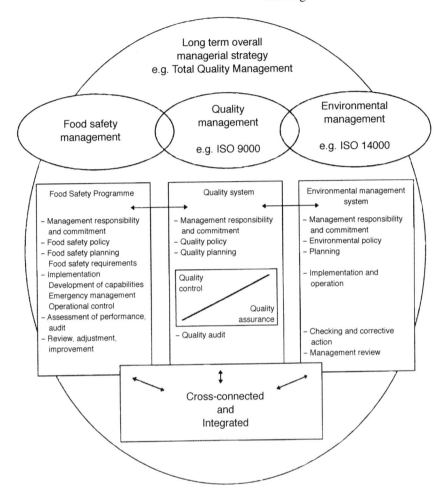

Fig. 12.2 An integrated approach to the management of key issues.

- Environmental impact (risk of damaging the environment during production through pollution, or inefficient use of energy or natural resources)
- Security (risk of being subjected to criminal activities during production).

Harmonisation of approaches and systems in these four key areas plus food safety, and their integration into an overall managerial strategy, are the only means for long-term success.

Because it incorporates the same concepts as TQM and quality or environment management systems, the food safety programme can be readily integrated with these programmes, where they exist. It should not be established independently, and must be part of the quality management system that addresses activities, procedures and processes closely related to food safety. Nevertheless the development of a complete food safety programme will show that a company has taken a comprehensive and consistent approach to dealing with food safety issues, its goals are clearly established and communicated, and operational activities are effective and meet food safety requirements.

This integrated approach resolves the problem of integrating HACCP with ISO 9000 quality systems. Present attempts have taken two main directions: expanding the HACCP methodology to incorporate some managerial activities organised and documented according to ISO 9000 requirements, or using the quality system to manage the HACCP system. Neither route is entirely satisfactory, because to keep HACCP strong and practical it should not be misused. Second, because food safety is crucial to accomplishing an organisation's strategy, its proper management should receive full consideration and it should not be considered a subset of another management system.

For practical implementation, there needs to be correspondence and compatibility of the food safety programme with environment management systems (ISO 14000) and quality systems (ISO 9000) (Fig. 12.2). ISO 9000 standards are under revision, to be brought in line with the ISO 14000 series. Therefore the food safety programme and the two other management systems may be easily integrated in the future. But HACCP needs to maintain its pragmatic value to establish operational controls and in this context the use of risk assessment is the novelty. It can be viewed as the 'Ariane thread' ensuring a consistent progression between the food safety programme and public health (Fig. 12.3).

12.5 HACCP revisited: introduction of risk assessment techniques

The future challenge facing managers and food business operators is to establish a clear link between operational controls and public health requirements, based on risk assessment, while keeping the practical nature of HACCP. Even though they incorporate the best scientific information, technical know-how and expertise, most current HACCP analyses are mainly qualitative. These HACCP studies have facilitated understanding and ownership by the workforce of the HACCP methodology and programmes. The practical strength of the system probably accounts for its current worldwide acceptance.

Today, however, the increasing complexity of food safety requires a better understanding of how the processing steps, their control and inherent variability and the possibility of failure interact to affect the safety of the food produced. This clearly demands revisiting the Hazard Analysis stage of the HACCP methodology. This would be improved by the introduction of a more quantitative, probability-based approach to evaluate the reliability of processes and align process controls to public health requirements. This is risk assessment.

The opportunity for incorporating risk assessment techniques into HACCP could deliver several benefits. Risk assessment offers identification of relevant hazards, a quantitative appraisal of the likely level of hazard(s) in food, while

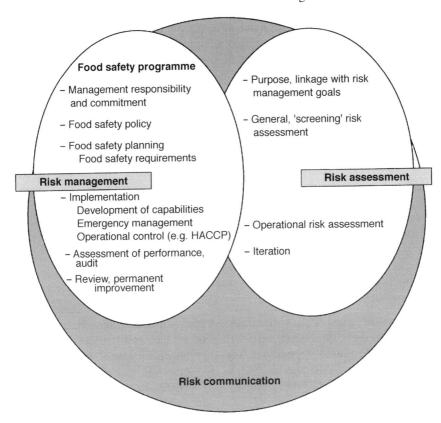

Fig. 12.3 Integrating risk analysis with the food safety programme.

taking into account the variability of raw materials and manufacturing. The better the determination of the level of hazard, the more product/process control requirements can be targeted, all refinements that are missing from the current HACCP system. Key advantages of the quantitative approach would be the ability to link the HACCP plan and controls to public health impact and to measure the level of confidence that the managers (and the evaluators) may have in the operational results.[4]

The usefulness of risk assessment techniques is not limited to improving operational HACCP plans. Risk and especially risk assessment should be considered in the preparation and planning of a food safety programme (Fig. 12.3). Risk assessment may differ in its scope and range depending only on where it is used. In preparation and planning, risk assessment needs to be conducted in a general, 'screening' manner to identify significant hazards and establish priorities for action, so that overall food safety requirements consistent with public health expectations are developed. In the development of a HACCP plan, its purpose is to introduce a probabilistic, quantitative approach based on hazards and their control, taking into account process variability, so that critical control points (CCPs) can be more accurately identified and specified. This requires an operational, process-oriented approach to risk assessment. In other words, a food safety programme can be meaningful only if it incorporates risk assessment activities prior to establishing operational controls.

The following sections will consider the evolving approach towards quantitative risk assessment and how this might be combined with HACCP.

12.5.1 The evolving approach to risk assessment

Risk assessment consists of hazard identification, exposure assessment, hazard characterisation (including dose–response assessment) and risk characterisation.[29] Within this framework, quantitative risk assessment involves the development and use of dynamic risk models and covers the stages of characterisation of the system, model development, analysis, risk estimation and description, and risk assessment.

Characterisation of the system to be studied
This includes[30]

- finding out whether particular microorganisms in a food may be associated with adverse health effects and determining the factors that affect their ability to be transmitted (conditions for survival, persistence, multiplication, inactivation or destruction; vectors and potential spread) and to cause adverse effects in the host (virulence, pathogenicity factors and their variation and evolution);
- characterisation of the source of materials and the process, including product flows (e.g. manufacturing and distribution within the food chain and specific stages in a process); determination of factors that may potentially go wrong and the relationship between the hazard, process, product contamination, and the level of hazard in the product;
- characterisation of the population according to its sensitivity to the hazard and its likely exposure to the product; identification of more sensitive sub-populations and their characteristics, factors that influence susceptibility to the hazard and the severity of disease, as well as consumption patterns and habits;
- characterisation of the disease, involving determination of the different outcomes and the dose–response relationship.

This information should provide an insight into the inherent variability of the different factors, their importance and the likely distribution of their values.

Development of a conceptual risk model
The risk model should integrate and structure the information mentioned above. It should cover both the main variables and their variation – statistical (probability distribution of values, conditions, individuals) and dynamic (evolution of microbial populations, of process parameters, such as pH or temperature). It may incorporate mathematical sub-models, such as microbial

predictive models and dose–response models, to provide direction for the analytical phase of the risk assessment.

Analysis
This stage refers to consideration of information from the exposure assessment and dose–response data. Today, a particularly promising tool for this is the 'Monte Carlo simulation'.[31] It can be used to simulate the interactions of the pathogen, the food, a population of consumers and exposure scenarios. Each probability distribution within the model is randomly sampled to reproduce the shape of the distribution and produce a large number of possible scenarios ('iterations' or 'trials'). The simulation determines the combined impact of the probability distribution of variables on the probability distribution of the possible outcomes. Therefore it represents a distribution of risk, based on combining the probability of values occurring. At present a variety of commercially available software products facilitate the calculations.

Risk estimation and description
This refers to risk characterisation and integrates key aspects of the analysis to provide an allocation of risk and a description of the factors that have the greatest impact on the risk. It includes a quantitative measure of the relationship between process targets and their variability and the overall performance of the system with regard to risk. Statistical techniques such as rank correlation (e.g. Spearman rank correlation, tornado diagrams) and sensitivity analysis have proved particularly useful.[32]

Unrestricted/restricted risk assessment
Risk modelling and analysis may first be used to characterise an existing situation or process and provide a baseline. This is 'unrestricted' risk assessment. Using this baseline it is possible to change the input parameters to take account of values resulting from different control interventions or measures and then observe the changes in the risk estimate. This is the 'restricted' risk assessment, which allows for evaluation and comparison of the effectiveness of control strategies and measures. Restricted risk assessment can also provide an objective input into analyses of cost-effectiveness,[20] allowing managers to make informed decisions about practical changes.

12.5.2 Incorporating quantitative risk assessment techniques into HACCP
Several investigators are presently demonstrating that it is possible to develop risk assessment models, and to use them to refine the hazard analysis stage of HACCP studies.[32–35] The process may appear rather sophisticated and demanding in terms of information and data, time and statistical expertise. However, a quantitative, risk-based approach represents the only way to realise the full potential of a food safety programme. The advantages warrant the increased work and any adjustments indicated can be implemented incrementally.

This process starts with a description of the structure of the system, including raw materials, final product and intended use, process steps and product flow and data on current product and process specifications.

The hazard analysis stage begins with hazard identification, a qualitative procedure aimed at identifying which microbial hazards are relevant to the product and process. Microbiological knowledge is required for this and a more informed, systematic procedure is desirable.[36] Hazard identification may be improved by the government-led collection and collation of information (e.g. research results, epidemiological studies), development of global risk assessments by public health authorities, aimed at assessing and ranking the food-borne health risks in a population, and communication of the information to business organisations and operators. Good communication will ensure the active exchange of information on the characteristics of realistic hazards, in particular quantitative changes in risk associated with the variability of microbial populations (e.g. resistance to external factors such as temperature).

The next activity in hazard analysis involves determining the conditions leading to the presence, contamination, survival or growth of each hazard and its impact on the level in the final product or percentage of non-conforming products. Here, one approach to introducing quantitative, probabilistic techniques may be the use of reliability tools such as Event Tree and Fault Tree analyses. Detailed descriptions of these tools may be found in the literature.[37-39] An Event Tree is a diagram illustrating the consequences of an event chain (where an event is a deviation in a manufacturing process). A Fault Tree diagram describes the causes of the deviation. Combining the two will allow systematic description of circumstances under which a system could fail and understanding the effects of chains of events, expressed in terms of frequency or probability. Once the diagrams have been outlined, the next task is to evaluate evidence on the probability of each event (i.e. the 'risk'). This may be given qualitatively, using expressions such as 'low', 'medium', or 'high' risk, or by a numerical ranking system, using for instance 10 for high risk, down to 1 for low risk. This is suggested by Failure Modes, Effects and Criticality Analysis (FMECA), another reliability tool currently used by industry. FMECA is an extension of Failure Modes and Effects Analysis (FMEA) from which HACCP originated. It is amazing that these tools seldom appeared in the classical descriptions of HACCP, which focus on general or specific hygienic practices. Introducing quantitative approaches into HACCP may be viewed as simply re-sourcing it!

Semi-quantitative evaluation provides a means of assessing the impact of failures and prioritising problems and may prompt the development of more advanced quantitative techniques. When more precise risk assessments are necessary, more accurate quantitative information needs to be sought, to determine the distribution of the probabilities of each adverse event occurring (probability distribution function). Software packages for stochastic simulation (e.g. Monte Carlo) give probability estimates and may provide a means of identifying potential critical points in a complex system[38] allowing better alignment of system capability.

Where a desired outcome has been identified (e.g. attainment of a food safety objective), another approach may be used to consider the effects of variations in materials and process specifications, or hazards, and their variability at all stages. This approach determines the impact of variations on the occurrence of non-conforming product.[40] To do this requires collecting additional information on the level and statistical distribution of microorganisms in the raw materials and at different process steps and in the final product, and next on how this is affected by the distribution of parameter values unique to each process stage (e.g. duration of lag phase, sensitivity to thermal processes, pH variation, time/temperature variation in a thermal process, etc.). This information is used to construct a risk assessment model, using a flow diagram for mapping the process and adding the parameters and their variability to the model. Sub-models may be utilised to refine the approach, such as microbial predictive or lethality models or heat transfer and other process models. The latter should take account of the variability of product dimensions and thermophysical properties, including product temperature at start of cooking and cooking conditions.[41]

After defining the features and variability in the basic model, it becomes possible to use simulation, e.g. Monte Carlo, to determine the impact of the distribution of variables on the predicted outcome. The main advantage of introducing a probabilistic approach is that it provides a rational and transparent way to address variability. Use of a qualitative approach tends to consider average or mean values, based on experience, or will default to the worst-case scenarios. Both approaches are unsatisfactory, the first because a system based on mean values may fail when confronted with extreme circumstances and the second because it may be over-conservative.[4] An approach based on probability considers the whole range of distribution of values, their chances of occurrence, and how they impact on overall variability in the system. This allows interventions to be directed towards reducing the variability and elimination of high-risk scenarios (e.g. poor microbiological quality of raw materials). It also allows the accurate establishment of critical process limits in order to reduce the risk of unsatisfactory performance and ensure conformance with specifications. With good knowledge, process parameters may be used to set processes closer to the edge and maintain safety levels (e.g. using lower heating temperatures in minimally processed foods) and to predict the impact of process changes on the risks of making non-conforming products (e.g. from changes in the quality of raw materials, product dimension or heating conditions, etc.).[41]

Incrementally it will become possible to refine and extend modelling activities to give a complete risk model covering consumer sensitivity and dose–response modelling.[42] Using this type of model, the results of the preceding simulations (e.g. level of hazard or rate of failure) can be correlated with their public health outcomes (e.g. probability of infection, probability of disease occurring). This is particularly important because it can be used to gauge the technical performance of an industrial system analytically and not only from a public health perspective.

It allows businesses to effectively apportion the impact on public health of process design and the stages used in processing and make informed decisions on management of product safety. Where trade-offs with other considerations can be foreseen, the management of compromises, based on a quantitative risk assessment, can balance public health requirements against other considerations such as technical feasibility, market necessities or cost. In the absence of quantitative information about risk, such trade-offs are nothing more than 'a matter of gut feeling'.[43] For informed and justifiable decisions, the whole process needs to be supported by effective communication within the company and its trading partners. Within the food industry, movement towards this approach is a must. The pace at which changes can be implemented, and the final success, depend only on the commitment of senior management of food businesses to improving food safety.

12.6 Summary

Food safety is a basic demand of the consumers. It can be considered as the price of admission to market in the sense that no other feature on which companies compete, such as satisfaction, service, nutrition, innovation, quality and cost, can be valued in the marketplace unless there is customer confidence in the safety of the food. In addition, trading conditions and legislation require food businesses to demonstrate their commitment to food safety issues.

The proposed approach to determining food safety requirements should ideally start with the development of a national food safety plan. It should be based on a government-led risk analysis process, identifying public health based food safety objectives, such as maximum contaminant levels and the level of consumer protection to be achieved. In response, the food industry should organise itself to provide greater evidence that procedures to ensure food safety are present and adequately managed. Integrated food safety management programmes should be widely developed and linked to management of other key issues such as quality and environmental impact within the long-term management strategy. In the context of the food safety programme, better control should be exercised over industrial processes to increase their reliability and the relevance of controls to ensuring public health goals. To that aim, HACCP should evolve, to include quantitative risk assessment techniques at the hazard analysis stage.

Whereas these principles and their application will probably not pose much of a problem to large food businesses, which have the necessary resources and expertise, it has to be appreciated that these changes would increase pressure on smaller, less developed businesses (e.g. the small and medium-sized enterprises, so-called SMEs). They would have unique needs for specific assistance and guidance. This should be provided by governmental authorities and professional organisations, and interpretation of the risk-based HACCP system should be kept flexible to allow them to apply it. To benefit all food producers and consumers, it is essential in particular that scientists from academia, government, professional organisations and industry work together to provide the necessary information, advice and technical support.

The increasing complexity of food safety and the significant changes occurring in the global economy present a unique opportunity and challenge. Going beyond HACCP towards a risk-based food safety management programme will be crucial for companies wishing to move from a regional or national scale to an international one. It is likely that only companies that recognise this need will be successful on the international marketplace during the twenty-first century.

12.7 References

1. THEYS J, 'La société vulnérable', in *La Société Vulnérable: Evaluer et maîtriser les Risques*, eds FABIANI J L and THEYS J, Paris, Presses de l'Ecole Normale Supérieure, 1987.
2. SANDMAN P M, 'Definition of risk: managing the outrage, not just the hazard', in *Regulating Risk: the Science and Politics of Risk*, eds BURKE T A, TRAN N L, ROEMER J S and HENRY C L, Washington DC, ILSI Press, 1993.
3. BAUMAN H E, 'HACCP concept, development and appreciation', *Food Technology*, 1990 **45**(5) 156–8.
4. BUCHANAN R L and WHITING R C, 'Risk assessment: a means for linking HACCP plans and public health', *J Food Protec*, 1998 **61**(11) 1531–4.
5. MAYES T and JOUVE J L, 'HACCP and risk analysis: a pragmatic way forward' (in preparation).
6. WHO, *Guidelines for strengthening a National Food Safety Programme*, Geneva, WHO/FNU/FOS/96.2, 1996.
7. JOUVE J L, 'Principles of food safety legislation', *Food Control*, 1998 **9**(2/3) 75–81.
8. WHO-FAO, *Application of Risk Analysis to Food Safety Issues*, Report of a joint Expert Consultation, Geneva, WHO/FNU/FOS/95.3, 1995.
9. NRC (National Research Council, Institute of Medicine), *Ensuring Safe Food from Production to Consumption*, Washington DC, National Academy Press, 1998.
10. FAO/WHO, *Risk Management and Food Safety*, Report of a joint expert Consultation, Rome, FAO Food and Nutrition Paper No. 65, 1997.
11. AHL A S, ACREE J A, GIPSON P S, McDOWELL R M, MILLER L and McELVAINE M D, 'Standardization of nomenclature for animal health risk analysis', *Rev. Sci. Tech. Off. Int. Epiz.*, 1995 **14**(4) 913–23.
12. GRAY G M, 'The case for improving risk characterization', in E V OHANIAN and J A MOORE (ed), 'Risk characterization: a bridge to informed decision making', *Fundam. Appl. Toxicol*, 1997 **39** 81–8.

13 Presidential/Congressional Commission on Risk Assessment and Risk Management, *Risk Assessment and Risk Management in Regulatory Decision Making*, Final Report, Vol. 2, 1997, http://www.riskworld.com/.
14 HATHAWAY S C, 'Risk assessment procedures used by the Codex Alimentarius Commission and its subsidiary and advisory bodies', Joint FAO/WHO Food Standard Programme, 20th session of the Codex Alimentarius Commission, Geneva, 1993.
15 HATHAWAY S C and COOK R L, 'A regulatory perspective on the potential use of microbiological risk assessment in international trade', *Int. J Food Microbiol.*, 1997 **36** 127–33.
16 ROSE J B, HAAS C N and GERBA C P, 'Linking microbiological criteria with quantitative risk assessment', *J Food Safety*, 1995 **15** 121–32.
17 HAVELAAR A, SLOB W, TEUNIS P and JOUVE J L, 'A public health basis for food safety objectives', int. conf. *Risk Analysis*, Rotterdam, SRA-Europe, 1999.
18 JOUVE J L, 'Establishment of food safety objectives', 2nd int. conf. *Food Safety and HACCP*, Bilthoven, Bastiaanse Com. bv., 1998.
19 VAN SCHOTHORST M (Secretary of the ICMSF), 'Principles for the establishment of microbiological food safety objectives and related control measures', *Food Control*, 1998 **9** 379–84.
20 MORALES R A and McDOWELL A M, 'Risk assessment and economic analysis for managing risks to human health from pathogenic micro-organisms in the food supply', *J Food Protec*, 1998 **61**(11) 1567–70.
21 FULKS F T, 'Total quality management', *Food Technology*, 1991 **45**(6) 96–101.
22 GOLOMSKI W A, 'Total quality management and the food industry, why is it important?', *Food Technology*, 1993 **47**(5) 74–9.
23 JURAN J M, GRYNA F M and BINGMAN R S, *Quality Control Handbook*, 3rd edn, New York, McGraw Hill, 1979.
24 NFPA (National Food Processors Association), 'HACCP and Total Quality Management, winning concepts for the 90s', *J Food Prot*, 1992 **55** 459–62.
25 SASHKIN M and KISER R H, *Putting Total Quality Management to Work*, San Francisco, Berrett-Koelher, 1993.
26 WEBB N B and MARSDEN J L, 'Relationship of the HACCP system to Total Quality Management', in *HACCP in Meat, Poultry and Fish Processing*, eds PIERSON A M and DUTSON T R, London, Blackie Academic and Professional, 1995.
27 JOHANSSON H J, McHUGH P, PENDLEBURY A J and WHEELER W A, *Business Process Reengineering*, Chichester, John Wiley and Sons, 1993.
28 JOUVE J L, STRINGER M F and BAIRD-PARKER A C, *Food Safety Management Tools*, Brussels, ILSI-Europe, 1998.
29 CAC (Codex Alimentarius Commission), 'Draft principles and guidelines for the conduct of microbiological risk assessment', Alinorm 99/13A, Appendix II, Rome, CAC, 1998.

30 ILSI (International Life Science Institute), 'A conceptual framework to assess the risk of disease following exposure to pathogens', *Risk Analysis*, 1996 **16** 841–8.
31 VOSE D, *Quantitative Risk Analysis: a Guide to Monte Carlo Simulation Modeling*, Chichester, John Wiley and Sons, 1996.
32 CASSIN M H, LAMMERDING A M, TODD E C D, ROSS W and McCOLL R S, 'Quantitative risk assessment for *Escherichia coli* O157:H7 in ground beef hamburgers', *International J Food Microbiol*, 1998 **41** 21–44.
33 MARKS H and COLEMAN M, 'Establishing distribution of micro-organisms in food products', *J Food Protec*, 1998 **61**(11) 1535–40.
34 SERRA J A, DOMENECH E, ESRICHE I and MARTORELL S, 'Risk assessment and critical control points from the production perspective', *Int J Food Microbiol*, 1999 **46** 9–26.
35 WHITING R C and BUCHANAN R L, 'Development of a quantitative risk assessment model for *Salmonella enteritidis* in pasteurized liquid eggs', *Int. J Food Microbiol*, 1997 **36** 11–125.
36 NOTERMANS S, ZWIETERING M H and MEAD G C, 'The HACCP concept: identification of potentially hazardous micro-organisms', *Food Microbiol*, 1994 **11** 203–12.
37 AVEN T, *Reliability and Risk Analysis*, Barking, Elsevier Science, 1992.
38 ROBERTS T, AHL A and McDOWELL R, 'Risk assessment for foodborne microbial hazards', in *Tracking Foodborne Pathogens from Farm to Table*, eds ROBERTS T, JENSEN H and UNNEWEHR L, Washington DC, USDA, Economic Research Service, 1995.
39 VILLEMEUR A, *Sureté de Fonctionnement des Systèmes industriels*, Paris, Eyrolles, 1988.
40 MILLER A J, WHITING R C and SMITH J L, 'Use of risk assessment to reduce listeriosis incidence', *Food Technol*, 1997 **61**(4) 100–3.
41 BILLON C M P, BROWN M H and DAVIES K W, 'Quantitative risk assessment', int. conf. on *Predictive Microbiology*, Quimper, ADRIA, 1997.
42 CASSIN M H, PAOLI G M and LAMMERDING A M, 'Simulation modeling for microbial risk assessment', *J Food Protec*, 1998 **61**(11) 1560–6.
43 BERNSTEIN P L, *Against the Gods: the Remarkable Story of Risk*, New York, John Wiley and Sons, 1996.

Index

accommodation, animal 39, 65
accredited auditing services 289
acridine orange direct count (AODC) 218, 219, 220
action plan 261–2
adenosine triphosphate (ATP) 219–20, 224
advisory graded scale 215–16
aerobic plate counts *see* total viable counts
Aeromonas 84, 116
air chilling 131, 138, 139
all-in, all-out practice 39, 65
American Meat Institute 29
American Meat Science Association 218
analytical methods 140–2, 171–3
animal feeds 40, 55, 67, 71–2, 293
animal rearing
   hazard analysis 38–43
   see also farm production
antibiotics 40–2
Arcobacters 166
area, sampling 86–7
assessment 283–4
attributes sampling 216–17
audit 199, 273–91
   auditing HACCP systems 284–6
   benchmarks for auditing 277–86
   future trends 288–90
   ground rules 278–84

   assessment and scoring 283–4
   follow-up 284
   format 281–2
   frequency 284
   preparation 280–1
   scope 279
   standards 279–80
   what auditor should look for 287–8

Bacillus anthracis 169
Bacillus cereus 169
Baclight stain 218, 219
Barr, John 4
barriers to implementation 197–8
bedding 39
beef carcasses 95, 209
   dressing 96–102
beef cattle 45, 69–73
benchmarks for auditing 277–86
biosecure housing 39
birds, wild 57
birth 39, 49–51
bovine mastitis 58–9, 162, 165
bovine spongiform encephalopathy (BSE) 40, 170–1, 293
breeding cattle 47–9
British Egg Industry Council 64
Brucella melitensis 169

calibration of control system 261

caliciviruses 171
Campylobacter coli 165
Campylobacter jejuni 123–5, 164–6
   in animals 163
   in man 165–6
carcass breaking processes 110–13
carcass cooling 107–10
carcass dressing 96–104
cattle 45–60
   beef cattle 45, 69–73
   dairy 45–60, 61, 69–73
   farm HACCP 47–57
   production stages 46
   salmonellosis 159
certification audits 288–9
characterisation of system 312
check sheets 281–2
chemical decontamination methods 148–9
chemical hazards 246
chilling/cooling
   carcass cooling 107–10
   monitoring 211–12
   offals 104–7
   poultry 131–2, 137, 138, 139
chlorinated-water sprays 134, 136, 137
chlorine 149
cleaning
   carcasses 96, 99–104
   equipment used for carcass breaking 110–13
   poultry processing plants 132–3, 135–6
cleaning-in-place systems 146
cleanliness of animals 39–40
Clostridium botulinum 168–9
Clostridium perfringens 123–5, 168
Codex Alimentarius 6–7, 30
   principles for validation and verification 177–8, 231–2, 233–5
Codex Alimentarius Commission 6–7, 11–12
colibacillary diarrhoea 161–2
coliforms 84, 141–2
comparison 259
competent authorities 235–6
conceptual risk model 312–13
confidence in management 25
consistency 22
contagious mastitis 58, 59
control points (CPs) 135
   see also Critical Control Points
control system calibration 261
cooked meat products 240–1
cooling *see* chilling/cooling

corrective actions 185–6, 261
   monitoring systems and 212–13, 214
   problems in implementing 224
Corynebacterium pseudotuberculosis 169
crates, poultry 129–30, 135–6, 137–8, 139, 145
Creutzfeldt-Jakob disease (CJD) 170–1
Critical Control Points (CCPs) 44, 181–2, 205, 239
   and GMPs 203–4
   poultry 134–42
      control measures at different stages of processing 135–7
      establishment and operation 137–9
      microbiological testing of products and equipment 140–2
      validation and verification 140
   problems in establishing 223
   records 260
   red meat 82–3, 90–1
   US 29
critical limits 182–3, 184–5, 206–8, 260
   determination of 207–8, 209
   establishment of 206–7
critical points 21
Cryptosporidium parvum 169–70
Cyclospora spp. 170

dairy cattle 45–60, 61, 69–73
data collection and analysis 83–7
deboning 143–5
decision making 242–3
decision tree 212, 214
decontamination of carcasses 99–104, 147–9
defeathering 130–1, 136, 146
dehairing of carcasses 95–6
design of equipment 112–13
deviations 261
dioxin 293
direct epifluorescent filter technique (DEFT) 218
directives, EU 13–20
   review of 22–5
disinfection 132–3
DNA chip technology 171
documentation *see* recordkeeping
dressing, carcass 96–104
drinking water 52, 54–5, 73
due diligence 32–3

Echinococcus granulosus 170
effectiveness of HACCP on the farm 67–76

monitoring *see* monitoring
eggs 64
enablers 289–90
enforcement 29
  ease of 25
enteric colibacillosis 161–2
Enterobacteriaceae 141–2
environmental management 308–10
environmental mastitis 58, 59
equipment
  cleaning carcass breaking equipment 110–13
  poultry processing 127, 130–1, 132–3
    CCPs 134–7
    development of more hygienic equipment 145–6
    microbiological testing 140–2
Escherichia coli (*E. coli*) 84, 161–4
  carcass breaking 111–12
  carcass dressing 96–7, 98
  cooling offals 106, 107
  disease in animals 161–2
  disease in man 162–4
  E. coli O157 4–6, 30
  FSIS verification system 216–17
  microbiological criteria 115–16
  outbreak in Scotland 3–4, 8–9, 14, 293
  Pacific Northwest outbreak 29
  poultry 134, 140, 141–2
European Food Safety Inspection Service (EFSIS) 289
European Foundation for Quality Management (EFQM) 289–90
European Union (EU) 11–26, 30, 137
  fishery products 20–2
  food policy and HACCP 12–14
  meat hygiene legislation and HACCP 14–20
    fresh meat hygiene directives 17–19
    general food hygiene directive 15–17
    meat products 20
    minced meat and meat preparations 19–20
  and its member states 11
  review of directives 22–5
  verification 235–6
EUROVOL project 145–6
Event Tree analysis 314
evisceration 215
  poultry 131, 136–7, 138, 139, 146
  red meat 96, 98–9
excision 85–6
experience-based audits 281

exposure assessment 237
external audits 280
  HACCP plans 286

faecal contamination 58, 141–2
  see also Escherichia coli
Failure Modes and Effects Analysis (FMEA) 296, 314
Failure Modes, Effects and Criticality Analysis (FMECA) 314
farm production 37–79
  effectiveness of HACCP 67–76
  HACCP plans for cattle 45–60, 61, 69–73
  HACCP plans for a pig unit 65–7
  HACCP plans for a poultry unit 62–5
  HACCP plans for sheep 61–2, 69–73
  hazard analysis in animal rearing 38–43
  setting up the HACCP system 43–5
farm quality assurance programmes 74–5
farm staff 56
Fault Tree analysis 314
Federal Register (FR) 27–8
feed, animal 40, 55, 67, 71–2, 293
fishery products 20–2
fitness for human consumption 17
flow-chart 129
flow cytometry 172
flow diagram 260
follow–up 284
Food and Drug Administration (FDA) 27
Food MicroModel (FMM) 244–5
Food Safety Initiative (FSI) 31
Food Safety and Inspection Service (FSIS) 27, 28–9, 233–4
  verification system 216–18
food safety management 6–7, 293–319
  future trends 295–7
  introduction of risk assessment techniques into HACCP 298, 310–16
  review of food safety issues 190–1, 302–3
  risk-based food safety strategy 297, 298–301
  vulnerability of food systems 293–5
  way forward 297–8
Food Safety Objectives (FSOs) 233, 301, 305–6
food safety policy 302

Food Safety Programme (FSP) 297–8, 301–10
  integration with other management programmes 308–10
  managed programme 302–3
  stages and components 303–8
    implementation 306–8
    performance assessment 308
    planning 304–6
    preparation 303–4
    review, adjustment and improvement 308
foodstuffs, animals 40, 55, 67, 71–2, 293
foreign bodies 246
format, audit 281–2
fresh meat hygiene directives 17–19

game 18–19
General Agreement on Tariffs and Trade (GATT) 12
general food hygiene directive 15–17
Giardia duodenalis 169
Giardia lamblia 169
giblets 142–3
global policy 11–12
goats 60
Good Manufacturing Practices (GMPs) 188, 242
  poultry processing 132–3
  prerequisite programme 203–4, 205
  small plants 113–15
grazing land 51, 69–70
ground rules for an audit 278–84
growth promotion techniques 41–2
Guillain-Barré syndrome 165
gut, portions of the 104, 105

HACCP 4, 274
  auditing HACCP systems 284–6
  EU and 23–4
  explanation of reasons for 190
  and food safety 6–7
  implementation *see* implementation
  and quality systems 275–7
  and risk assessment 237–8
    incorporation of risk assessment techniques 298, 310–16
  setting up the system for farm production 43–5
  stages in construction of system 82
  strength and limitations 295–7
  successful implementation 8–9
  US 28–31

HACCP plans
  documentation and explanation of 259–61
  farm production
    for cattle 45–60, 61, 69–73
    for pigs 65–7
    for poultry 62–5, 76
    for sheep 61–2, 69–73
  scope of for processing poultry 127
HACCP team 127–8, 191
  implementation sub-teams 193
  involvement in validation and verification 263–4
haemolytic uraemic syndrome (HUS) 5, 163
haemorrhagic colitis 5
hazard characterisation 237
hazard identification 179–81, 237
  poultry processing 129–32
HAZOP (Hazard Analysis and Operability) 295–6
heat treatment 31–2
hide 225
horizontal controls 15–16
hormones, steroid 41
housing, animal 39, 65
Hygiene Assessment System (HAS) 215–16
hygiene information 261

immunological techniques 172
impedimetry 172
implementation of HACCP 8–9, 177–201
  allocation of resources 193
  auditing and review 199
  differences between large and small businesses 188–9
  elements requiring implementation 178–87
    CCPs 181–2
    corrective actions 185–6
    hazard identification 179–81
    monitoring 183–5
    recordkeeping 186–7
    targets and critical limits 182–3, 184–5
    verification 187
  explanation of reasons for HACCP 190
  general approach for processing red meat 87–92
  implementation process 187–8
  measuring performance of the plan 198
  planning for implementation 191–3

processing poultry 127–32
review of food safety issues 190–1
selecting teams and activities 193
tackling barriers 197–8
training 193–5
transferring ownership to production personnel 195–7
US 30–1
implementation team *see* HACCP team
implemented HACCP plans, validation of 264
information
  assembly for verification 258–9
  production and hygiene information 261
  requirements for validation and verification 248–9
  sources for validation 253, 254
integrated approach to management 308–10
intensive pig production systems 66
internal audits 280
  HACCP plans 285–6
International Commission on Microbiological Specifications for Foods 149, 205
irradiation 32
ISO 9000 system 273–4, 275–6, 277, 278, 284–5, 288–9

kill steps 31–2

large businesses 188–9
legislation *see* regulation
limitation of exposure, principle of 300–1
line speed 225
Listeria 84, 116
  monocytogenes 30–1, 123–5, 167–8
  rapid methods 220–1
log mean numbers of bacteria 84–5
  microbiological criteria for red meat 116

management
  confidence in 25
  integrated approach 308–10
  involvement in validation and verification 263
  responsibilities for training 194
markers 252
mastitis, bovine 58–9, 162, 165
material-related hazards 179
measurement, monitoring by 211–12

meat preparations 19–20
meat products 20
medications 40–2, 68, 70–1
Medicines (Restrictions on the Administration of Veterinary Medicinal Products) Regulations 1994 41
'Mega-Reg' 29, 30
microbial adenosine triphosphate (mATP) test 219–20, 224
microbiological hazards 157–76, 180–1
  analytical methods 171–3
  bovine mastitis 58, 59
  Campylobacter jejuni 123–5, 164–6
  Clostridium botulinum 168–9
  Clostridium perfringens 123–5, 168
  decontamination of carcasses 99–104, 147–9
  E. coli see Escherichia coli
  future trends 172–4
  HACCP and risk assessment 237–8
  Listeria see Listeria
  in milk 58
  online monitoring and the use of microbiological data 224
  parasites 53, 169–70
  poultry 123–7, 146–7, 147–9
    testing of products and equipment 140–2
  predictive modelling 146–7, 238, 244–5
  red meat 81–3
    criteria 115–16
    data collection and analysis 83–7
  Salmonella see Salmonella
  Staphylococcus aureus 123–5, 131, 167
  TSEs 170–1
  validation and 239–41, 244–6, 252
  verification 213–18
    rapid methods 218–21
  viruses 171
  Yersinia enterocolitica 166
microbiological testing 140–2, 171–3
milk production 45–60, 61
  see also dairy cattle
milking 59, 60, 61
minced meat 19–20
Ministry of Agriculture, Fisheries and Food (MAFF) 64, 215
modified atmosphere packaging 128
molecular typing 173
monitoring 21, 203–30, 243–4, 259
  determination of critical limits 207–8, 209

326   Index

monitoring *continued*
  establishing CCPs 205
  establishing criteria 206–7
  feedback and improvement 224
  future trends 224–6
  HACCP prerequisite programmes 203–5
  identifying problem areas 223–4
  implementation and 183–5
  need for 206
  setting up monitoring systems 208–13
    corrective actions 212–13
    function of monitoring 208
    monitoring procedures 208–12
  validation *see* validation
  verification *see* verification
Monte Carlo simulation 313
Mycobacterium paratuberculosis 169

National Advisory Committee on
    Microbiological Criteria for Foods
    (US) (NACMCF) 149, 234
National Fisheries Institute (US) 29
national food safety plans 298
National Marine Fisheries Service (US)
    (NMFS) 28
necessity of regulation 22–3
new food vehicles 173
new HACCP studies, validation of 264
Norwalk-like viruses (NLV) 171
nucleic acid based assays 172–3

occupational safety 308–10
offals: collection and cooling 104–7
online monitoring 218–21, 224
online visible monitoring system 210–11
operations 89, 90
optimisation, principle of 301
organic acids 148
over wintering outside 53–4
ownership of HACCP, transferring 195–7

packaging 128
parasites 53, 169–70
parturition 39, 49–51
Pasteurella spp. 169
pasteurisation 31–2, 102
Pathogen Modelling Program 244–5
pathogens
  poultry 123–7
  rapid methods for 220–1
  red meat 81–3
  see also microbiological hazards; *and
    under individual names*

pepperoni 206, 207, 208
performance assessment 308
performance indicators 198, 308
personal equipment 110–13
pH, monitoring 211
physical decontamination methods 147–8
physical hazards 246
pig carcasses
  carcass cooling 110
  carcass dressing 103–4
  cleaning and dehairing 95–6
  visual monitoring 210–11
pigs 159
  HACCP plans for a pig unit 65–7
planning
  Food Safety Programme 304–6
  HACCP plans *see* HACCP plans
    for implementation 191–3
plant management *see* management
plant sanitation programmes 132–3
polishing 95–6
portioning 143–5
portions of the gut 104, 105
poultry 123–53
  decontamination of carcasses 147–9
  development of more hygienic
    processing equipment 145–6
  establishing CCPs 134–42
  EU regulation 18
  giblets 142–3
  HACCP plans for a poultry unit 62–5,
    76
  hazard analysis in slaughter process
    127–33
  microbiological hazards 123–7, 146–7,
    147–9
  portioning and deboning operations
    143–5
  Salmonella 61–4, 123–5, 135, 142,
    159–60
  stages in processing operation 124
predictive modelling 146–7, 238, 244–5
predressing treatment processes 94–6
pre-HACCP requirements 113, 114
prerequisite programmes 203–5
prescription 24–5
preserved products 247
procedures, auditing and 287–8
process flow-chart 129
process-related hazards 179
processes
  general approach to HACCP
    implementation 87–92

reassessment of 223
Produce Safety Initiative (PSI) 31
product recalls 30–1
production information 261
production personnel
　talking to during audit 288
　training 194–5
　transferring ownership of HACCP to 195–7
prophylaxis 40–1
proportionality 22–3, 24–5
provisional operating procedures 90
Pseudomonas spp. 125–6

Quality Assurance (QA) systems 66, 83, 232–3
　auditing HACCP-based QA systems 273–91
　decision making within a HACCP-based system 242–3
　HACCP and quality systems 275–7
　and implementation of HACCP 183–5
　integrated approach to management 308–10
questionnaires, audit 282

rabbit 18–19
rapid methods 218–21, 225
raw materials 16, 260
recalls, product 30–1
reception of stock 40, 92–3
recordkeeping 22, 186–7, 198, 199
　on the farm 67, 68
　of HACCP system and its implementation 186
　monitoring and 224
　performance of HACCP system 186–7
　verification 259–61
red meat 81–122
　carcass breaking and equipment cleaning 110–13
　carcass cooling 107–10
　carcass dressing 96–104
　collection and cooling of offals 104–7
　EU regulation 17
　HACCP implementation 87–92
　microbiological criteria 115–16
　microbiological data collection and analysis 83–7
　slaughter and predressing treatment processes 94–6
　smaller plants 113–15
　stock reception 92–3

regulation
　animal health and human health 75
　EU 11–26
　necessity of 22–3
　US 27–33
regulatory audits 289
reporting
　validation 267, 268
　verification 261–2, 267, 268
residues 68, 69–73
resources, allocation of 193
restricted risk assessment 313
results 289–90
review
　of food safety issues 190–1, 302–3
　FSP 308
　implementation 199
　verification 259
Richmond Committee Report 64
rinse sampling 141
risk, perception of 294
risk analysis 237, 299–300, 306, 313
risk assessment 299–300, 306, 310–16
　evolving approach 312–13
　HACCP and 237–8
　incorporation into HACCP 298, 313–16
　integration with FSP 310, 311
　unrestricted and restricted 131
risk-based food safety strategy 297, 298–301
　principles for implementation 300–1
risk characterisation 237–8, 313
risk communication 299, 300
risk management 237, 299
risk model, conceptual 312–13
rodents 55, 65
Rome, Treaty of 12–13

Salmonella 40, 158–61
　pigs 66, 159
　poultry 62–5, 123–5, 135, 142, 159–60
　red meat 115–16
　salmonellosis in animals 159–60
　salmonellosis in man 160–1
　US 30
　verification 217–18
　　rapid methods 220–1
sampling of carcasses and other raw products 85–6, 140–1, 216–17
sampling plans
　for validation 265
　for verification 265–6

Sanitary and Phytosanitary (SPS) Agreement 12
sanitation programmes, plant 132–3
Sanitation Standard Operating Procedures (SSOPs) 29, 204, 205, 220
scalding 130, 136, 138, 139, 145–6, 147
scope of audit 279
scoring system for audits 283–4
Scotland 3, 7, 8–9, 14, 293
scrapie 170
security 308–10
selective media 171–2
self-regulation 24–5
sheep 159
  carcass dressing 102–3
  HACCP plan 60, 62–3, 69–73
shelf-life 140
singeing 95–6
skills
  validation 250–1
  verification 256
skinning 96–7, 99, 100
slaughter 7
  changes in slaughter technology 225–6
  information from slaughter plant 67, 68, 74
  poultry 127–33
  and pre-dressing treatment for red meat 94–6
small businesses/plants 113–15, 188–9
specifications 260
spoilage bacteria 245–6
  poultry 123–7
  see also microbiological hazards
spray chilling 109–10
Standard Operating Procedures (SOPs) 91, 204, 205
standards 279–80
Staphylococcus aureus 123–5, 131, 167
steam decontamination 102, 147–8
steroid hormones 41
sticking 94
stock reception 40, 92–3
strategic medication 40–1
straw bedding 39
stunning 94, 130
summer mastitis 58–9
supply chain
  coverage of HACCP study 241
  verification and stages of 262–3
surface adhesion immunofluorescent (SAIF) technique 220–1
surface drying 109

swabbing 85–6, 141
Swann committee report 42
system, characterisation of 312
systematic colibacillosis 162
systems audits 278
  auditing HACCP systems 284–6

taint in milk 59–60
targets 182–3, 184–5
teams
  HACCP *see* HACCP team
  validation 251
  verification 256–7
teat canal 58
technical audits 278
technology, changes in 225–6
temperature controls 16
therapeutic antibiotics 40–1
third party auditing services 289
thrombotic thrombocytopaenic purpura (TTP) 5, 163
Total Quality Management (TQM) 275, 277
total viable counts (TVCs) (aerobic plate counts) 84, 116
  verification 214–16, 218–20
tour of plant 287
Toxoplasma gondii 170
trade 74, 75
training 189, 193–5, 307
  continuing 195
  at the factory floor level 194–5
  management responsibilities 194
transmissible spongiform encephalopathies (TSEs) 170–1
transportation 7
  poultry 64, 129–30
Trichinella spiralis 170
trimming 99–100, 101
trisodium phosphate (TSP) 148–9

ultraviolet (UV) light 148
United States (US) 27–33
  development of HACCP 28–30
  due diligence 32–3
  implementation of HACCP 30–1
  kill steps 31–2
  regulatory background 27–8
United States Department of Agriculture (USDA) 27, 28, 38, 82
  microbiological criteria 115–16
unloading of poultry 130
unrestricted risk assessment 313

vaccines 42–3, 64
vacuum cleaning 99–100
validation 21–2, 206, 222–3, 231–72
  background to 238–41
  function of 222
  HACCP and risk assessment 237–8
  implemented HACCP plans 264
  information requirements 248–9
  involvement of HACCP team 263–4
  involvement of plant management 263
  microbiological and other hazards 244–7
  nature of process 249–50
  new HACCP studies 264
  outputs 267, 268
  people responsible for 251
  poultry processing 140
  principles of Codex Alimentarius 233–5
  procedures 222–3, 251–3
  reassessment 223
  sampling plans 265
  skills needed 250–1
  sources of information 253, 254
  when it should be done 251
variables sampling 217
verification 21–2, 206, 213–21, 231–72
  action plan 261–2
  additional requirements for the meat industry 262–3
  assembly of information 258–9
  background to 238–41
  comparison and review, control and monitoring 259
  decision making 242–3
  documentation and explanation of HACCP plan 259–61

EU legislation 235–6
function of 213
HACCP and risk assessment 237–8
and implementation 187
information requirements 248–9
involvement of HACCP team 263–4
involvement of plant management 263
nature of process 253–5
outputs 267, 268
people responsible for 256–7
poultry processing 140
principles of Codex Alimentarius 233–5
procedures involved 257–8
rapid methods 218–21
reporting 261–2, 267, 268
sampling plans 265–6
skills needed 256
stages 258–61
systems 213–18
when it should be done 255–6
vertical controls 15, 16–20
viruses 171
visceral organs 104, 105
visitors, farm 56–7
visual observations 210–11
vulnerability of food systems 293–5

washing
  carcasses 101–2, 147
  stock reception 92–3
water, drinking 52, 54–5, 73
water immersion chilling 131, 137, 147
wild birds 57
World Trade Organisation (WTO) 12, 233

Yersinia enterocolitica 166